Astronomers are confronted with an increasing inflow of data from different modern facilities, like spacecraft, radio telescopes, CCD receivers, and Schmidt telescopes. Statistical analysis of this data can give a more accurate interpretation. This book gives a complete analysis of the mathematical methods available and puts them in an astromomical context. Subjects covered in this volume include data treatment, combining and weighting data; curve fitting; treatment of small samples, bootstrapping and jacknife for instance; modelling and interference; errors, both accidental and systematic; bias and treatment of missing data and special types of data like missing data. This book stems from a meeting held in September 1989 at the International Stellar Data Centre, in Strasbourg.

Errors, bias and uncertainties in astronomy

Errors, bias and uncertainties in astronomy

EDITORS

C. Jaschek
Strasbourg Observatory, Strasbourg, France

and

F. Murtagh
European Southern Observatory, Garching, West Germany

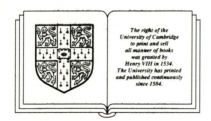

The right of the
University of Cambridge
to print and sell
all manner of books
was granted by
Henry VIII in 1534.
The University has printed
and published continuously
since 1584.

CAMBRIDGE UNIVERSITY PRESS

Cambridge

New York   Port Chester

Melbourne   Sydney

Published by the Press Syndicate of the University of Cambridge
The Pitt Building, Trumpington Street, Cambridge CB2 1RP
32 East 57th Street, New York, NY 10022, USA
10 Stamford Road, Oakleigh, Melbourne 3166, Australia

First published 1990

Printed in Great Britain at the University Press, Cambridge

British Library cataloguing in publication data available

Library of Congress cataloguing in publication data available

ISBN 0 521 39300 0 hardback

CONTENTS

Contents

Introduction

As explained in the introductory speeches, the idea of this meeting arose at the Munich colloquium on "Astronomy from Large Databases" in 1987. It was helped into life by an active Scientific Organising Committee, the members of which were H. Eichhorn, E. Feigelson, A. Heck, M. Huber, C. Jaschek (Chairperson), F. Murtagh, P. Nobelis, B. Ripley and P. Seidelmann.

A large gathering of the participants was held at the Observatory on Sunday evening. The Colloquium opened at 9.00 on Monday, September 11, with welcome addresses by Prof. D. Bernard, Vice-President of the Louis Pasteur University of Strasbourg; by Dr. D. Egret, Deputy Director of Strasbourg Observatory; and by Prof. C. Jaschek on behalf of the Scientific Organising Committee. The audience comprised over ninety scientists from twenty countries.

The programme was subdivided into six sessions, entitled "Data Treatment", "Fitting of Data", "Small Samples", "Modelling and Inference", "Bias" and "Special Topics".

Thanks are due to the chairmen of the different sessions, which were respectively Drs. Murtagh, Usowicz, Worley, Nousek, Bijaoui and Upgren.

Each session consisted of one or two invited lectures followed by one or two contributed papers. The remainder of the papers listed in the programme were presented as posters, which were on exhibit for two days. All papers, whose authors were present at the Colloquium, are reproduced in these proceedings. The meeting closed with a general discussion chaired by Dr. M. Huber.

The local organisation was handled very efficiently by the Local Organising Committee, composed of D. Egret (chairperson), A. Acker, P. Dubois, M. Hamm, G. Jasniewicz, C. Jaschek and B. Traut.

The work of the Committee was greatly helped by the collaboration of a number of volunteers from Strasbourg Observatory: Ms. Wagner, Bruneau and Langenbacher, and Messrs. Didelon, Hohnadel and Messmer.

Sponsoring and financial support was received from the Commission Nationale des Etudes Spatiales (CNES) and from the Institut National des Sciences de l'Univers (INSU). We express our thanks to both institutes.

Thanks are also due to the Louis Pasteur University for the reception given at the Planetarium, and to the Municipality which organised a reception at the Town Hall.

Finally we express our thanks to the Université des Sciences Humaines which permitted the use of its buildings for this meeting.

As for the text of these proceedings, papers are reproduced such as they were presented by the authors. Discussions were summarized by one of us (F.M.) on the basis of the notes provided by the participants. It is hoped that no-one's views have been seriously misrepresented.

A certain number of papers, the authors of which were prevented from attending the meeting, have been excluded. Most of these will be published in the *Bulletin d'Information du Centre de Données Stellaires*, No. 38.

It is a great pleasure to thank Mrs. M. Hamm for the preparation of the manuscripts for printing.

C. JASCHEK   F. MURTAGH

Welcome Address
by
André Heck
Director, Strasbourg Astronomical Observatory

Ladies and Gentlemen, good morning.

It is an honour and a real pleasure for me to add a few welcoming words to those of the Vice-President of our University.

As you may know, Strasbourg Observatory was the host in 1983 of the first specific colloquium on what we could call *statistical astronomy*. This meeting was entitled *Statistical Methods in Astronomy* and was organized with the support of the European Space Agency[1].

A few years later, in 1987, Fionn Murtagh and I organized a follow-up meeting at the Space Telescope – European Coordinating Facility in Garching. This colloquium, entitled *Astronomy from Large Databases*, had a strong methodological component and the proceedings were published by the European Southern Observatory[2].

Then, Fionn and I thought it was important to go on with the series and we proposed the holding of another meeting in Strasbourg on the present theme that you all know.

I am very pleased to see that this idea has now been substantiated in a promising meeting, thanks to Carlos Jaschek as head of the Scientific Organizing Committee, and to Daniel Egret as head of the Local Organizing Committee.

To conclude, let me wish each of you a successful and fruitful meeting. I hope also that you will take some of your time to enjoy life in this very pleasant city.

---

[1]Rolfe, E.J. (ed.) 1983, *Statistical Methods in Astronomy*, European Space Agency Special Publication ESA SP-201, xviii + 262p.

[2]Murtagh, F. and Heck, A. (eds.) 1988, *Astronomy from Large Databases – Scientific Objectives and Methodological Approaches*, ESO Conference and Workshop Proceedings, No. 28, xiv + 512p.

Opening remarks

C. JASCHEK

Observatoire Astronomique, Centre de Données Stellaires,
Strasbourg, France

An organizer of a meeting can do preposterous things, like leaving an opening for himself right at the start of the meeting. Please be patient. I shall not talk more than 65 minutes or perhaps somewhat less.

As you know, this is the second meeting on Statistics in Astronomy here, the first was held in 1983. Then there was a meeting at Garching end of 1987, attended by that many colleagues that it because clear that statistics was becoming popular. In a conversation with Messrs Heck and Murtagh the idea of a third meeting at Strasbourg came up. Thanks to an excellent scientific organizing Committee the meeting took shape rather quickly so that we have the third meeting in 1989. This acceleration testifies that an increasing number of colleagues is now aware of the usefulness of statistics for astronomy in general. In reality one can say that statistics was born for astronomy -it was that successful that other sciences adapted statistics for their own needs, so that after a bright start in astronomy with Legendre, Laplace and Gauss -to name just a few- statistics developed away from astronomy, and it took one century that statistics came back -this time we hope to stay with astronomy forever.

What was the starting point of statistics ? The treatment of "errors and uncertainties" - i.e. two of the words of the title of this meeting. So let us look a little bit into the reasons why errors occupy such a central position in the minds of astronomers.

We all learned that all measures in physics are beset with uncertainties and that errors are a fact of life. And textbooks also tell us that errors can be divided into accidental and systematic ones, a distinction due to D. Bernouilli. Usually in textbooks one gets one example of a well behaved error which can be split easily into the two components, and that is it !

We are then told that accidental errors can be dealt with by means of a convenient statistical treatment. Textbooks are usually very short on "systematic errors", even the old textbooks. A real step toward clarification came from the work of Eichhorn and Cole (1985) how showed that systematic errors come from incomplete modelisation. Consider the case that you are observing the pressure and the volume of gases. You arrive at the conclusion that all different gases obey an relation of the type p.v. = cte, but you also observe that this constant shifts unpredictably by small amounts i.e. you have a systematic error. Once you learn that one should really include a third parameter, the temperature, the systematic parameter disappears. This is clear and convincing, except that if by chance the dependence upon the new parameter is not very strong, one may certainly claim that the modelisation is OK and that we have simply a larger-than-expected accidental error (which includes a systematic component). The recipe is therefore simple, in principle : vary the experimental conditions as much as you can, to find out if something happens - i.e. if you can split off a systematic error.

Unfortunately as you know astronomers have some particular difficulties of their own on top of the ones any physicist has. This comes from the fact that astronomy, as different from physics, has its objects out of touch. That is, no astronomers can hope to perceive objects outside the solar system by anything else than by electromagnetic radiation or by high energy particles. We are observers, not experimentators and this is one fundamental difference. The second difference is that in astronomy everything is variable in time : positions, proper motions, colors, magnetic fields and so on. Therefore a repetition of measurements makes only sense if one knows before- hand that the time scale of the variations is much longer than the interval over which we observe, and this sometimes is very tricky.

How can we then hope to split off systematic components ? We can do it only in an indirect way, namely by modeling phenomena down to the last detail. Then one must calculate a-priori the accidental errors to be expected and if these differ from the observed ones one may conclude about the existence of systematic errors. The whole difficulty lies in the model - what means "modeling down to the last detail" ? Each epoch has a classical model -for stellar radiation it used to be the black body, now it is a model atmosphere, in non-thermal equilibrium. But when we compare our predictions to observed fluxes, we find clearly systematic differences, which mean that the models are not yet good enough for such an approach. On the other hand when one compares flux measures of very similar stars one finds also differences and these are larger than what one could expect from the errors inherent in a flux measurement, so one suspects systematic errors, but where do they come from ??

Another better way is to look for a completely different procedure to determine the parameter, for instance using another instrument or another technique. This is fine, as long as one measures strictly the same parameter, for instance positions with meridian circles or photographic zenith tubes. It is well known that even such a technique does not permit to eliminate all systematic errors ; HIPPARCOS was the best way to get rid of much of the systematic errors. A case which may come up in this context is that when we change instruments we do not measure exactly the same parameter, but the parameter plus something else. For instance in very high precision radial velocity measurement one measures the stars motion down to 10 ms^{-1}, a fact which makes these measurements prone to very small perturbations on or beyond the stellar surface. So one can demonstrate and correct the systematic errors of the "old" radial velocity technique, but one is left with the problem of the systematic errors of the new technique.

The problem changes if one does not determine an observational parameter like positions or radial velocity, but a physical parameter like the temperature of the star. Here each technique measures slightly different things and complete modelisation becomes even more necessary to compare things.

You may wonder why I insist so much on an apparently trivial point - the fact is however that one discovers over the years that as a rule of thumb accidental errors are twice as large as observers indicated, and systematic errors may be five times larger than indicated. This you might say is an exceedingly pessimistic outlook, but unhappily it is also a realistic one. One should also note that the preocccupation with errors is nothing recent. We find remarks on error treatment in Ptolemy, Al Biruni, Indian authors of the XII century and in Kepler, just to name a few of them. They all dealt in varying shades with experimental design problems, selection of data and stochastic treatment of data (see for instance Sheynin, 1974).

Obviously astronomical complications rise their head again when one speaks of bias. All astronomical observations are biased in the fundamental sense that we are bound to one single observational base to carry out all observations of the Universe : whatever we do, we cannot escape this limitation and look for instance at our galaxy from 10 Kpc distance, or a star (besides the sun) from one AU distance. This a very obvious fact, so obvious that we tend to forget it.

It is like trying to as certain the characteristics of a whole nation from the observation of one small hamlet of some hundred inhabitants. Obviously we can only arrive at some credible results for the whole population if we use the assumption of homogeneity of the whole population, which astronomers enshrine into the "principle of homogeneity". Apparently when called principle it acquires a status of respectability that the word assumption apparently does not have.

But this very fact of our limited view point (both in space and time) raises apparently some psychological resistances in the mind of astronomers and as a consequence of this studies related to this effect creep only slowly into the mind of astronomers.

Take for instance such an obvious effect as the Malmquist bias, discovered for stellar astronomy in the 1920's - it took fifthy years until extragalactic astronomers rediscovered it. A somewhat similar fact appears when one considers galaxies of low surface brightness ; they disappear quickly into the background unless they are very nearby. At large distances we see only big things, but of course that does not imply that the smaller ones do not exist. However this effect is seldomly discussed.

What we get as a consequence is a curious fact - the average density of the Universe (visible matter only) is

$$3,5 \times 10^{-27} \text{ gr cm}^{-3} \quad \text{within } 100 \text{ Kpc}$$
$$1,3 \times 10^{-28} \quad \text{within } 300 \text{ Kpc}$$
$$1,8 \times 10^{-29} \quad \text{within } 800 \text{ Kpc}$$

implying either that we live in a high density region or that our vision is biased.

I think that I did my share of talking and I propose now to pass to the concrete things.

THANKS AND LET US HAVE AN ENJOYABLE MEETING

***

BIBLIOGRAPHY

EICHHORN H. & COLE C.S., 1985. Celestial Mechanics, **37**, 263

JASCHEK C., 1989. Data in Astronomy (Cambridge University Press)

SHENIN O.B., 1974. Arch. Hist. Ex. Sc. Vol. **11**, 97

STIGLER S.M., 1986. *"The history of statistics. The measurement of uncertainty before 1900"*. (Belknap Press of Harvard University Press)

List of Participants

This list includes all those associated with the meeting. The names of co-authors who were unable to attend the conference are shown with small letters.

ACKER A.	Observatoire Astronomique, 11, rue de l'Université, F - 67000 Strasbourg, France
ADORF H.M.	Space Telescope - European Coordinating Facility, Karl Schwarzschild Strasse 2, D-8046, Garching, FRG
Akritas M.G.	Pennsylvania State University, Department of Statistics, 525 Davey Laboratory, University Park, PA 16802, USA
ALFARO E.J.	Instituto de Astrofisica de Andalucia, Apdo 2144, E - Granada 18 080, Spain
ANSARI S.G.	Observatoire Astronomique, Centre de Données Stellaires, 11, rue de l'Université, F - 67000 Strasbourg, France
ARENOU F.	C N R S, Observatoire de Paris, Section de Meudon, 5, Place Janssen, F - 92195 Meudon Cédex, France
Babu G.J.	Pennsylvania State University, Department of Statistics, 525 Davey Laboratory, University Park, PA 16802, USA
Balona L.A.	South African Astronomical Observatory, P.O. Box 9, Observatory, 7935 South Africa
Barone F.	Universita di Napoli, Dipartimento di Scienze Fisiche, Mostra d'Oltremare, Pad. 20, I - 80125 Napoli, Italy
BENEST D.	Observatoire de la Côte d'Azur, Observatoire de Nice, PB 139, F - 06003 Nice Cedex, France
BHAVSAR S.P.	Department of Physics and Astronomy, University of Kentucky, Lexington KY 40506, USA
BIJAOUI A.	Observatoire de la Côte d'Azur, Observatoire de Nice, BP 139, F - 06003 Nice Cedex, France
Bonoli C.	Osservatorio Astronomico, Vicolo Dell'Osservatorio 5, I - 35100 Padova, Italy
Bonatto C.	Universidade Federal de Rio Grande do Sul, Instituto de Fisica, Departamento de Astronomia, Avenida Bento Gonçalves 9500, BR - 90 099 Porto Alegre RS, Brazil
Bougeard M.L.	Université Paris X, IUT, 1, chemin Desvallières, F - 92410 ; C N R S, Observatoire de Paris, 61, avenue de l'Observatoire, F - 75014, France
BRANHAM R.	C.R.I.C.Y.T., C.C. 131, RA - 5500 Mendoza, Argentina
BRIOT D.	Observatoire de Paris, 61, avenue de l'Observatoire, F - 75014 Paris, France

BRONIATOWSKI P.	I R M A, 7, rue René Descartes, F - 67000 Strasbourg, France
BROSCHE P.	Observatorium Hoher List, Universitätssternwarte Bonn, D -5568 Daun, FRG
Bucciarelli B.	Space Science Telescope Institute, 3700 San Martin Drive, Baltimore, MD 21218, USA
CABRERA-CANO J.	Depto de Fisica Atomica, Molecular and Nuclear, Facultad de fisica, Apdo 1065, E - Sevilla 41080, Spain
CHEN BING	Purple Mountain Observatory, Nanjing, China
CHOLONIEWSKI J.	Astronomical Observatory of the Warsaw University, Aleje Uzazdowskie 4, PI - 00 478 Warzawa, Poland
COLE S.C.	U.S. Naval Observatory, 34th Mass. Avenue N.W., Washington 20392 5100, USA
Comeron F.	Departament de Fisica de l'Atmosfera, Astronomia i Astrofisica, Universitat de Barcelona, Diagonal, 647, E - 08028 Barcelona, Spain
CREZE M.	Observatoire de Besançon, 41, avenue de l'Observatoire, F - 2500 Besançon, France
CUYPERS J.	Koninklijke Sterrenwacht van Belgie, Ringlaan 3, B - 1180 Brussel, Belgium
DEMERS Ch.	Astronomisches Institut, Ruhr Universität Bochum, Steinhügel, 105, D - 5810 Witten, FRG
DENIZMAN L.	Observatoire Astronomique, Centre de Données Stellaires, 11, rue de l'Université, F - 67000 Strasbourg, France
DESBAT L.	Groupe d'Astrophysique de Grenoble, CERMO, BP 53 X, F - 3841 Grenoble, France
Dick W.R.	Zentralinstitut für Astrophysik, Rosa-Luxemburg Strasse 17a, Potsdam 1591, DDR
DUBOIS P.	Observatoire Astronomique, Centre de Données Stellaires, 11, rue de l'Université, F - 67000 Strasbourg, France
Ducati J.R.	Instituto de Fisica U F R G S, av. Ben. Gonçalves 9500, BR 91500 Porto Alegre, Brazil
DUCOURANT C.	Observatoire de Floirac, BP 89, F - 33270 Floirac, France
EGRET D.	Observatoire Astronomique, Centre de Données Stellaires, 11, rue de l'Université, F - 67000 Strasbourg, France
EICHHORN H.	Department of Astronomy, 211, Space Sciences Building, University of Florida, Gainesville, Florida 32611, USA
Falco E.E.	Harvard-Smithsonian Center for Astrophysics, 60 Garden Street, Cambridge, MA 02138, USA
FASANO G.	Osservatorio Astronomico, Vicolo Dell'Osservatorio 5, I - 35100 Padova, Italy
FEIGELSON E.	Department of Astronomy, Pennsylvania State University, University Park, PA 16802, USA
FIGUERAS F.	Dpt Fisica de l'Atmosfera i Astrofisica, Avda Diagonal 647, E - 08028 Barcelona, Spain
FITZGERALD M.P.	Department of Physics, University of Waterloo, Waterloo ON N2L 3G1, Canada
Garcia B.	Observatoire Astronomique, Centre de Données, 11, rue de l'Université, F - 67000 Strasbourg, France
Giudicelli M.	Observatoire de la Côte d'Azur, Observatoire de Nice, BP 139, F - 06003 Nice Cédex, France
GLIESE W.	Astronomisches Rechen-Institut, Moenchhofstrasse 12-14, D - 6900 Heidelberg, FRG

Gomez A.E.	C N R S, Observatoire de Paris, Section de Meudon, 5, Place Janssen, F - 92195 Meudon Principal, France
GRAFAREND E.W	Geodätisches Institut, Universität Stuttgart, D - 7000 Stuttgart, FRG
Grenier S.	C N R S, Observatoire de Paris, Section de Meudon, 5, place Janssen, F - 92195 Meudon Principal, France
GUARINOS J.	Observatoire Astronomique, Centre de Données Stellaires, 11, rue de l'Université, F - 67000 Strasbourg, France
HAUCK B.	Institut d'Astronomie de l'Université de Lausanne, CH - 1290 Chavannes des Bois, Switzerland
Harris G.L.H.	Department of Physics, University of Waterloo, Waterloo ON N2L 3G1, Canada
HECK A.	Observatoire Astronomique, 11, rue de l'Université, F - 67000 Strasbourg, France
HEJNA L.	Astronomical Institut of Czechoslovakia, Academy of Sciences, 251 65 Ondrejov, Czechoslovakia
HENDRY M.A.	Department of Physics and Astronomy, Mathematics Building, University of Glasgow, Glasgow G12 8QQ, Scotland, U.K.
HENSBERGER H.	Observatoire Royal de Belgique, Ringlaan 3, B - 1180 Bruxelles, Belgique
HERNANDEZ M.	Dpt Fisica de la Atmosfera, Astronomia y Astrofisica, Diagonal 647, E - 08028 Barcelona, Spain
Hirte S.	Zentralinstitut für Astrophysik, Rosa-Luxemburg Strasse 17a, Potsdam 1591, DDR
HOGEVEN S.J.	Astronomical Institute, University of Amsterdam, Roeterssraat 15, NL - 1018 WB Amsterdam, The Netherlands
HUBER M.C.E.	ESA/ESTEC, Space Science department, P.O. Box 299, NL - 2200 AG Noordwijk, The Netherlands
ISOBE T.	Pennsylvania State University, Department of Astronomy, 525 Davey Laboratory, University Park PA 16802, USA
Jahreiss H.	Astronomisches Rechen-Institut, Moenchhofstrasse 12-14, D - 6900 Heidelberg, FRG
JAKIMIEC M.	Zachodnia 11/8, PL - 53 643 Wroclaw, Pologne
JANKOV S.	Astronomical Observatory, Volgina 7, YU - Beograd, Yougoslavia
JASCHEK C.	Observatoire Astronomique, Centre de Données Stellaires, 11, rue de l'Université, F - 67000 Strasbourg, France
JASCHEK M.	Observatoire Astronomique, Centre de Données Stellaires, 11, rue de l'Université, F - 67000 Strasbourg, France
JASNIEWICZ G.	Observatoire Astronomique, Centre de Données Stellaires, 11, rue de l'Université, F - 67000 Strasbourg, France
JORDI C.	Dpt de Fisica de l'Atmosfera, Astronomia i Astrofisica, Avda Diagonal 647, E - 080028 Barcelona, Spain
JUPP P.E.	Department of Mathematical Sciences, University of St Andrews, North Haugh, St Andrews KY16 9SS, UK
Jylänne E.J.	Department of Physics, University of Waterloo, Waterloo ON N2L 3G1, Canada
KIANG T.	Dublin Institute for Advanced Studies, School of Cosmic Physics, Dunsik Observatory, IRL - Dublin 15, Irland
Kristian J.	Observatories of the Carnegie Institution of Washington, 813 Santa Barbara Street, Pasadena CA 91101, USA
KURTZ M.	Harvard-Smithsonian Center for Astrophysics, 60, Garden Street, Cambridge, MA 02138, USA

Lattanzi M.G.	Space Science Telescope Institute, 3700 San Martin Drive, Baltimore, MD 21218, USA
Lentes F.T.	Observatorium Hoher List, Universitätssternwarte Bonn, D - 5568 Daun, FRG
Llacer J.	Engineering Division, Lawrence Berkeley Laboratory, Berkeley, CA 94720, USA
LOPEZ-GARCIA A.	Observatorio Astronomico, Universidad de Valencia, Avda Blasco Ibanez 13, E - 46010 Valencia, Spain
LOPEZ-MACHI R.	Observatorio Astronomico, Universidad de Valencia, Avda Blasco Ibanez 13, E - 46010 Valencia, Spain
LOPEZ-ORTI J.	Observatorio Astronomico, Universidad de Valencia, Avda Blasco Ibanez 13, E - 46010 Valencia, Spain
MALKOV O. Yu	Observatoire Astronomique, Centre de Données Stellaires, 11, rue de l'Université, F - 67000 Strasbourg, France
MARSHALL H.L.	Space Sciences Laboratory, University of California, Berkeley CA 94720, USA
MERMILLIOD J.C.	Institut d'Astronomie de l'Université de Lausanne, CH - 1290 Chavannes des Bois, Switzerland
Milano L.	Dipartimento di Scienze Fisiche, Pad. 19, Mostra Oltremare, I - 80125 Napoli, Italy
MORENO M.	Dept Fisica de l'Atmosfera, Astronomia i Astrofisica, Diagonal 647, E - 08028 Barcelona, Spain
MORRISON H.	Carnegie Observatories, 813 Santa Barbara Street, Pasadena, CA 91101, USA
MURTAGH F.	ST - ECF, Karl Schwarzschild Strasse 2, D - 8046 Garching, FRG
NOBELIS P.	Institut de Mathematique, 7, rue Descartes, F - 67000 Strasbourg, France
NOUSEK J.	Department of Astronomy, Pennsylvania State University, University Park, PA 16802, USA
NUNEZ J.	Departamento de Fisica de la Atmosfera, Astronomia y Astrofisica, Universidad de Barcelona, Diagonal 647, E - 08028 Barcelona, Spain
PELT J.	Institute of Astrophysics and Atmospheric Physics, Tartu Toravere 202441, Estonian, SSR
PFLEIDERER J.	Institut für Astronomie der Universität Innsbrück, Technikerstrasse 15, A - 6020 Innsbrück, Austria
Polvorinos A.	Universidad de Sevilla, E - Granada, Spain
RAPAPORT M.	Observatoire de Bordeaux, BP 89, F - 33270 Floirac, France
RUSSO G.	Universita di Napoli, Dipartimento di Scienze Fisiche, Mostra d'Oltremare, Pad. 20, I - 80125 Napoli, Italy
Santoro E.J.	US Naval Observatory, 34 & Massachusetts Avenue NW, Washington, DC 20390, USA
SCARGLE J.D.	Mail Stop 245 - 3, NASA - AMES Research Center, Moffet Field, California 94035, USA
SCHILBACH E.	Zentralinstitut für Astrophysik, Rosa-Luxemburg Strasse 17a, Potsdam 1591, DDR
Schmidt R.E.	U.S. Naval Observatory, 34 & Massachusetts Avenue NW, Washington DC 20390, USA
SCHOENMAKER A.A.	Kapteyn Observatory, Mensingheweg 20, NL KA Roden, The Netherlands
Scholz R.D.	Zentralinstitut für Astrophysik, Rosa-Luxemburg Strasse 17a, Potsdam 1591, DDR

SEIDELMANN P.K.	US Naval Observatory, 34 & Massachusetts Ave. NW, Washington DC 20390, USA
SIMMONS J.F.L.	Department of Physics and Astronomy, Glasgow University, Glasgow G12 8QQ, UK
SOUBIRAN C.	M.A.M.A, Observatoire de Paris, 61, Avenue de l'Observatoire, F - 75014 Paris, France
Stasinska G.	D A E C, Observatoire de Paris, Section de Meudon, 5, place Janssen, F - 92195 Meudon - Principal, France
STAVREV K.	Department of Astronomy, Bulgarian Academy of Sciences, Boul. Lenin 72, BG - 1784 Sofia, Bulgaria
TAFF L.G.	Space Science Telescope Institute, 3700 San Martin Drive, Baltimore, MD 21218, USA
TEJERO F.C.	Departamento de Fisica de la Atmosfera, Astronomia y Astrofisica, Diagonal 647, E - 08028 Barcelona, Spain
TORRA J.	Departamento de Fisica de l'Atmosfera, Astronomia y Astrofisica, Diagonal 647, E - 08028 Barcelona, Spain
UPGREN A.R.	Van Vleck Observatory, Wesleyan University, Middletown, CT 06457, USA
USOWICZ J.	Torun Radio Astronomy Observatory, UL. Chopina 22/18, PL - 87100, Poland
Verschueren W.	Theoretical Mechanics and Astrophysics, University of Antwerp (RUCA), Belgium
Weis E.W.	Van Vleck Observatory, Wesleyan University, Middletown, CT 06457, USA
Welsh A.H.	Department of Statistics, The Faculties, Australian National University, Australia
WORLEY C.E.	U.S. Naval Observatory, 34 & Massachusetts Ave. NW, Washington D.C. 20392, USA

DATA TREATMENT

Astronomical Time Series Analysis : Modeling of Chaotic and Random Processes

SCARGLE J.D.

Theoretical Studies Branch, Space Science Division, Mail Stop 245-3
NASA–Ames Research Center, Moffet Field, California, USA 94035

Astronomical data are frequently in the form of time series, and their analysis is carried out with time series analysis techniques. This is natural if one is explicitly studying the time dependence of physical quantities, but is also applicable when the observations are repeated over time just to reduce the errors, uncertainties and/or bias in the measurement of a quantity known or assumed to be constant.

Observational time series always have a stochastic element -- to some degree the variations are disordered and therefore unpredictable. First, there are always the random observational errors. Second, in addition to such noise there may be intrinsic randomness, in the form of unpredictable variations in the source above and beyond the observational errors. Such variability is rarely completely random but rather is correlated to some extent; indeed the correlation structure may be of great interest. The apparently stochastic fluctuations in the brightness of certain stars and the quasi-stellar objects are examples of such astrophysically interesting variability.

Recent developments in nonlinear dynamics have led to the realization that there is a third source of apparent disorder, namely **chaos**. Even very simple physical systems can produce time series which have a general appearance identical to those produced by random processes and yet are completely deterministic (i.e. the future evolution of the system can be exactly calculated from its initial values).

This new paradigm will have a significant impact on the way astronomers analyze and interpret time series data. In some cases what appear to be random observational errors may in fact be deterministic chaos. While to my knowledge no examples of this have been identified, chaos occurs in almost all nonlinear systems for at least some range of physical parameters. Hence chaotic processes are certainly operating in detectors and other aspects of astronomical data acquisition. If such processes are identified and modeled, their effects can be removed from the data to yield improved determination of the relevant physical quantities. In other cases the intrinsic variations in the astronomical source may be chaotic (as opposed to random in the conventional sense) and this discovery can lead to improved understanding of the underlying physical processes.

This paper describes techniques which can be used to study both errors and intrinsic variations in time series data. In either case the primary objective is detection of chaotic or conventional random processes, and the subsequent separation and/or modeling of these processes if they are found to be present.

Since it is relatively new I introduce chaos with four examples demonstrating its essential features. I then outline a few mathematical results to set the stage for the specific data analysis techniques which are the major subject of this paper.

THE NATURE OF CHAOS

Many astronomical studies, notably in celestial mechanics, can be phrased as initial value problems. One specifies the dynamical laws and initial conditions and then solves for the system's evolution forward from the initial state. The laws are typically differential equations which have a unique solution passing through the initial state. Within this framework all dynamical evolution is necessarily deterministic -- if the initial conditions are reproduced the system behavior is identically repeated.

Classically, while the time dependence of these solutions may be difficult to calculate exactly, the general behavior can almost always be described in the following simple terms.

- There are transients which . . .
    ◦ depend on the initial conditions, and
    ◦ decay away as time progresses.
- The solution approaches a state which may be:
    ◦ a steady state,
    ◦ periodic, or
    ◦ quasi-periodic.

While the possibility of another, completely different evolution (namely disordered, aperiodic behavior) has been known since the time of Poincare, the revolutionary discovery of recent work is that such apparently random behavior is nearly ubiquitous and appears in even very simple nonlinear systems. Stochastic behavior is not, as was previously assumed, limited to complex systems with many degrees of freedom. The reader should consult works such as [1]-[14] for a systematic treatment of chaos; [15] and [16] discuss interesting connections between fractals, time series, and random walks. Here we demonstrate the main features of chaos using examples.

Example 1: Quasiperiodic Functions

The first is **not** really chaos but nevertheless exemplifies a principle that lies at its heart: a simple deterministic process can generate time series that look disordered. A *quasiperiodic function* depends on several arguments and is periodic in each of them. Figure 1(a) depicts such a function, the sum of two sinusoids with incommensurate periods. This process appears rather disordered but its hidden simplicity is revealed by the power spectrum, Figure 1(c), consisting of two delta functions. Figure 1(b) will be explained later.

Example 2: The Logistic Equation

The next example deserves considerable study, as it is archetypal of chaos. It is a seemingly trivial nonlinear system that behaves in an extraordinarily complex way. Most of the features of this example apply to chaotic processes in general.

Assume that the evolution of observable **X**, measured at discrete times **n=0,1,2, ...** , is given by the following simple nonlinear rule:

$$X_{n+1} = \lambda \, X_n(1 - X_n) \qquad 0 < \lambda < 4. \qquad (1)$$

This relation specifies each new value as a function of the previous one -- and therefore determines the complete history, for each choice of the initial value X_0.

The behavior of this system for typical values of X_0 is as follows: For small values of λ, **X** quickly approaches a constant. As λ is increased there is suddenly a value for which the asymptotic behavior is a periodic oscillation between two values. As λ increases further the period suddenly becomes twice as long and **X** oscillates between 4 values. These events as λ increases are known as *period doublings*, or *bifurcations*, and the character of the evolution changes qualitatively at these points. The period doubling continues, producing oscillation between 8 values, then 16, *ad infinitum.*

Eventually the behavior becomes chaotic, followed by a complicated series of returns to periodic behavior. Finally, for the λ =4, the largest value for which **X** does not diverge to infinity, we obtain the uncorrelated ("white") chaotic time series shown in Figure 2(a). Actually two time series are shown with the same λ and only a tiny difference in the initial value. The rapid divergence of the two time series (after about 40 steps) is an example of *sensitivity to initial conditions*; this phenomenon is fundamental to chaos and will be discussed further.

Because simple equations usually generate simple behavior, this complex time dependence is surprising. Indeed, we have only hinted at the full complexity of this system. Of particular beauty are the often reproduced **bifurcation diagrams** which show the complex relationship between the values of λ and the resulting values of **X**. The reader is urged to conduct numerical experiments with and consult the large literature (e.g. [1]-[3]) on eq. (1), the *logistic equation.*

We will study this example using an important tool for analyzing chaotic data, namely the *return map* -- a plot of X_{n+1} vs. X_n. If the underlying process is chaotic, rather than random, this plot may reveal that the apparently complex process is actually simple. The return map for the data of Figure 2(a) is extraordinarily simple, since by equation (1) all the points must fall on a parabola [Figure 2(b)]. Part of the reason for a flood of interest in this subject is the popularity of the dream that one's messy, apparently random data will collapse to a dramatically simple relationship as in this idealistic (but theoretically important) example.

There are practical problems even if a simple chaotic process does underlie the observed variability. Measurement errors typically spread out the points in the return map and can thereby obscure its structure. Further in more complicated problems the return map may have to be examined in a higher dimensional space [e.g. (X_n, X_{n+1}, X_{n+2})].

The power spectrum in Figure 2(c) is characteristic of chaos, namely "broad band" -- i.e., it has power at a range of frequencies. Sometimes it is said that the spectrum is continuous, a correct description of the power averaged over an ensemble of initial values. The spectrum of any realization of the logistic equation is actually discrete and consists of a forest of spectral lines. Only in the average is the power a smooth and continuous function of frequency -- here a flat ("white") spectrum.

The probability distribution for the initial values used in this computation is the so-called *invariant measure*. The assumption that the process is stationary implies that all of the X_n, including the initial value X_0, have the same distribution function. This distribution is called the invariant measure because it does not change under the transformation $n \longrightarrow n+1$. Numerical determinations are easy, and simply involve collecting a histogram of the values generated by the map (1). Exact invariant measures exist for only a few special cases.

What is the origin of the randomness of processes like the logistic process? This is a tricky question. The following three sources of randomness act together:

• The initial values are randomly distributed, as discussed above. But this is obviously not responsible for the disorder in a single realization -- which corresponds to a single initial value.

• Each iteration of equation (1) produces a stretching and subsequent folding of the **X**-axis, as in the kneading of taffy. This mixing action produces disorder and also amplifies errors.

• The initial value itself can lead to randomness, as we will see in the next sample chaotic process.

Example 3: The Bernoulli Shift

The following evolution formula can be derived from (1) by a simple change of variable [5]:

$$Y_{n+1} = (2 \ Y_n) \ \text{mod} \ 1 \ . \qquad (2)$$

This is again a map of the interval (0,1) onto itself, and consists of a twofold stretch followed by a fold in half. This transformation is called the *Bernoulli shift* because it discards the most significant binary digit of the previous iterate and shifts the remaining digits one place to the left. Rational initial values thus lead to periodic solutions. Those represented by a truncated digit string lead to convergence to zero after a finite number of iterations. The more dense and therefore more characteristic irrationals lead to chaos, because of the randomness of the digits. There is much discussion of this deceptively simple concept in the context of computation and complexity theory (see, e.g. [4]-[6] and [39]).

Thus eq. (2) does not generate randomness but simply reveals that which is contained in the initial value. Such processing of external randomness is called *homoplectic* behavior in [6] .

Example 4: An Automaton Process

There are dynamical systems in which the evolution from even trivially simple initial conditions is disordered, for example the cellular automata discussed by Wolfram [6]. Since the initial conditions are too simple to contain the information necessary to specify the resulting random process, he argues that the randomness is **created** by the rule which determines the evolution. Wolfram calls such behavior *autoplectic*.

The following example is a minor modification of the one discussed in [6]. Introduce a new independent variable **m** (meant to represent space) and denote the dependent variable $Y_{n,m}$. Let **Y** take on only the two values 0 or 1, and postulate the dynamical evolution rule

$$Y_{n+1,m} = [\ Y_{n,m-1} + D(Y_{n,m}, Y_{n,m+1})\ ]\ \mathrm{mod}\ 2\ , \qquad (3)$$

·where $D=0$ if both of its arguments are 1 and $D=1$ otherwise.

The evolution from the initial state $Y_{0,0}=1$ and $Y_{0,m}=0$ for m not 0 can be depicted by a diagram in which $Y=1$ is a black square and $Y=0$ is white (Figure 3). Consider the values along the center column, i.e. the $Y_{n,0}$. Figure 4(a) is a plot of the running sum

$$X_n = \sum_{k=0}^{n} (\ Y_{k,0} - .5\ ), \qquad (4)$$

which looks quite random but is a deterministic process since the rule which generates it involves no randomness. Moreover, this process starts from a trivially simple initial state; hence the randomness is not external but is generated purely by the recurrence equation.

The return map for the time series data is shown in Figure 4(b). The peculiar structure of the map is due to the fact that X can increase or decrease by only one quantized increment at a time. Note the gap in the structure near the top of the panel, caused by the vagaries of the evolution. The power spectrum, shown in Figure 4(c), is again broad band. Note, however, that we here cannot appeal to an averaging procedure to show that the spectrum is continuous, because the initial condition for this process is not random.

The above examples are all in discrete time, appropriate for problems with inherently discrete dynamics. But there are also situations in which discrete maps can be derived from continuous equations such as differential or delay equations [8]-[13].

The existence of deterministic systems which look disordered has several implications for data analysis:

- time series data which seem random may be chaotic
- chaos and randomness can be present in the same data
- new models and analysis tools are needed to:
  - detect the presence of chaos and randomness
  - study the structure of chaotic and random dynamics
  - separate chaos from randomness if both are present

We now consolidate the lessons of the examples into a rough, picture of what a chaotic process is. This is not meant to be a mathematical definition of chaos. There is a good deal of confusion on the relationship between deterministic chaos and randomness, partly because there is no agreed on definition of chaos. But chaotic processes are considered by most to be **the special case of stationary random processes in which the initial value is a random variable in the usual sense but future values are deterministic functions of the initial value**. Thus the elements of a chaotic process are highly dependent random variables -- the opposite extreme to the variables of a completely random process, which are independent of each other. This dependence is explicitly expressed in a recurrence equation, such as equations (1)-(3) above. This "map" has a stretch-and-fold character which gives rise to great sensitivity of the future states on the initial values.

A few words about predictability are in order. Chaotic and random time series, in extreme cases, are so mixed up that consecutive values seem unrelated to each other. Such disorder would normally prevent prediction of future values, say based on observations of past values. Nevertheless, the underlying order in a chaotic process can be used to make predictions [20],[21]. Unfortunately, extreme sensitivity to initial conditions places severe limits on how far into the future such predictions are good.

In the next three sections we will see that a discernible distinction between randomness and chaos occurs in the system's state space, rather than in the appearance of the time series.

THEORETICAL BASIS OF TIME SERIES ANALYSIS

In nonlinear dynamics a physical system is almost always described in its state space -- the coordinates of which are a complete set of physical variables (e.g. position and momentum of all constituent particles).   A state of the system is represented by a point in this multidimensional space, often called *phase space*.   Evolution of the system in time is represented by a trajectory in this space beginning at the point corresponding to the initial state. This geometrical view has been useful in dynamics for some time [7].

Unfortunately a real system's complete state can rarely be observed, because it is impractical to measure all of the relevant variables.   For example, an astronomer studying variable star dynamics can measure only brightness and spectral changes, not the position and momentum of the fluid internal to the star.

This problem is solved by an extraordinarily mathematical result [22]-[24]: analysis of a time series in just one variable can reveal the structure of the trajectories in the full multivariate state space.   This variable need not be a true dynamical variable. Further, one does not need prior knowledge of the dimension of the state space.

For there exists another multidimensional space, called the *embedding space*, the coordinates of which are easily determined from the data on the single observed variable.   The trajectories in this fake state space are topologically the same as the trajectories in the real state space (see Fig. 5).   Mathematically, there is a smooth mapping from the full state space to the simpler (and observable) embedding space that preserves the topology of the trajectories. Reference [24], a recent overview of this theory, contains a simplified discussion of the difficult mathematics in [23].

The coordinates of the embedding space are the observed variable evaluated at a set of **N** "lagged" times:

$$X = (X_n, X_{n+k}, X_{n+2k}, \ldots, X_{n+(N-1)k}). \qquad (5)$$

It is not necessary that the lags be multiples of a constant as shown here, and this equation could be generalized to allow arbitrary lags. Although in principle (i.e. with an infinite amount of noise-free data) the value of the lag does not matter, in practice one must determine a good value for the lag as well as the dimension **N** of this space. How to do this is the subject of a large and growing literature (e.g. [27]-[35]).

Another useful result is that stationary processes have a general and simple representation consisting of separate random, chaotic, and deterministic parts. The first two parts have the same simple form, the *moving average* -- a practical and well studied time-domain model that can be relatively easily fit to time series data.

The *Wold Decomposition Theorem* [25] states that any stationary process can be decomposed into the sum of a purely random and a purely deterministic process, and further the random part is in the form of a moving average -- the convolution of a fixed pulse shape with a white noise process. Because chaotic processes are a special case of random processes, the Wold representation applies to them too. Some simple analysis shows that [17],[18] a stationary process **Y** can be uniquely represented in the form

$$Y = R*C + X*B + V , \qquad (6)$$

where **C** and **B** are constant filters (usually interpretable physically as pulse shapes), **R** is a white noise process, **X** is a white chaotic process. Process **V** is rigidly deterministic, the same in every realization even if the initial conditions are different. In addition it can be shown [18] that **X** almost everywhere satisfies a recurrence equation (or return map) of the form $X_{n+1} = f(X_n, X_{n-1}, \ldots)$.

A DECONVOLUTION PROCEDURE

The basic problem of estimating the parameters of these moving averages is similar to the usual deconvolution problem for random processes [26]. Assume that we have a purely chaotic process, so the random and rigidly deterministic parts can be ignored, with the model:

$$Y = X * B . \tag{7}$$

From time series data on **Y** it is desired to estimate the filter **B** and the uncorrelated chaotic process **X**; the recurrence function **f** can be estimated by constructing a return map for **X**.

This problem (and the solution I propose) is similar to standard deconvolution. The preliminary goal is to find the inverse of **B**, as convolving it with Eq. (7) would recover **X**. In the analogous situation for random processes, we seek **A** which makes **A*Y** maximally random; here we seek maximal chaos. This is done by maximizing some measure of the degree of chaos with respect to the parameters of **A**, which, as in [18], is represented as a two-sided autoregressive filter. The only significant difference between this technique and that routinely used for deconvolution of random processes is the penalty function in the model fitting.

A large variety of measures of the degree of chaos of a time series have been discussed, for example ([27] and references indicated): Kolmogorov entropy [28]-[29], metric entropy [30], topological entropy [30], mutual information [31] and other information measures [32], Lyapunov characteristic exponents [33], fractal dimension [34], Hausdorff dimension, and the natural measure. In this context, small dimension means large degree of chaos, as can be seen from the examples above. Thus we minimize a dimension estimate to maximize the degree of chaos.

Details of this procedure and tests on simulated data are given in [18]. Here I will briefly summarize the results of such simulations. The parameters of low-order autoregressive models were recovered quite accurately, although noise added to the data to simulate observational errors caused moderate degradation of the accuracy.

The method was found to be surprisingly successful at deconvolution of *random* processes, still using the penalty function tailored to chaos. Since conventional techniques are incapable of determining the phase character of random processes, special methods have been developed [36],[37]. Modeling with the penalty functions defined here readily distinguishes a pulse from its time reverse, and recovers the correct values of the parameters of noncausal, nonsymmetric moving averages -- normally considered very a difficult problem.

Chaos can be distinguished from randomness by inspection of the state-space plots: the former produces one dimensional distributions of points and the latter space-filling two dimensional ones. In higher dimensional embedding spaces, similar results hold. The method can not only distinguish randomness from chaos but can separate them in some simple cases in which both are mixed together (see Figure 6). The technique should be used not only to model chaotic processes, but in modeling any stochastic processes. It is superior to, and should be used instead of, the L_1 techniques proposed in [36] even for the kinds of problems for which that technique was developed.

Some practical problems are still being investigated. For example, an important roadblock in some applications is the inappropriateness of the assumption that the data sampling interval is equal to the recurrence interval in the underlying physical process. The latter is rarely known when data are being taken; even if it is, it may not be possible to sample at the corresponding interval.

SUMMARY

Chaotic time series appear random but are generated by simple deterministic rules. Power spectrum analysis is not a diagnostic for chaos, but the return map is. Remarkably this new tool is effective for random process too. The dynamics of a physical system in its full state space can be studied by analyzing a single scalar time series, and stationary random and chaotic processes can both be generally represented as filtered uncorrelated processes. These results lead to a practical deconvolution technique for modeling chaos and randomness, detecting their presence in data, and distinguishing and separating them when both are present.

ACKNOWLEDGEMENTS: I am grateful to Phyllis Scargle for stimulating discussions, to David Donoho and Gary Villere for various suggestions, and to Enders Robinson for suggesting that I write [38], which led to some of the work reported here.

REFERENCES

[1] H.G. Schuster, *Deterministic Chaos*, VCH Publishers, New York, 1988.

[2] R. M. May, Simple Mathematical Models with Very Complicated Dynamics, *Nature*, **261**, 459-467 (1976).

[3] C. Grebogi, E. Ott, and J. Yorke, Chaotic Attractors in Crisis, *Physical Review Letters*, **48**,1507-1510 (1982).

[4] J. Ford, How Random is a Coin Toss?, *Physics Today*, **36**, 40 (1983).

[5] J. Ford, Chaos: Solving the Unsolvable, Predicting the Unpredictable!, in *Chaotic Dynamics and Fractals*, eds. M. F. Barnsley and S. G. Demko, Academic Press, New York, 1986.

[6] S. Wolfram, Origins of Randomness in Physical Systems, *Physical Review Letters*, **55**, 449-452 (1985).

[7] R. H. Abraham and C. D. Shaw, *Dynamics--The Geometry of Behavior, Part 2: Chaotic Behavior*, Aerial Press, Santa Cruz, 1983.

[8] A. J. Lichtenberg and M. A. Lieberman, *Regular and Stochastic Motion*, Springer-Verlag, New York, 1983.

[9] H. Bai-Lin, *Chaos* , World Scientific Pub. Co., Singapore, 1984.

[10] P. Berge, Y. Pomeau, and C. Vidal, *Order within Chaos: Towards a deterministic approach to turbulence*, John Wiley & Sons, New York, 1984.

[11] J. M. T. Thompson and H. B. Stewart, *Nonlinear Dynamics and Chaos*, John Wiley and Sons, New York, 1986.

[12] P. W. Milonni, M.-L. Shih, and J. R. Ackerhalt, *Chaos in Laser-Matter Interactions* , World Scientific Pub. Co., Singapore, 1987.

[13] F. C. Moon, *Chaotic Vibrations: An Introduction for Applied Scientists and Engineers* , John Wiley & Sons, New York, 1987.

[14] H. Bai-Lin, *Directions in Chaos, Vol. 2*, World Scientific Pub. Co., Singapore, 1988.

[15] B. B. Mandelbrot, *The Fractal Geometry of Nature*, W.H. Freeman and Company, New York, 1983.

[16] J. Feder, *Fractals*, Plenum Press, New York, 1988.

[17] J. Scargle, A Chaos Representation Theorem, preprint.

[18] J. Scargle, Studies in Astronomical Time Series Analysis. IV: Modeling Chaotic and Random Processes, preprint, submitted to *Astrophysical Journal*.

[19] R. H. Abraham, Is there Chaos without Noise?, in *Chaos, Fractals, and Dynamics* , eds. P. Fischer and W. Smith, Marcel Dekker, Inc., New York, 1985.

[20] D. Farmer and J. Sidorowich, Predicting Chaotic Time Series, *Physical Review Letters*, **59**, 845 (1987).

[21] D. Farmer and J. Sidorowich, Exploiting Chaos to Predict the Future and Reduce Noise, preprint, 1988.

[22] N. Packard, J. Crutchfield, J. Farmer, and R. Shaw, Geometry from a Time Series, *Physical Review Letters*, **45**, 712 (1980).

[23] F. Takens, in *Dynamical Systems and Turbulence*, Vol. 898 of *Lecture Notes in Mathematics*, eds. D. A. Rand and L. S. Young, Springer, Berlin, 366, 1981.

[24] N. Gershenfeld, An Experimentalist's Introduction to the Observation of Dynamical Systems, in [14].

[25] H. Wold, *A Study in the Analysis of Stationary Time Series* , Almqvst and Wiksell, Uppsala, 1938.

[26] E. Robinson, Recursive Decomposition of Stochastic Processes, in *Econometric Model Building*, ed. H. Wold, North-Holland, Amsterdam, 111-168, 1967.

[27] D. Farmer, E. Ott, and J. Yorke, The Dimension of Chaotic Attractors, *Physica* **7D**, 153-180 (1983).

[28] P. Grassberger and I. Procaccia, Characterization of Strange Attractors, *Physical Review Letters*, **50**, 346-349 (1983).

[29] P. Grassberger and I. Procaccia, Estimation of the Kolmogorov Entropy from a Chaotic Signal, *Physical Review A*, **28**, 2591-2593 (1983).

[30] J. P. Crutchfield and N. H. Packard, Symbolic Dynamics of Noisy chaos, *Physica* **7D**, 201-223 (1983).

[31] A. M. Fraser and H. L. Swinney, Independent Coordinates for Strange Attractors from Mutual Information, *Physical Review A*, **33**, 1134-1140 (1986).

[32] J.-P. Eckmann and D. Ruelle, Ergodic Theory of Chaos and Strange Attractors, *Reviews of Modern Physics*, **57**, Part I, 617-655 (1985).

[33] J.-P. Eckmann, S. O. Kamphorst, D. Ruelle, and S. Ciliberto, Liapunov Exponents from Time Series, *Physical Review A*, **34**, 4971-4979 (1986).

[34] H. Froehling, J. P. Crutchfield, D. Farmer, N. H. Packard and R. Shaw, On Determining the Dimension of Chaotic Flows, *Physica* **3D**, 605-617 (1981).

[35] A. N. Kolmogorov, A New Invariant for Transitive Dynamical Systems, *Dokl. Akad. Nauk SSSR*, **119**, 861-864 (1958).

[36] J. Scargle, Studies in Astronomical Time Series Analysis. I. Modeling Random Processes in the Time Domain, *Astrophys. J.Supp.*, **45**, 1-71 (1981).

[37] D. Donoho, On Minimum Entropy Deconvolution, in *Applied Time Series Analysis II*, ed. D. Findley, Academic Press, New York, 1981.

[38] J. Scargle, An Introduction to Chaotic and Random Time Series Analysis, to appear in the *International Journal of Imaging Systems and Technology*.

[39] J. Crutchfield and K. Young, Inferring Statistical Complexity, *Phys. Rev. Lett.*, **63**, 105-108 (1989).

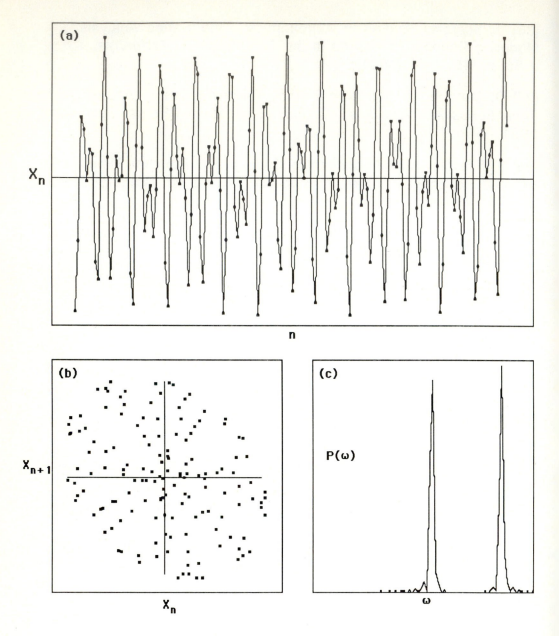

Figure 1

Figure 1: The quasiperiodic process $X_n = \sin(n) + \sin(5n/\pi)$.
    (a) the time series;
    (b) the return map;
    (c) the power spectrum (arbitrary scales).

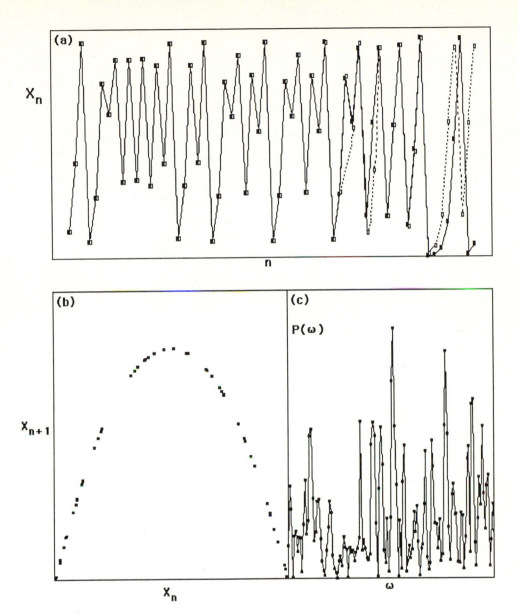

Figure 2

Figure 2: The logistic process for I = 4.
      (a) Time series for two very close initial values:
          filled boxes: X_0 = .123
          open boxes: X_0 = .123+10^{-14} ;
      (b) the return map;
      (c) the power spectrum (arbitrary scales).

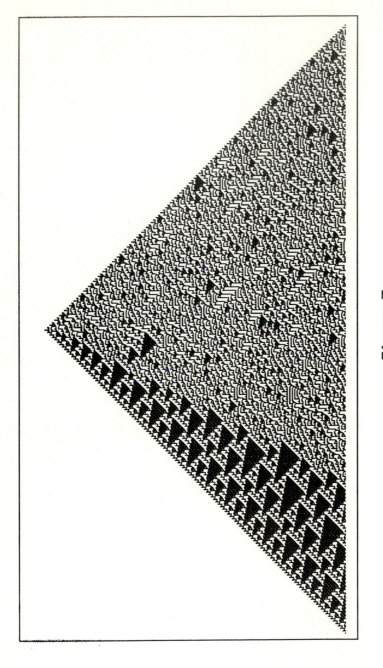

Figure 3

Figure 3: Graph for cellular automaton as described in the text.

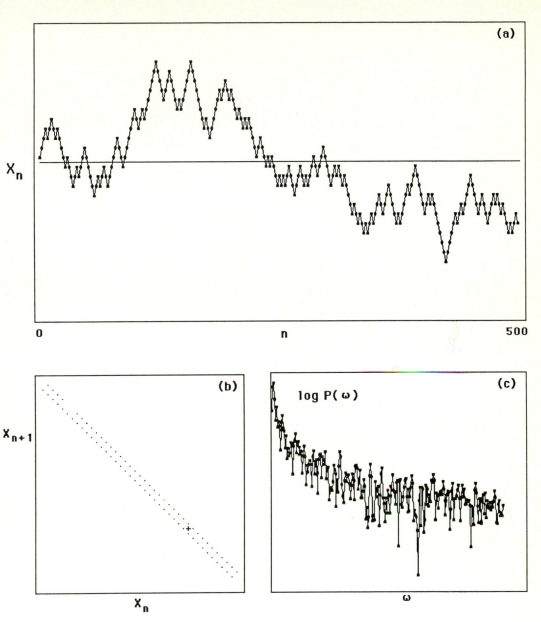

X_n

0   n   500

X_{n+1}

X_n

log P(ω)

ω

Figure 4

Figure 4: Deterministic process (derived from the cellular automaton of Fig. 3) which looks random due to autoplectic generation of disorder.
       (a) the time series;
       (b) the return map;
       (c) the power spectrum (arbitrary scales).

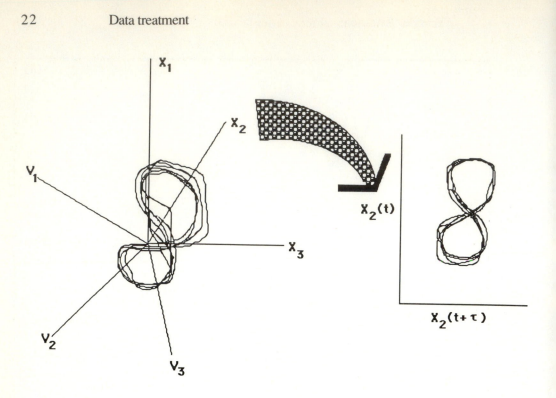

Figure 5

Figure 5: The full dynamical state space (left) and trajectories which it is not practical to observe, embedded in the space (right) where the coordinates are the directly observable time series of one of the dynamical variables evaluated at times which differ by a constant lag.

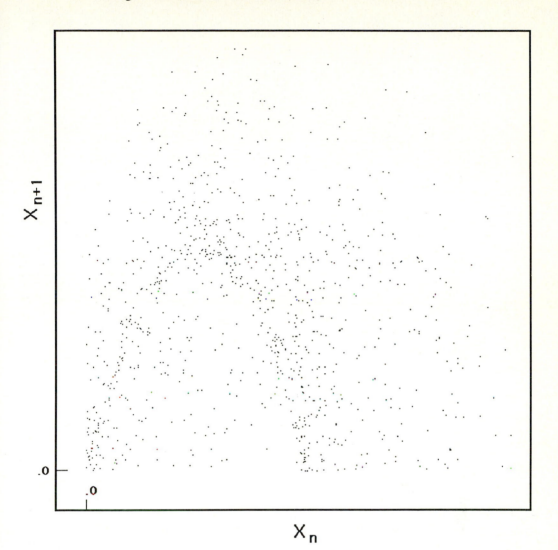

Figure 6

Figure 6: Underlying process determined by deconvolving a mixture of chaos and randomness. The randomness produces the general scatter in this plot, and the chaos produces the parabolic shape characteristic of the logistic map that was used to generate the chaotic part of the data. The parameters of the model were recovered very accurately.

Linear Regression : Which method should be used ?

T. ISOBE – E.D. FEIGELSON
Department of Astronomy, Pennsylvania State University, Pennsylvania, U.S.A.

M.G. AKRITAS – G.J. BABU
Department of Statistics, Pennsylvania State University, Pennsylvania, U.S.A.

SUMMARY

Five methods for obtaining linear regression fits to bivariate data with unknown measurement errors are discussed: ordinary least squares (OLS) regression of y on x and x on y, the bisector of the two OLS lines, orthogonal regression and reduced major axis regression. The applicability of the procedures is discussed with respect to their mathematical properties, the nature of the data , and the scientific purpose of the regression. The standard OLS regression is appropriate for exploratory and predictive purposes, but we favor orthogonal regression under many circumstances such as comparing data with theory.

INTRODUCTION

In astronomy, it is often of interest to describe the relationship between two variables by fitting a straight line on collected data of the form (x_i, y_i), i = 1, 2, ..., n. By far, the most common way of doing so is the method of ordinary least squares (OLS) linear regression where x is called the independent variable and y the dependent variable. OLS, however, does not treat the two variables symmetrically, since OLS assumes that the independent variable has no errors at all and the dependent variable has random error with a constant variance.

In many astronomical problems, these assumptions may not hold: both variables may have errors due to measurement and/or intrinsic scatter, and the labeling of one variable as dependent and the other as independent may be arbitrary. Some aspects of this problem were realized and discussed in the astronomical literature in relation to measurement errors (e.g. Seares 1944; Deeming 1968). Astronomers, however, do not always know how the errors in each variable are divided between instrumental and intrinsic effects.

If the independent variable is non-stochastic (no intrinsic scatter), one may use OLS for prediction as well as in establishing the functional relation between the two variables. However, astronomers are generally interested in obtaining functional relationships between the two variables when the independent variable is stochastic. In such cases OLS may not be an appropriate method to use in the analysis; instead one should use a method which treats the variables symmetrically.

SUMMARY OF MATHEMATICS

There are three methods available which treat the variables symmetrically: OLS bisector (Pierce and Tully 1988), orthogonal regression (OR; Pearson 1901), and reduced major axis regression (Strömberg 1940; Kermack and Haldane 1950). The OLS bisector regression is the bisector of the two OLS (x/y and y/x) lines. The OR minimizes the square deviations perpendicular to the line. The reduced major axis regression minimizes the product of the x and y deviations.

Theoretical values of the slopes estimated by these methods can be written as functions of the population standard deviation of x (σ_x), that of y (σ_y), and the correlation coefficient (ρ). Under the assumption of $\rho \neq 0$

OLS(y/x) $\beta_1 = \rho\sigma_y / \sigma_x,$

OLS(x/y) $\beta_2 = \sigma_y / \rho\sigma_x,$

OLS bisector $\beta_3 = \dfrac{\rho}{1+\rho^2} \left\{ \dfrac{\sigma_y^2 - \sigma_x^2}{\sigma_x\sigma_y} + [(\dfrac{\sigma_x}{\sigma_y})^2 + (\rho^2 + \rho^{-2}) + (\dfrac{\sigma_y}{\sigma_x})^2]^{1/2} \right\},$

Orthogonal $\beta_4 = \dfrac{1}{2\rho\sigma_x\sigma_y} \left\{ \sigma_y^2 - \sigma_x^2 + [(\sigma_y^2 - \sigma_x^2)^2 + 4\rho^2\sigma_x^2\sigma_y^2]^{1/2} \right\},$

Reduced Major Axis $\beta_5 = \dfrac{\sigma_y}{\sigma_x}.$

We also provide analytical expressions for the uncertainties of the slopes and intercepts for each method; in some cases these formulas were not previously published.

The representation clearly reveals problems with some of these estimates. First, when ρ is small (i.e. the data exhibit large scatter), a small variation in the estimation of correlation greatly affects the estimators of OLS(x/y) and OLS bisector.

Second, since the slope of the reduced axis does not depend on the correlation coefficient and entirely dependent on the marginal variances of the variables, it cannot help us in understanding the relation between x and y. The same value of the slope is obtained whether x and y are independent or they are strongly dependent on each other.

Third, if $\sigma_x = \sigma_y$, then the OLS bisector, OR, and reduced major axis regressions always gives slope of unity independent of the true relation between x and y.

These limitations are due to lack of knowledge of relative contributions of intrinsic and measurement errors to the total variances. Without more information concerning the nature of the errors, the true population slope cannot be reliably estimated.

GUIDELINES FOR ASTRONOMERS

We recommend that astronomers examine all of these fits for their linear regression problems. If the slopes differ by a wide margin with regard to the slope errors provided here, caution is advised in arriving at scientific conclusions. In such cases, the following guidelines may be used for choosing the preferred method:

(i) If the goal is to predict the y value of a new object from the relation found in a pre-existing sample, then OLS(y/x) should be used.

(ii) If the goal is to estimate the underlying functional relation between the variables, as needed for comparing data with an astrophysical theory, then one of the regression lines between the OLS extremes should be used. We recommend *orthogonal regression provided the variables are scale-free* (e.g. logarithmic variables or ratios of scaled variables). We prefer OR because, in addition to being symmetric in both variables, it has a clear mathematical and physical interpretation (e.g. it always passes through the centroid of the data points and is the axis of the minimum moment of inertia). If the variables must have scaled units [e.g. luminosity in linear units of erg s^{-1} rather than log units of log(erg s^{-1})], then OR is inappropriate and we recommend use of the OLS bisector. Though its mathematical properties are less well established, it should be a reasonable measure of the functional relationship except when the OLS(x/y) slope becomes very large. We also note that our numerical simulations suggest that the analytical error estimate tends to be too small. Of the three symmetric regression lines, we believe the *least reliable* is the reduced major axis, known to astronomers as Strömberg's (1940) "impartial" line. We, and researchers in other fields, believe it has less merit because it depends entirely on the variances of the x and y variables and not on the correlation between the variables .

(iii) Whatever method is adopted, we can unequivocally state that the derived regression coefficients should be accompanied by their appropriate error estimates.

We emphasize that this entire discussion refers to data in which the scientist cannot distinguish intrinsic dispersion of the observed objects from errors due to the measurement process. If such distinction is possible, then weighted or "errors-in-variable" methods should be used .

REFERENCES

Deeming, T. J. 1968, *Vistas Astron.*, **10**, 125.

Kermack, K., and Haldane, J. B. S. 1950, *Biometrika*, **37**, 30.

Pearson, K. 1901, *Phil. Mag. Ser. 6*, **2**, 559.

Pierce, M. J., and Tully, R. B. 1988, *Astrophys. J.*, **330**, 579.

Seares, F. H. 1944, *Astrophys. J.*, **100**, 253.

Strömberg, G. 1940, *Astrophys. J.*, **92**, 156.

ADDRESSES

Michael G. Akritas and Gutti Jogesh Babu : Department of Statistics, Penn State University, 219 Pond Lab., University Park, PA 16802, USA.

Eric D. Feigelson : Department of Astronomy, Penn State University, 525 Davey Lab., University Park, PA 16802, USA.

Takashi Isobe : Center for Space Research, Massachusetts Institute of Technology, 77 Massachusetts Ave., Cambridge, MA 02139, USA.

The Combination of Astronomical Data of Different Precision ; An Interim Report

A.R. UPGREN – E.W. WEIS

Van Vleck Observatory, U.S.A.

This is a preliminary report on an ongoing program which seeks to combine parallaxes for a star made using different methods. It is common in lists of stars such as the Catalogue of Nearby Stars (Gliese 1969, Jahreiss and Gliese 1989) to provide the best parallax for each star derived from each of as many as three methods; trigonometric, spectroscopic and photometric. In truth, a star has but one parallax, and the best single representation of it should be only a matter of combining the results from the various techniques into a single weighted mean value. In practice this has almost never been done, because a realistic weighting scheme which makes the best use of the data is very difficult to derive. Among the problems encountered in such a task are the fact that the errors in some of the methods are distance dependent and others are not. An assessment of a realistic set of relative weights between the spectral classification systems in widespread use or between the various contributors of photometric data is also a nearly impossible effort. Nevertheless, one of the goals of this investigation is to determine the best single value of the parallax to use for a star for which a multiplicity of distance methods is available. In this brief paper, we report on two developments which are but two small steps toward this eventual goal.

In the first of these, measures have been made on two machines for parallax and proper motion of all measurable stars appearing on 13 fields located near the center of the Hyades star cluster. The two sets of measures were made of the same photographic plates but otherwise are totally independent of each other. One of the machines is a conventional two-screw Mann machine located at the Van Vleck Observatory and the other is the PDS microdensitometer of the Yale Observatory.

The two sets of measures have been analyzed to determine the relative and absolute precision in terms of the internal and external parallax errors. A preliminary

description of this project has been given previously
(Upgren et al. 1988). Now the analysis has been extended
to include all of the stars measured, field stars as well
as cluster members. The full report is being prepared for
publication elsewhere; here we report only the primary
results and conclusions.

The analysis shows that the internal and external
errors of the parallaxes found from the PDS measures are,
on average, about half those derived from the Mann measures
of precisely the same set of stars. In both sets of data,
the internal errors represent the external errors well,
indicating that the Van Vleck parallaxes can almost be
characterized by a single standard error. For the stars
within the optimum ranges in magnitude and locations in the
field, the standard parallax errors for the PDS and Mann
measures are about 4 and 8 milliarcseconds (mas),
respectively; these increase to more than 5 and 10 mas for
other areas within the unvignetted region of the field.
The results concur with earlier ones of Upgren (1973) Lutz
and Hanson (1979) van Altena (1988) and Upgren et al.
(1988). The low coma in the Van Vleck refractor optical
system found by Russell (1976) is confirmed. For the
exposure times typical of the astrometric program (3 to 5
minutes), the useful range in apparent magnitude extends
from about 9 to about 13. The weighted mean of the
absolute parallaxes of 24 near-certain members of the
cluster in this magnitude range is given in Table 1 along
with the distance moduli and that of Hanson (1980) which
summarizes the distance determinations from a number of
techniques. The agreement between our distance and that of
Hanson is very close and suggests that the parallaxes of
these stars are not affected by any significant systematic
error.

Table 1 - Parallaxes in mas and Distance Modulus of Hyades

	Mann	PDS	Combined
Weighted mean absolute parallax	21.71 ± 1.76	22.02 ± 1.18	21.92 ± 1.03
Mod., m-M	3.32 ± 0.18	3.29 ± 0.12	3.30 ± 0.10
Hanson (1980) mod.			3.30

Photometry is being obtained for all of the members
and as many of the field stars as possible in order to
provide a database useful for combining high-precision
trigonometric parallaxes with photometrically derived

parallaxes on as many photometric systems and indices as possible. Two other sets of data are more available now and the remainder of this paper describes them and some initial results.

The second result examined in this preliminary report concerns the combination of photometric data. Two color indices comprise between them the large majority of photometric data presently available for the nearby stars. These are the B-V and the R-I indices, and the latter are mostly on the Kron system. The absolute magnitudes and thus also the distances of stars should, in principle, be more precisely derived from weighted means of these parameters found from the two indices than from either one alone, providing no systematic difference is present between them. It is evident that some correlation between the colors is present, regardless of the source of the data, but this does not vitiate the conclusion. The situation is analogous to the combination of parallaxes derived from measures made along the axes aligned with right ascension and declination into a single value, despite a substantial correlation between the two separate parallaxes which are usually made from the same photographic plate material. In a study of the parallaxes determined at four observatories, Lutz and Upgren (1980) found no systematic differences between the parallax results from the two measures and concluded that the measures of parallax in right ascension and declination are measures of the same quantity although its variance has been underestimated by all four observatories by different amounts.

Two samples were chosen to examine the question of the combination of absolute magnitudes from the two color indices. The first was taken from a homogeneous photometric sample of 73 very probable members of the Praesepe star cluster (Upgren, Weis and DeLuca 1979, Weis 1981). All of the observations were made at the Kitt Peak National Observatory using a single telescope, photometer and set of filters. The second was taken from a sample of very heterogeneous photometry of the entire set of nearby K and M dwarfs appearing in the five lists compiled by Vyssotsky and his colleagues. Although 70 sources in the literature include photometry of at least one of the stars, most data comes from about a half dozen individual contributors. Such photometry can be expected to show very little correlation since it was made from both hemispheres using a variety of telescopes, photocells and filter sets.

The combination of the observations into a single value for the absolute visual magnitude, M_v, for a star was made after weighting them by the inverse squares of the

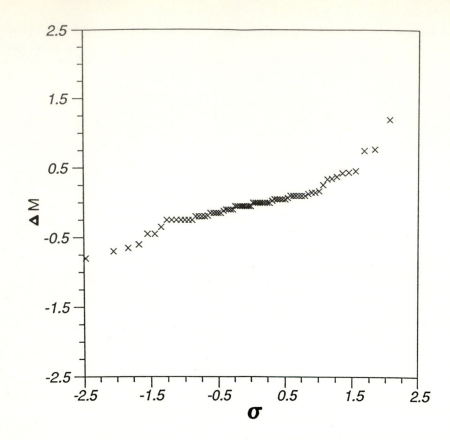

Figure 1. Probit plot for 73 probable members of the
Praesepe star cluster. The mean and dispersion are -0.035
and ±0.335, respectively.

slopes of the main sequences at the point where it fell on
the main sequence in the color magnitude diagram. The
individual errors in the photometric data were not taken
into account because they are not usually given and are
hard to determine. But the slopes of the main sequences
undoubtedly affect the errors in the absolute magnitudes
because a given error in a color translates into one in M_V
in proportion to the slope of the sequence at the color of
the star. The slopes were estimated from color magnitude
diagrams of stars with s.e.M_V < ±0.30 mag. taken from the
review by Gliese, Jahreiss and Upgren (1986). They are
given in Table 2 below. The entries in the table indicate
the variations in M_V for a variation of one whole magnitude
in the color indices.

     The differences between the absolute magnitudes
produced from each of the two colors was found to follow
closely a Gaussian Distribution for each of the two
samples. Figures 1 and 2 indicate this; they are

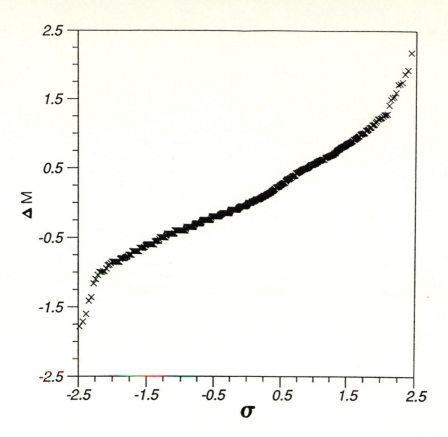

Figure 2.   Probit plot for 978 nearby K and M dwarf stars.
The mean and dispersion are +0.036 and ±2.246, respectively.

Table 2 - Slopes in M_V of Main Sequences

Range in B-V	Slope	Range in R-I	Slope
0.00 to 1.25	5.0	0.20 to 0.31	15.0
1.26 to 1.30	7.0	0.32 to 0.36	12.5
1.31 to 1.35	9.0	0.37 to 0.46	8.0
1.36 to 1.40	11.0	0.47 to 0.96	6.0
1.41 to 1.70	13.0	0.97 to 1.14	5.5
		1.15 to 1.70	8.0

cumulative probability plots, or probit plots, in which a
Gaussian Distribution would be represented by a straight
line.   The inner 80% of the data in both figures is nearly
linear, indicating a Gaussian Distribution is a valid
assumption in both cases.   However, the dispersion is much
greater for the field stars than for the cluster members.

This may be due in part to the much more heterogeneous observational material available for the field stars. It is more likely a result of the fact that the field star group contains a mixture of stars of all ages and chemical compositions found in the entire disk population whereas these properties are much more uniform among the cluster stars. Although different degrees of correlation appear in the different data, the normality apparent in both figures indicates that a single weighted mean value for the photoelectrically determined absolute magnitude is a better representation of its true value than one derived from a single index alone. The parallaxes of the field stars are too uncertain to make a test of this last point, but the distance of the Praesepe cluster is well known. The true absolute parallaxes of its member stars is + 6 ± 1 mas and their absolute magnitudes are well known. The residuals from the weighted mean absolute magnitudes are, as they should be, smaller than those from either index alone, indicating that they appear to be the best photoelectric data to use.

The authors acknowledge the support of the National Science Foundation through research grant AST-8610424 and its supplements.

REFERENCES

Gliese, W. 1969, Veroff. Astron. Inst. Heidelberg, No. 22.
Gliese, W., Jahreiss, H. and Upgren, A.R. 1986, "The Galaxy and the Solar System", R. Smoluchowski, J.N. Bahcall and M.S. Matthews, eds., publ. University of Arizona Press, Tucson, p. 13.
Hanson, R.B. 1980, IAU Symposium No. 85, "Star Clusters", J.E. Hesser ed., publ. D. Reidel, Dordrecht, p. 71.
Hanson, R.B. and Lutz, T.E. 1983, Mon. Not. Royal Astron. Soc. **202**, 201.
Jahreiss, H. and Gliese, W. 1989, These Proceedings.
Lutz, T.E. and Upgren, A.R. 1980, Astron. J. **85**, 1390.
Russell, J.L. 1976, Thesis, University of Pittsburgh.
Upgren, A.R. 1983, Astron. J. **78**, 79.
Upgren, A.R., Weis, E.W., Fu, H.-H. and Lee, J.T. 1988, IAU Symposium No. 133, "Mapping the Sky – Past Heritage and Future Directions", S. Debarbat, J.A. Eddy, H.K. Eichhorn and A.R. Upgren eds., publ. Kluwer, Dordrecht, p. 469.
Upgren, A.R., Weis, E.W. and DeLuca, E.E. 1979, Astron. J. **84**, 1586.
Van Altena, W.F., Lee, J.T. and Hoffleit, E.D. 1989, "General Catalogue of Trigonometric Stellar Parallaxes", Yale University Observatory, New Haven.
Weis, E.W. 1981, Publ. Astron. Soc. Pacific **93**, 437.

A Possible Re-Reduction Scheme For The Guide Star Catalog

L.G. TAFF, B. BUCCIARELLI, and M.G. LATTANZI*

Space Telescope Science Institute

1 SUMMARY

The constructors of the Guide Star Catalog for the Hubble Space Telescope had to overcome two different sources of systematics. They faced the traditional problem of constructing an all-sky catalog, of overcoming the bifurcation of quality reference catalogs at the celestial equator. In addition, because the source material for the Guide Star Catalog (GSC) is large-scale Schmidt plates, they faced the heretofore intractable problem of astrometric reduction of Schmidt plates. The latter problem has been completely solved (Taff in press in Astron. J. 98, 1989; Taff, Lattanzi, and Bucciarelli submitted to Astrophys. J., 1989) but not in a fashion applicable to the entire celestial sphere. The reason relates to the first difficulty of GSC construction, the non-uniformity between the northern and southern hemispheres. The technique we have developed depends on a high density, quality catalog for its success. While the AGK3 is barely adequate, the SAOC is not. Therefore, we have invented another solution to the Schmidt plate astrometric reduction problem that, while it does not work as well as our first method, is still substantially better than all its other competitors and can be applied over the entire celestial sphere. We present this method in this paper and results of its application in the northern hemisphere.

2 MOTIVATION

The reason why all other workers have failed to successfully reduce large-scale Schmidt plates is that Schmidt plate modelling is dominated by systematics. Carrying over the habits developed with astrographs, where the gnomonic projection is a good approximation and the optical system supports an almost linear model between standard coordinates and measured coordinates, can not work because of the patterns of distortions induced in Schmidt plates mainly by the curved plateholders. Hence, plate modelling per se must be completely abandoned and a method found to correct the systematic deformations impressed onto the plate. We have invented such a method (although this is theoretically unnecessary as the distortions could be mapped at the telescope with a calibrated grid). We have applied it to a large sample of Schmidt plates and show the results here.

* Affiliated with the Astrophysics Division, SSD of ESA; on leave from Torino Observatory.

3 METHOD

If you accept the hypothesis that there exists a mean pattern of distortions on all the Schmidt plates taken with the same instrument, then how can you discover them ex post facto? One method would be to use a very high density, high precision catalog that occupies the 0.01 steradian of the sky a 6.5 deg × 6.5 deg Schmidt plate does. Then after reducing a plate of this area of the sky in the usual fashion, one could construct a map of the residuals. Unfortunately, no such catalog exists.

An alternative method is to use a very high quality catalog which is sparse. Now one would reduce many plates taken with the same instrument, compute the residuals for the few reference stars on each plate, and sum the results (in a plate-based coordinate system) over all plates. If a clear signal emerges from such an integration, then the hypothesis that there is an average pattern of distortions is not only proved, but the means to correct the pattern is at hand. Smoothing the residual map with a moving average is advisable too. Such a mask is shown in Fig. 1 for the Palomar Schmidt.

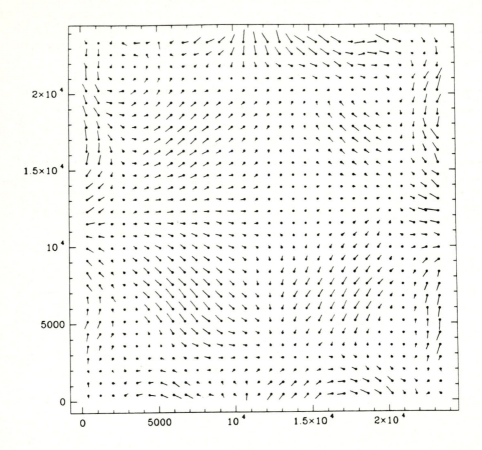

Figure 1. Smoothed mean mask of vector residuals for northern hemisphere GSC plates. Axes labels are pixels. Note multiple swirls. The largest symbol represents 0".8.

3.1 Details

The details of construction of this mask are as follows. First we reduced 180 plates using AGK3R stars as reference points. As there are only approximately 40 such stars per plate, this is an insufficient number of stars to support a complicated model. Hence, we pre-corrected for the standard cubic terms of Schmidt departures from a gnomonic projection and then followed with a full quadratic polynomial in x and y. Neither magnitude nor color terms were included. The actual model is immaterial, it is only a platform to use for the construction of the residual maps. Almost any reasonable model will do; what is of paramount importance is that its coefficients be very well-determined. Hence, the precision of the positions in the reference star catalog are much more important than the number of reference stars (once there is a sufficient number). Using a denser, but less precise catalog, leaves more noise in each plate's residuals. Then these must be smoothed out too by the integration over all the plates in addition to the plate-to-plate deviations from the mean mask. More explicitly, the plate-to-plate deviations from the mean must arise from two different sources; different physical circumstances for the exposure of a plate (such as temperature, zenith distance, humidity, hour angle, and so on) and systematic differences imparted to a particular plate's residuals as a consequence of poorly determined plate model constants. Even a perfect reference catalog (i.e., no random errors in its positions) would not eliminate the latter contribution because of plate measurement errors whereas an invariant instrument and physical circumstances would eliminate the former.

Having reduced each of the 180 plates with the AGK3R, we need an independent catalog to construct the residuals. The best other catalog at our disposal is the set of Carlsberg Meridian Circle Catalogues La Palma (Nos. 1, 2, and 3) As they are nearly coeval with the Schmidt plates of the GSC, the lack of precision proper motions is irrelevant. The density of these catalogs is about 1.0 stars per square degree. Hence, after adding up 180 plates we have about 7,000 individual pairs of residuals. These were binned into a 30 × 30 grid across the plate and average values computed per bin. Not only are the distortions crystal clear in these maps, they are already clear in the individual residuals. The application of a moving average technique is very effective in smoothing out the remaining noise.

3.2 Correction

Having proved the existence of a pattern, the proof that a mean mask can correct them is obtained by taking another 20 plates, from the same instrument but not used to construct the mask, reducing them with our cubic pre-corrected quadratic AGK3R model, and then applying the mask with its sign reversed. Once again Carlsberg stars are used as an absolute external reference. Residuals can be computed after the AGK3R reduction alone and after the AGK3R plus mask reduction. The nature and the amplitude of the improvement made thereby is shown in Fig. 2.

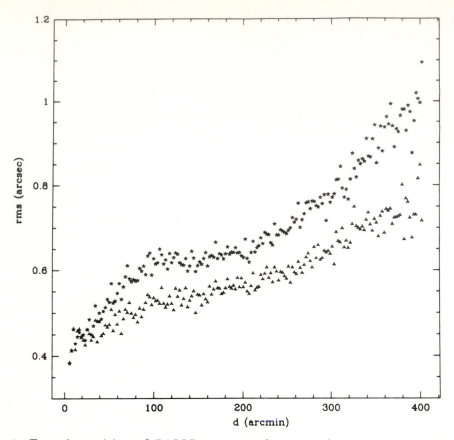

Figure 2. Error in position of CAMC stars as a function of separation on plate. Triangles are this paper, asterisks are the current GSC.

4 THE GSC

The literature on the GSC refers to its performance specification of 1/3" relative accuracy over 30' for Hubble Space Telescope pointing. We have computed the error in angular distance between pairs of Carlsberg stars on our 180 plate sample using GSC values for their positions and using our values (i.e., AGK3R plus mask). In both cases the distance between Carlsberg stars using Carlsberg positions is the reference value.

Figure 2 shows a clear diminishment in the reduction of the systematics across the plate (i.e., our curve is flatter). It also demonstrates that an improvement in precision has been made too. Indeed, the only relevant issue with previous all-plate reductions of Schmidt plates has been accuracy. Until the systematic effects are much reduced, discussing the precision of the results is meaningless. Finally, our technique can easily, and equally, be extended to the southern hemisphere for the SRS can be used in lieu of the AGK3R. Neither the SAOC nor the CPC is a comparable substitute for the AGK3. The GSC is worse in the south than the north and our preliminary results show an even more marked improvement there.

Principal Component Analysis of Galaxian Data

P. BROSCHE – F. Th. LENTES

Sternwarte der Universität Bonn, GFR

Summary: The comparison of the treatment of old and recent data reveals a remarkable stability of the method. Results from recent data are provided for spirals, SO-galaxies and ellipticals. They show a certain similarity even between these three classes.

If we want to treat the obversable parameters of galaxies as a manifold in the mathematical sense, we first encounter the problems connected with the morphological type: there is no continuous definition of this parameter; but one can define the "stage" as a number and one can believe that the discrete character of the presently given stages is only a matter of observational "resolution". Further, we have to recognize that there is perhaps more than one manifold because the transition from the ellipticals via the SO to the spiral galaxies has been questioned (Biermann and Shapiro, 1979). And with regard to more quantitative parameters as size or luminosity we are not sure whether some extremes (dwarfs, giants) are connected with the normal manifold or not. Therefore in what follows we consider normal galaxies and we separately treat the 3 above mentioned morphological ranges.

We start with a certain number m of observational parameters (in practice m is around half a dozen). It appeared from the beginning that there are empirical relations and correlations, hence the dimension p of the manifold (the number of its independent parameters) is certainly smaller than m. From a general point of view, the question of the true dimension of the various manifolds of galaxies is the first which has to be answered. In practice, this question is intimately related with the next one: if p < m, every observed or derived quantity must be representable as function of m independent parameters (be it observables or derived quantities) and the second question is the one for this functional dependence. The prerequisite for any solution of the two questions is a statistical model. So far it turned out that the assumption of a linear dependence between all parameters was sufficient. An extension towards nonlinear dependence is not a severe problem.

The most important result of the first study (Brosche, 1973) and of nearly all further studies was the dimension number p = 2. For the present investigation, we used data from the Reference Catalogues RC1 (A. and G. de Vaucouleurs, 1964) and RC2 (A. and G. de Vaucouleurs, Corwin, 1976). Our kinematical data for ellipticals mainly stem from Davies et al. (1983), for SO-galaxies from Dressler and Sandage (1983) and for spirals from Whitmore (1984 and private communication). The variant of principal component analysis used here was described by Lentes (1983).

Every physical diagramme (e.g. a two-colour-diagramme) must contain the zero points and the scales of its axes. In our case, we must consider the averages and the dispersions of the observables (Table 1). They will not agree for any two samples. In order to provide a fair comparison of the results for two samples, we have to subject our results to an affine transformation of the independent arguments such that the resulting gradients of the observables agree as good as possible (one could make two specific observables to agree exactly and check the remaining m-p ones, but we prefer a compromise transformation since there are no a priori distinguished observables). Naked-eye comparisons (including rotations and reflections as special affine transformations) with the results of other colleagues have always shown a good qualitative agreement. Here we want to present a more quantitative comparison of the eldest results (Brosche, 1973) with the work of Whitmore (1984), both, because this paper is a recent one based on homogeneous data and because the author claimed that the old data are completely obsolete. Dr. Whitmore kindly provided his data so that we could obtain results with dispersions and averages. In such a way and with a least squares requirement for the gradients we produced Figure 1.

It shows that there is much more than a random coincidence between the oldest and a recent study. In our view, this astonishing stability stems from the fact that our method uses all mutual m (m-1)/2 connections between the observed variables. The resulting compromise has a correspondingly large stiffness.

In the following Figures 2,3,4 we present the results for the 3 morphological categories. For the spirals and SO we can estimate the inclination from the apparent figure; in such a way we can transform the observed amplitude of the rotation curve into an absolute value in the main plane of the galaxy. The difficult observations of the rotation of ellipticals cannot be corrected so easily. Therefore the gradients of the observed rotational velocities and of the apparent axis ratios both are influenced by the unknown inclination i of the galaxy's rotation axis. At worst, the

whole second dimension of the ellipticals could be nothing else but an inclination effect, that is, an independent parameter which does not belong to the galaxies themselves. We have found the following way-out of the dilemma:
For an isotropic distribution of the rotation axes, the space average of sin i has a dispersion σ (lg(sin i)) = 0.183. We obtained a dispersion of the observed rotational velocities σ (lg v') = 0.271 ± 0.045. This is significantly larger than σ (lg(sin i)) and it follows a dispersion of σ (lg v) = 0.200 where v = v'/sin i. If the observed ellipticities ϵ = (a-b)/a are produced by oblate ellipticals of a true ellipticity q = 0.5, we can derive individual inclinations and thereby an 'observed' value of σ (lg (sin i)) = 0.100 ± 0.016 which, in turn would lead to σ (lg v) = 0.252. In each case we conclude that the gradient of the observed rotational velocities is dominated by the variation of the true rotational velocities and not by a variation of inclinations. Moreover, the gradient of the ellipticity is almost perpendicular on the gradient of the rotational velocities and the communality of the ellipticities with the two dimensions is small – so perhaps the ellipticity itself is rather an inclination measure.

We do not know how much our observed cutting outs of the total manifolds are affected by selection effects. The fact that the limits of our sample are defined by such effects does not automatically mean that our gradients are biased. Our strongest observational demand is the knowledge of rotational velocities. This could be influenced by the value of the surface brightness. So if our gradients are biased then most probably they are somewhat to small in the direction of the surface brightness gradient (in the data themselves a cutoff in that direction cannot be seen). In general, it seems remarkable that the qualitative behaviour of those parameters which are common to all 3 classes is quite similar. This is an empirical argument for their mutual connection. A real proof of this conjecture is not possible since not all variables are identical.

References

Biermann, P., Shapiro, St. L.: 1979, Astrophys. J. 230, L33
Brosche, P.: 1973. Astron. Astrophys. 23, 234
Davies, R.L., Efstathiou, G., Fall, M., Illingworth, G., Schechter, P.L.: 1983, Astrophys. J. 266, 41
de Vaucouleurs, G., de Vaucouleurs, A.: 1964, Reference Catalogue of Bright Galaxies, University Press, Austin
de Vaucouleurs, G., de Vaucouleurs, A., Corwin, J.G. Jr.: 1976, Second References Catalogue of Bright Galaxies, University Press Austin
Dressler, A., Sandage, A.: 1983, Astrophys. J. 265 664

Lentes, F. Th.: 1983, Proc. Statistical Methods in
    Astronomy, Symp., Strasbourg, ESA SP-201
Whitmore, B.C.: 1984, Astrophys. J. 278, 61

Adress of the authors

Observatorium Hoher List der Universitäts-Sternwarte Bonn
D-5568 Daun, R.F. Allemagne

Table 1   Averages and dispersions of the observed variables for our
         data on 3 classes of galaxies.

Variable	E	SO	S
T	–	–	2.7 ± 1.4
U–V ⎫	1.40 ± 0.15	1.52 ± 0.17	–
B–V ⎬ [mag]	–	–	0.68 ± 0.11
M_B ⎭	-21.15 ± 1.36	-20.65 ± 0.97	-21.47 ± 1.10
$\lg(R_{25}/\text{kpc})$	1.24 ± 0.31	1.17 ± 0.19	1.32 ± 0.24
$\lg(R_o/\text{kpc})$	–	–	0.57 ± 0.26
$\lg(v/\text{km s}^{-1})$	1.67 ± 0.27	2.14 ± 0.22	2.37 ± 0.13
$\lg(\Sigma/\text{km s}^{-1})$	2.33 ± 0.16	2.32 ± 0.12	2.11 ± 0.17
$\lg \varepsilon$	-0.64 ± 0.14	-0.40 ± 0.07	–
Mg	0.31 ± 0.09	–	–

U–V,B–V	corrected colour index ⎫	in	
M_B	absolute luminosity ⎬	magnitudines	
R_{25}	photometric radius in kpc at 25^m /(arcsec)^2		
v	characteristic rotational velocities in km/s		
	for ellipticals: observed maximal value;		
	for SO: observed value at surface brightness $21^m\!.5$/(arcsec)^2		
	corrected for inclination and integration along line of sight		
	(mainly edge-on galaxies)		
	for spirals: maximum rotation corrected for inclination		
Σ	central velocity dispersion in km/s		
T	type parameter = stage number		
Mg	Mg_2-index at 5178 Å in magnitudines		
ε	apparent ellipticity = (a–b)/a (of the bulge for SO galaxies)		
R_o	a characteristic length scale of the inner rotation curve		

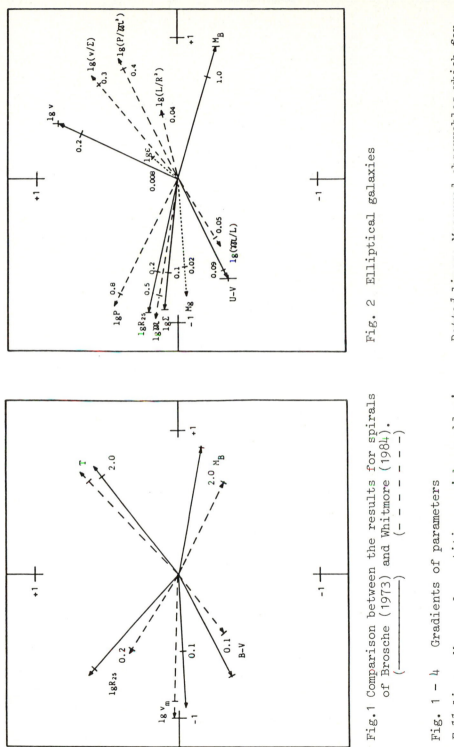

Fig.1 Comparison between the results for spirals
of Brosche (1973) and Whitmore (1984).
(————) (— — — —)

Fig. 2 Elliptical galaxies

Fig. 1 – 4 Gradients of parameters

Full lines: Measured quantities or 'observables'
which were used primarily in the factor analysis.

Dotted lines: Measured observables which for
some reasons were only secondarily embedded
into the gradients of the primary observables.

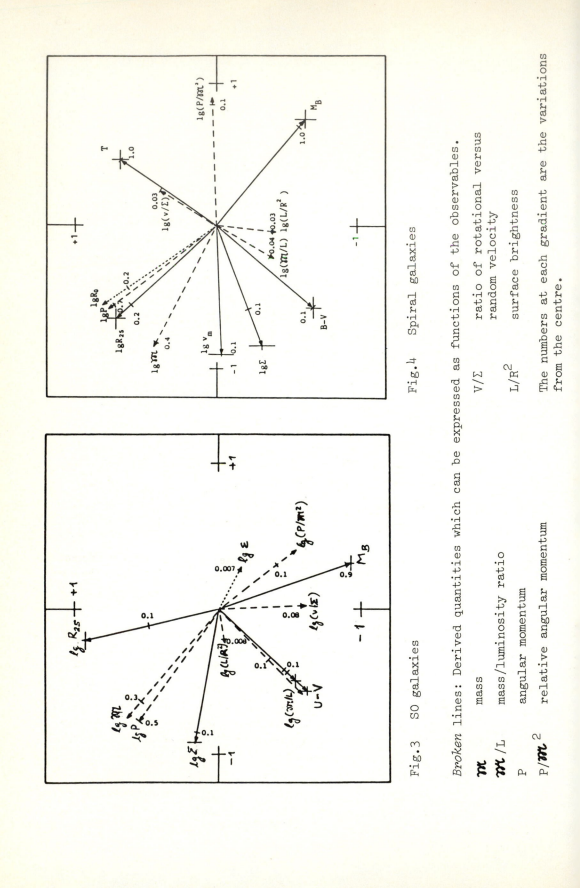

Fig.3 SO galaxies

Fig.4 Spiral galaxies

Broken lines: Derived quantities which can be expressed as functions of the observables.

$\mathcal{m}$	mass	V/Σ	ratio of rotational versus random velocity
$\mathcal{m}/L$	mass/luminosity ratio	L/R^2	surface brightness
P	angular momentum		
$P/\mathcal{m}^2$	relative angular momentum		

The numbers at each gradient are the variations from the centre.

DISCUSSION

As with the discussions following other sessions, these remarks and questions are based on material provided by the speakers. As far as possible they are in chronological order, although this cannot be guaranteed. The following is formatted in such a way that the paragraph following the question is the addressee's answer.

J. Usowicz to J.D. Scargle: Can your approach be generalized to higher dimensional dynamical systems?

Yes, the generalization to higher dimensional systems is straightforward in principle. In practice there are problems due to the limited amount of data and connected with the difficulty of displaying high dimensional data. All of my examples have been 2-dimensional for simplicity, especially in the display of the phase space.

S.P. Bhavsar to J.D. Scargle: In a real astronomical situation you may be sampling only a small part of this "phase" orbit. Could you please address this.

Yes, this is an important problem. In many astronomical contexts the data are so limited that the coverage of the chaotic attractor in phase space is quite incomplete. Unless one can obtain more data it may be impossible to obtain a clear picture of the trajectories in phase (state) space.

C.S. Cole to J.D. Scargle: To find a particular chaotic process do you need to look for that particular chaotic process?

No. The form of the recurrence equation $x_{n+1} = f(x_n)$ does not have to be assumed, but is in fact estimated in the procedure. On the other hand, I did assume that the order of the process (e.g. the number of autoregressive parameters) is known, but I am sure there are ways to estimate the order too.

M.A. Hendry to J.D. Scargle: Is it possible to use the penalty function you have developed as a means of discriminating between chaotic and random processes, in their definition?

The method distinguishes random and chaotic processes by means of their different appearance in phase space – random processes are "area filling" (2D) while chaotic processes are "curve hugging" (1D). More generally, a chaotic process yields a phase diagram which has lower dimensionality than the space in which the data are plotted.

S.J. Hogenveen to J.D. Scargle: You showed the procedure you developed works for an observable which results from a random process *and* a chaotic process. What happens when your observable is the result of two (independent) chaotic processes?

This is a very interesting question, which can also be asked for the case of two (independent) random processes. As discussed in reference [36] of paper I of my astronomical time series papers it is a curious result of the Wold theorem that such a composite *process* can be represented *exactly* by a single stationary process, which is in some sense an average of the two. It is presumably possible to find methods to separate such processes.

J. Nousek to J.D. Scargle: The distinction between chaos and random noise is the existence of an iterative expression to predict the next value. Can your deconvolution lead to determination of the expression and, if so, can you predict the next value in the series?

Yes, the form of the functional relationship between successive values is determined, can can be used to make predictions. However, the phenomenon called *sensitivity to initial conditions* severely limits such prediction. The point is that in a chaotic process, even very tiny differences between two initial states are amplified (exponentially) as time progresses. Thus any error of measurement will be amplified, rendering prediction far into the future impossible.

H. Eichhorn to J.D. Scargle: What happens to a chaotic process if the original independent variable (time) is eliminated from the differential equations governing the process and replaced by another one? Will the process still be chaotic in the new independent variable?

Chaos is a property of the solution of a set of differential equations, and thus any correct manipulation of the equations cannot change the chaotic nature of the solution. Manipulations of the type you describe have proven useful in deriving more convenient forms of chaotic differential equations because the resulting phase space is simpler.

H.L. Marshall to J.D. Scargle: I understood that the random number generators used on computers are essentially deterministic (of necessity), so they must be actually chaotic number generators. Why wouldn't these data show up as deterministic in these numerical experiments?

You are correct that computer "random" number generators are chaotic – like all computer algorithms they are deterministic. I believe that good random number generators are designed so that they correspond to a chaotic process of very high dimension, and therefore the determinism would show up only with very careful analysis in a state space of high dimension.

E. Feigelson to J.D. Scargle: (1) Are "quasi-periodic oscillator" time series recently discovered in X-ray sources likely to be due to chaotic processes? (2) Would it be valuable for astronomers to become familiar with the extensive econometric literature, and associated commercial software packages, to study astronomical time series? What package(s) would you recommend in particular?

(1) It is certainly possible that the "QPO"s are chaotic, and various workers are investigating this question. (2) Yes, there is a large literature, principally in geophysics (seismic prospecting), econometrics, and speech analysis – astronomers will find this an abundant source of time series analysis methods. It is ironic that the autoregressive model was invented long ago (by Yule) to study sunspot numbers, and has rarely been seen in astronomy since. I do not recommend any software packages. In my opinion, the dangers of misusing a "black box" software package are great, and it is preferable to develop and understand your own software.

F. Murtagh to J.D. Scargle: Could you comment on the work being done at Los Alamos National Laboratory on multi-layer neural networks for analyzing chaotic phenomena?

I do not know the details of this work, but the non-linear dynamics group at Los Alamos is very good.

J.D. Scargle to T. Isobe: Can you explain the distinction you are making between "intrinsic errors" and "instrumental errors"?

The intrinsic error is the error due to the nature of objects. The instrumental error or measurement error is the error due to observation. If the latter does not exist, OLS can be used for both predictions and functional relations.

R. Branham to T. Isobe: One must distinguish between a model where it is unimportant whether one does a regression of y on x and x on y and one where a definite model with error-free coefficients is given. For the latter we do not want to do orthogonal regression.

I agree. For the latter case, the OLS should be used (assuming the independent variable is error-free).

A.A. Schoenmaker to A.R. Upgren: Was the Mann machine used completely manually?

Yes, by visually bisecting stars projected on a screen. One single (experienced) measurer made all of the hard measures on the Mann machine.

H.-M. Adorf to A.R. Upgren: A technical question: where you are reducing measurements from the PDS microdensitometer you ought to employ an intensity calibration. So I assume you are using such a calibration, aren't you?

Yes.

R. Branham to A.R. Upgren: Is the resolution of the Mann and PDS the same so that the larger scatter, 2 to 1, in the Mann can be attributed entirely to the difference between measuring visually and automatically?

The resolution may account for some of the scatter difference but centering ability has a larger role.

M. Kurtz to A.R. Upgren: Might the differences between the human and automatic measurements be due to the edges of the star images containing superior data, as they are not so saturated as the centers?

Saturation is rarely an influence in the precision of the measures except, perhaps, at the very faint end. Poor seeing and especially poor guiding are more influential. The ability of the automatic machine to determine a consistent midpoint from the edges seems greater than that of the eye.

C.S. Cole to A.R. Upgren: Did you allow for correlation of the two data sets when estimating the error of the combined solution for parallax?

Yes.

T. Kiang to A.R. Upgren (comment): As a "historical" episode with automatic reduction techniques, I may mention that about 30 years ago (*The Observatory*, circa 1958) I reduced the parallax plates of one star using an Adams Hilger microdensitometer, i.e. the distance between the star and reference scratch lines was measured with a ruler on the recording chart. In spite of the feared coarseness of the chart moving mechanism, the result compared favourably with that obtained by the "eye" measurement with a micrometer.

H.-M. Adorf to L.G. Taff: You've made a clear distinction between "they" (the Guide Star Catalog people) and "we" (TLB). Could you, please, briefly comment on the future, i.e. are there any signs of a merge between the two groups or an adoption of your superior method by the GSC people?

"We" are the authors of this paper. "They" are the original constructors of the Guide Star Catalog (i.e. Russell, Lasker, McLean, Sturch, et al.). "We" are only recently at STScI, did not participate in the original GSC, and have used newer materials not available to the original GSC group. The GSC people have been very helpful to us and supportive of our research. I am in charge of the Fine Guidance Sensors, so we cannot really merge but we do share thoughts, software, information, criticism, and so on. It is my impression that the GSC people, who retain responsibility for the GSC, will re-reduce it in the future, probably after HST launch.

A. Upgren to L.G. Taff: In Venezuela, Stock has achieved success in objective prism radial velocities using CTIO Schmidt material and a general third-order polynomial across the entire field. The distortions imposed by the prism can be evaluated and removed. He is now looking into positions from Schmidt plates – are you acquainted with this effort and can you comment on it? Would your result of systematic differences between widely separated regions within a single Schmidt field be consistent with high-precision differential work, or with a relatively small telescope such as the Curtis Schmidt?

I am aware of Stock's work. If polynomial models work for him, his plates must be small in area or not curved very much. Yes – in a small area high precision work is possible, even probable, if it is relative. You must examine the pattern of residuals first though, to see where they are stable. Over about $\frac{1}{3}$ of a $6^o.5 \times 6^o.5$ Schmidt plate we see steep residual gradients implying no continuity even for differential work. Subplate overlap (Taff, 1989, in press in *Astron. J.* **98**) solves the whole Schmidt plate astrometric reduction problem.

M.P. Fitzgerald to L.G. Taff: To obtain magnitudes from at least the Curtis Schmidt you also have to use some techniques, reducing errors by about a factor of two, given sufficient photoelectric standards.

Nice to hear, I am not surprised. A successful recalibration of the Guide Star Catalog magnitude system using CCD calibration fields is well underway.

H. Morrison to L.G. Taff: A comment on the precision of the Stock radial velocity results from Schmidt plates – this may depend on the individual Schmidt telescope he used (the Curtis Schmidt). Which Schmidt telescope(s) are your results for, and do the residuals across the plate vary from telescope to telescope?

Four Schmidt telescopes contributed to the Guide Star Catalog plate collection. Most plates are from Palomar or the UK (SERC) Schmidt. I can't remember the other two off-hand. The residual patterns definitely do vary from telescope to telescope; even for one "telescope" if some critical part, such as the filter or connector plate, is altered. As to Stock's results, perhaps his Schmidt plates are small area or not bent very much.

H.L. Marshall to L.G. Taff: It seems to me that what you've actually done is replace a model with 5–20 parameters based on optics modelling with one based on empirical modelling with approximately 900 parameters (per coordinate).

Not really. In the sense meant in astrometry, there exists no plate model for a Schmidt plate. No lower order polynomial model adequately represents the swirls seen in the residuals. Only an empirical determination of the residuals makes sense. We have no model per se – we have a trial model as a platform to use to construct the residual mask. Only the combination of the two has a meaning; a different model would yield a different residual mask but the sum of the two would still be the same.

M. Kurtz to L.G. Taff: Have you found any effects other than the telescope, such as emulsion or hour angle, on the slope of the masks?

We have not investigated this owing to lack of sufficient statistics (see response to Russo). I am sure that they exist.

G. Russo to L.G. Taff: You mentioned the assumption that the "mask" is independent of the telescope position. Did you make tests to verify this assumption?

No. One needs 50–100 plates (because of the sparseness of the CAMC stars and the need to get good signal-to-noise) to build up the mask. Even if we have zenith distance or hour angle information for each plate, there would likely be insufficient numbers per orientation to test for this dependence. It probably is there and limiting the success of the near mask adjustment.

J. Nousek to L.G. Taff: How are residuals calculated to apply corrections to measurements?

We pre-correct for the difference between the tangent and its arc. Then we fit a quadratic model based on the AGK3R or SRS. The residuals for the CAMC stars on the plate are calculated from this pre-corrected quadratic model. The residuals are binned in a plate-based coordinate system and summed over the plates.

J.D. Scargle to T. Kiang: (1) Do you need to take into account the magnitude of the errors in the times of occurrence of the flares? (2) Physically it is likely that the flares are not strictly uniformly distributed, but instead have a "dead time" – i.e. two flares cannot occur at exactly the same time. Using such a model would change your result.

(1) In this problem the timing error is small compared to the inter-arrival times. So, in a first approximation, it need not be. Conceivably, if one assumes some error distribution, one can work out the uncertainties in the final estimate of the random probability. (2) The effect of the "dead time" in this sense is not considered here. The "dead period" in this paper refers to the time when the telescope is not pointed at the star; hence it involves the question of "missing data" rather than that of "instrumental dead time".

C. Worley to T. Kiang: But YY Cen is a close binary. Have you considered that the flare activity may come from both components? In the case of UV Ceti, for example, many observers have assumed that only B was flaring. In fact, our parallax plates with the 152cm reflector show bright flares on *both* components.

Since one period requires two flares to define, two periods – one from each star – will require four flares just to define. With only 4 observations, there can be no statistical problem if we assume each star flares periodically.

FITTING OF DATA

PHYSICAL STATISTICS

L. G. TAFF

Space Telescope Science Institute

SUMMARY

This paper summarizes the motivational basis for a new branch of physics—physical statistics. The realm of physical statistics is the manipulation of data obtained from scientific experiments in order to provide point estimates of parameters in the underlying model. As a branch of physics it should be constructed in accordance with guidelines obtained from modern, successful physical theories. This severely limits the possibilities for the new principle of statistical estimation. The further requirement that verifiable predictions be made from this maxim decides between two contenders. Applications of the new rule are given in curve-fitting, differential orbit correction, and predictive contexts.

1 BACKGROUND

The statistical adjustment of scientific data had its genesis in astronomy. Classical astronomy is a field of interest to me; Taff (1981, 1985). Out of the necessity to deal with practical statistical problems in this field, a new perspective on statistical philosophy and technique has emerged. A formal statement of the new principle of statistical estimation may be found in Taff (1988a). Several successful applications of the new maxim have already been developed; Morton and Taff (1987), Taff (1988b, 1988c). This paper is mainly devoted to the underlying ideas and their impact rather than to physical abstraction or formal development.

2 MOTIVATION

If a physicist were going to construct a theory of statistical estimation, how would he do it? One method would be to examine successful physical theories and to abstract universal qualities from them. Moreover, if these effective theories are refinements of earlier versions, as quantum mechanics is with regard to classical mechanics, then the nature of the improvement might shed light on fruitful avenues to pursue. Finally, if there were an existing structure he intended to supplant, then its shortcomings should be addressed and rectified.

There is an existing theory of statistical estimation. It consists of two parts. One component is that branch of mathematics known as (mathematical) statistics. The other element is the hypothesis that mathematical statistics is a good model for physical reality. This composite hypothesis is quite complex for it not only contains the obvious information content of its statement, another aspect of the hypothesis is a trial set of associations. The primitives of mathematical statistics, for example events, sample spaces, probabilities, and set theory, must be identified with real physical occurrences or relations. Furthermore, while the language of mathematical statistics may be deliberately suggestive, and even derive from a desire to relate its content to physical reality, this does not by itself imply success. Concepts in this vein are such familiar ones as random error, systematic error, consistency, robustness, correlation, and weight.

Ultimately the physicist requires an experimental verification of the theory. Because the theory is inherently statistical in character, this is a non-trivial task to fulfill. Mathematical statistics does not strive to obtain the *correct* answer, it strives to compute the "best" answer (in some well-defined sense). Physics is different. If we postulate a relationship between the impressed force $\mathbf{F}$, the inertial mass m, and the resultant acceleration $\mathbf{a}$, then we want the right relationship. Either $\mathbf{F} = ma$, or $\mathbf{F} = 2ma$, or $\mathbf{F} = m^2a$. Nature will inform us of the correct formula. Were a statistician given data to determine the inertial mass dependence between force and acceleration, and obtained the result $(1.03 \pm 0.05)m^{0.98 \pm 0.03}$, this is no failing or even shortcoming of statistics. A statistician would correctly argue that if the physicist desired more precision, then the physicist should supply more or better quality data. Indeed, given certain plausible (but unverifiable) assumptions, the statistician can predict the probability of a deviation of a given size from the true dependence on m.

We see then that the appropriate test of *mathematical* statistics cojoined with the hypothesis that it is a good representation of physical reality, is a statistical one. Hundreds, perhaps thousands of sets of measurements should be performed. The resulting data should be statistically adjusted to (say) a power law form Am^B, "best" estimates for A and B calculated, and so on. Maybe some (weighted) averaging of these estimates should be used; alternatively all the data could be combined into one global fit. Adjustments via different branches of mathematical statistics could be tried yielding other "best" estimates for A and B. Does anyone really do all this? If not, then where, *exactly*, is the experimental evidence to support this theory? I think that neither statisticians nor physicists conduct this type of experiment. While clearly a shortcoming of its practitioners and not of the underlying theory, the reality is that there is a paucity of experimental evidence for mathematical statistics per se. As a physicist this woefully inadequate basis is a poor reason to accept mathematical statistics as the correct description of Nature. I understand that it might be and that someone could demonstrate this to all of us. However, $\mathbf{F}$ is equal to exactly ma and a theory of statistical estimation that deals with physical experiments should lead one to this conclusion (in appropriate circumstances).

This kind of unconscious error—of combining a branch of pure mathematics with a hypothesis concerning its applicability to reality—has occurred before in physics. For millennia there was an existing theory of space-time. One component was that branch of mathematics known as Euclidean geometry. The other element was the hypothesis that Euclidean geometry was a good model for physical reality. This composite hypothesis is quite complex for it not only contains the obvious information content of its statement, another aspect of the hypothesis is a trial set of associations. The primitives of geometry, for example points, lines, planes, and notions of congruence and similarity, must be identified with real physical objects or relations. Furthermore, while the language of Euclidean geometry may be deliberately suggestive, and even derive from a desire to relate its content to physical reality, this does not by itself imply success. Concepts in this vein are such familiar ones as triangle, volume, parallel lines, positive-definite intervals, and orthogonal coordinates.

Ultimately the physicist requires an experimental verification of this theory. Because the old theory is quite good at small speeds or over short arcs, and we were technically incapable of moving rapidly or very far for most of our history, testing it thoroughly was a non-trivial task to fulfill. With a post-1916 viewpoint the shortcomings of a product space-time, and especially of a flat space-time, are obvious. Once Newton saw the proverbial apple fall, he could have known that this model of physical reality was wrong (with post-1916 vision). Of course Newton did know that it was unsatisfying, he just did not know how to resolve his unease.

Now we all know that we live in a curved four-dimensional space-time continuum. Space and time are inextricably enmeshed; one cannot construct a physically correct product space-time. In addition, Einstein showed us how to relate the curvature of space-time, as represented by the metric tensor, to the local energy-matter density. The field equations of general relativity allow us to predict the (apparent) falling of the apple as well as new phenomena such as the "falling" of light toward the Sun. Real space-time is whatever Nature provided, not what was intuitively obvious nor reasonable nor plausible. The mensuration of space-time belongs to physics. *The analysis of the data belongs to physics too.* Mathematics may provide an excellent language to describe the process and the results, but a clumsy correct description of reality is better than an elegant incorrect one.

3 IS IT GOOD PHYSICS?

Aside from the issue of experimental evidence, which could theoretically be provided, does mathematical statistics plus the application to reality hypothesis share the qualities of a good physical theory? The answer is a resounding no. (Mathematical statistics need only be good mathematics but the combination must be good physics).

How does this composite fail? For one thing it is not coordinate-free in formulation. Mathematical statistics, no matter which branch is your favorite, is a set of hypotheses concerning the errors of observation and how to manipulate them. According to one branch we should take the sum of the squares of the errors and minimize the result. Another branch would have us maximize the likelihood of this particular occurrence or realization of the errors. Good physical theories should be cast in a coordinate-free fashion. Newton's second law of motion,

$$\mathbf{F} = m\mathbf{a}$$

is true in all (inertial) coordinate systems if it is true in one of them.

When mathematical statistics is utilized to analyze scientific data this issue is partially recognized in the application of the reality part of the theory. To make this point concrete consider the differential correction of a spacecraft orbit. In general some subset of the kinematic state—location plus velocity—is observed. An optical observatory relying on the detection of reflected sunlight would measure position on the sky. A radar installation would determine this plus topocentric distance. If the radar were appropriately equipped, then the radial velocity—the direction and magnitude of the speed along the line of sight—could be ascertained by the mechanism of the Doppler shift.

Clearly all these measures are topocentric; they are relative to the observer. Not only is the observer located on a rotating, and therefore non-inertial, platform, the observer is displaced from the force center. Hence, when we compare the data to the predictions of the theory, we transfer the orbit to the instantaneous vantage point of the observer, impart to it the observer's motions, and then make the comparison. Notice all the effort expended to provide a true comparison. Nobody would advocate the direct adjustment of non-inertial, topocentric data to inertial, force center predictions.

A good physical theory would have been expressed in a coordinate-free way to begin with. Angular direction or radial velocity are not invariant to coordinate transformations. They cannot serve as the basis of comparison in a physical theory of scientific data reduction. Only scalar or vector or, more generally, tensor quantities can fulfill this role. Thus, in the simplest orbital scenario where the only force is equivalent to point-mass gravitation, the energy (a scalar), the angular momentum (a vector), and the Laplace (-Runge-Lenz) vector must play a central role in the differential correction process (see below; realistic orbital scenarios have additional forces such as atmospheric drag, radiation pressure, non-spherically symmetric primaries, and third body perturbers. As I shall show, it is possible to incorporate dissipative and symmetry-breaking forces in a natural way too.)

There is another aspect to the "No!" by which mathematical statistics plus the reality applicability hypothesis fails as a good physical theory. This defect is revealed by considering how physical theories have been successfully refined and improved upon. Examine the following list: Ptolemy's epicycles, Cartesian vortices, phlogiston, centrifugal force, caloric fluid, the aether, electric fluids, absolute time and absolute space, absolute reference frame, simultaneity, absolute acceleration, action at a distance, particle-wave duality, an empty vacuum, perfect measurements and deterministic dynamics, and distinguishable identical particles. All these out-moded concepts share two things; they describe unobservables and they have been discarded as central or crucial elements of some branch of physics. Errors of observation are both unknown and unknowable. We know nothing about their random versus systematic components, their weights, their correlations or lack thereof, their distribution functions, and so on and so forth. Errors of observation undoubtedly occur, there is experimental proof of this. Beyond their existence the depth of our ignorance is bottomless.

The twin pillars of modern physics, relativity theory and quantum mechanics, have forcefully driven home the message that we should construct the foundations of physical theories on what we know. Moreover, if an operational definition of exactly how we know it can be provided, then so much the better. Having seen the error of our way, no physicist could accept the unknown and unknowable errors of observation as a centerpiece for a physical theory.

Finally, let me return to a theme mentioned in several of the preceding paragraphs. Physical theories are supposed to be correct (or nearly correct) descriptions of Nature. If electromagnetism is a better match to experiment then electricity and magnetism are, then it is electromagnetism we must accept. If the electroweak interaction is a better match to experiment than electromagnetism and the weak force are, then it is the electroweak field we must accept. In addition, physical theories must make new, verifiable predictions. Electromagnetism predicted the existence of electromagnetic waves. The electroweak theory predicted the existence of three massive intermediate vector bosons.

Mathematicians are more free. If the principle of statistical estimation known as maximum likelihood leads to more interesting mathematics than does the principle of least squares, then they can pursue the former over the latter. If non-Euclidean geometry is more appealing than Euclidean geometry, then delve into it. No one geometry is correct just as no one principle of statistical estimation of mathematical statistics is correct. While there may be more than one candidate for *the* principle of estimation of physical statistics, only one can be right.

4 THE PRINCIPLE OF ESTIMATION OF PHYSICAL STATISTICS

Given the guidance from successful, modern physical theories, what are reasonable candidates for a principle of statistical estimation within physical statistics? I believe that there are only two. The limitation to two possibilities is a combination of the hemming-in constraints of constructing a good physical theory and my lack of imagination. The discrimination between the candidates arises from the fact that this is physics. As such, one should be able to make predictions which can be experimentally tested. I can use only one of the contenders to make a (verifiable) prediction.

A formal statement of the principle exists in language appropriate to physicists (Taff, 1988a). A presentation in a curve-fitting context might be more accessible. Furthermore, illustrating the technique by way of a specific example, my point may be more easily grasped by a wider audience. Curve-fitting problems fall under the provenance of the new theory because determining the values of the coordinates is the result of a physical measurement process. Therefore, consider the following problem: A series of measurements of rectangular coordinates x, y are made. Both x and y are imperfectly determined. Denote the set of measures by $\{(x_n, y_n)\}$, $n = 1, 2, \ldots, N$. We also know that these points are to fit a circle, see Fig. 1. How can we perform the statistical adjustment of the data? (That this simple problem is already outside the realm of mathematical statistics is interesting but not critical.)

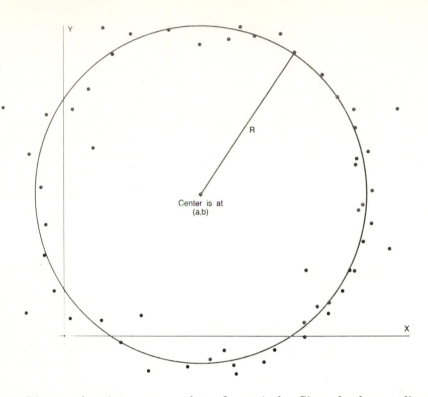

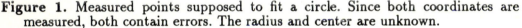

Figure 1. Measured points supposed to fit a circle. Since both coordinates are measured, both contain errors. The radius and center are unknown.

Clearly we must focus on the invariants of this or any other problem. Invariants are things we know, putative probability distributions are merely analytically convenient. A circle has one invariant appropriate to this data, that is it is the locus of points (in a plane) all of which are equidistant from a fixed point. If we denote the unknown coordinates of the fixed point by (a, b) and the unknown radius of the circle by R, then we can either compare each D_n,

$$D_n \equiv [(x_n - a)^2 + (y_n - b)^2]^{1/2}, \quad n = 1, 2, \ldots, N$$

directly to R or to each other D_m, $m = 1, 2, \ldots, N$, $m \neq n$. That is, we would seek the minimum of

$$T' = \sum_{n=1}^{N} (D_n - R)^2$$

with respect to a, b, and R or we would seek the minimum of

$$T = \frac{1}{2} \sum_{n,m=1}^{N} (D_n - D_m)^2$$

with respect to a, b, and R. The symmetry of a circle implies that an overall rotation of the coordinate system is undetectable. The use of a quadratic measure of discrepancy is a consequence of having to have an invariant with respect to all metrics. Clearly the inverse hyperbolic tangent of T or T' would satisfy this requirement too. Should Nature demand inverse hyperbolic tangents, then that is what we would have to accept.

If you actually examine the "normal equations" obtained by differentiating T' and T, then you can see that those arising from T' are easier to deal with. This appears to be a general feature in all problems. Moreover, $\partial T/\partial R$ vanishes identically because T is independent of R. This too is a general feature, an isolated relevant invariant always drops out of T. Thus, a special argument must be made to compute it if it is of interest (as it is for the circle problem as posed). There is physics in the cancellation of R, as will be shown in the next section, and this is a theoretically attractive aspect of T. The one prediction I can make also derives from T so it is my candidate for the correct choice. Thus, the new principle of statistical estimation is:

> To find the "best" estimates for a set of parameters, first enumerate the relevant invariants of the system. Next, form pairwise a symmetric quadratic measure of the discrepancies among the values of the independent invariants. Lastly, minimize this measure with respect to the sought-for parameters.

Perhaps a more complicated problem would be of interest. Again, imagine making pairs of measurements of rectangular coordinates x, y, say $\{(x_n, y_n)\}$, $n = 1, 2, \ldots, N$. Suppose this time we know that the points are to fit an ellipse (Fig. 2). Finally, assume that the major axis of the ellipse may have undergone a rotation by ψ and that the center of the ellipse may have been translated from the origin to the point (h, k). How do we find estimates for the semi-major axis a, eccentricity e, rotation angle ψ, and translation parameters h, k?

Once again we concentrate on what we know. One way to define an ellipse is the locus of points such that for every point on the curve the sum of the distances from two fixed points is a constant. The two fixed points are the foci. In the coordinate system in which the ellipse's equation is

$$\xi^2/a^2 + \eta^2/b^2 = 1, \qquad b^2 = a^2(1 - e)^2$$

the foci are at $(\pm ae, 0)$. After rotation by ψ, these values are mapped into $(\pm ae \cos \psi, \mp ae \sin \psi)$. When this is followed by the hypothesized translation they become $(\pm ae \cos \psi + h, \mp ae \sin \psi + k)$. Therefore, the principle instructs us to form

$$T(a, e, \psi, h, k) = \frac{1}{2} \sum_{n,m=1}^{N} [(D_n^+ + D_n^-) - (D_m^+ + D_m^-)]^2$$

and to find the minimum of T with respect to a, e, ψ, h, and k. The quantities $D_n^\pm$ are defined by

$$D_n^\pm = \left[(x_n \mp ae \cos \psi - h)^2 + (y_n \pm ae \sin \psi - k)^2\right]^{1/2}.$$

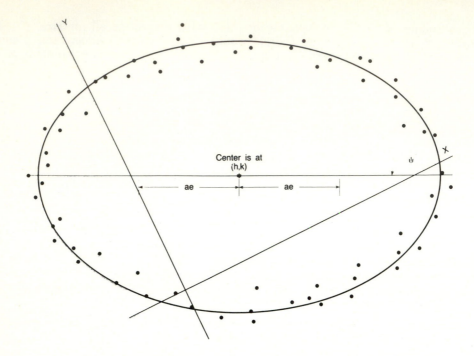

Figure 2. As for Fig. 1, but for an ellipse. Now there is the possibility of a rotation too.

As another example, we may define a hyperbola as the locus of points such that for every point on the curve the difference between the distances from two fixed points is a constant. Therefore, we would seek to minimize

$$\frac{1}{2} \sum_{n,m=1}^{N} [(D_n^+ - D_n^-) - (D_m^+ - D_m^-)]^2.$$

Finally, the ovals of Cassini (see Fig. 3) are the locus of points such that for every point on the curve the product of the distances from two fixed points is a constant. Hence, in this case we would seek to minimize

$$\frac{1}{2} \sum_{n,m=1}^{N} [(D_n^+ D_n^-) - (D_m^+ D_m^-)]^2.$$

None of these problems can be treated within the standard context of mathematical statistics.

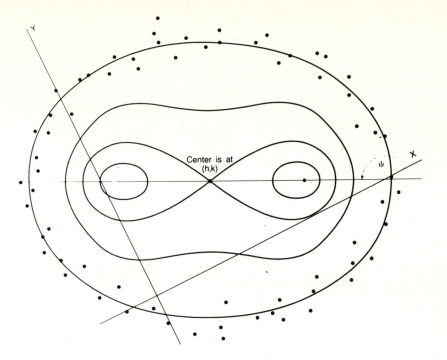

Figure 3. As for Fig. 2, but for the ovals of Cassini.

5 A SIMPLE PHYSICS EXAMPLE

Consider a ball of mass m thrown vertically off a building of height H. If the only force acting is a constant gravity downwards, acceleration equal to $g > 0$, then the equations of motion reduce to

$$F = dz^2/dt^2 = m\ddot{z} = -mg$$

where z measures height above the ground. Denoting the initial velocity of the ball by $\dot{z}(0) = v$, the solution for the height is

$$z = -\frac{1}{2}gt^2 + vt + H.$$

As posed this rectilinear motion problem has two invariants; one is the mass of the ball and the other is the energy of the ball. As this is not a collision problem, the mass of the ball is an irrelevant invariant. The energy E is the sum of the kinetic energy $K = (1/2)m\dot{z}^2$ and the potential energy V. We know the gradient of V since it is equal to the external force,

$$F = -dV/dz = -mg.$$

Potential energies are always uncertain to within an arbitrary, additive constant. I choose the zero of V to be at the ball's initial location,

$$V = mg(z - H).$$

(The customary zero level is on the ground.) Thus,

$$E = K + V = \frac{1}{2}m\dot{z}^2 + mg(z - H) = \frac{1}{2}mv^2$$

is a constant. Given, say, a sequence of observations of heights and times, $\{(z_n, t_n)\}$, $n = 1, 2, \ldots, N$, how are the parameters of the model to be found? Supposing that g is uncertain too, perhaps as a result of the presence of a nearby mountain, we are instructed to form T

$$T = \frac{1}{2}\sum_{p,q=1}^{N}(E_p - E_q)^2$$

and minimize it with respect to g, v, and H.

You can now see why the ball's mass m is an irrelevant invariant, it factors completely out. Note that E_n is a *partially random* variable;

$$E_n = \frac{1}{2}m\dot{z}^2(t_n) + mg(z_n - H)$$

where z_n is a measured value while $\dot{z}(t_n)$ is a computed quantity from the solution to the equations of motion, viz.

$$\dot{z}(t_n) = -gt_n + v.$$

Also, once again a parameter of interest cannot be directly determined. In every pair $E_p - E_q$ the height of the building cancels out by subtraction. Since this is the arbitrary additive constant in the potential energy, it should *not* be directly derivable. Even if the T' formula were used, then H would still not appear because of the translational invariance of T' (or T). Seeing $H - H = 0$ is more obvious than noting that $0 - 0 = 0$, thus my unconventional zero level.

The remainder of the problem is straightforward to work out. The derivatives $\partial T/\partial g$ and $\partial T/\partial v$ are set equal to zero and the resulting system of equations solved for g and v. These equations are different than those obtained by the least squares method. The formulas for g and v are *different* functions of the data in the two cases. How they differ offers more information concerning the divergence of physical statistics from mathematical statistics.

We can write T as

$$T = \frac{1}{2}\sum_{p,q=1}^{N}(E_p - E_q)^2$$

$$= \frac{1}{2}\sum_{p,q=1}^{N}\left\{\left[\frac{1}{2}m\dot{z}^2(t_p) + mg(z_p - H)\right] - \left[\frac{1}{2}m\dot{z}^2(t_q) + mg(z_q - H)\right]\right\}^2$$

$$= \frac{m^2}{8}\sum_{p,q=1}^{N}[g^2(t_p^2 - t_q^2) - 2gv(t_p - t_q) + 2g(z_p - z_q)]^2.$$

Note that T is fourth order in g and quadratic in v. Thus, we can expect that setting $\partial T/\partial g$ and $\partial T/\partial v$ equal to zero to search for the minimum of T will yield a set of cubic equations for g and v. In the mathematical statistics formulation, the normal equations are linear in the unknowns. This increase in complexity of the "normal equations" is a common feature of physical statistics. In this problem (as in every other one that I have worked out) the increased analytical difficulty is only partial, since the ball accelerates, $g \neq 0$, therefore, the system is really quadratic in g $(g > 0)$ and linear in v.

Suppose instead of a radar at the base of the building that measures height, it measures Doppler shift. Were such a radar at the base of the building, then the observational data would be $\{(\dot{z}_n, t_n)\}$. In this case the mathematical statistician would form the sum of the squares of the residuals R,

$$R = \sum_{n=1}^{B} [\dot{z} - \dot{z}(t_n)]^2 = \sum_{n=1}^{N} [\dot{z}_n - (-gt_n + v)]^2$$

and proceed to minimize R with respect to g and v (no H). I start from T but write it as ($\dot{z}$ measured now and z from the solution of the physics problem)

$$T = \frac{1}{2} \sum_{p,q=1}^{N} \left\{ \left[\frac{1}{2}m\dot{z}_p^2 + mg\left(-\frac{1}{2}gt_p^2 + vt_p + H - H\right) \right] \right.$$
$$\left. - \left[\frac{1}{2}m\dot{z}_q^2 + mg\left(-\frac{1}{2}gt_q^2 + vt_q + H - H\right) \right] \right\}^2$$

(no H either! ever!!). Once again m factors out,

$$T = \frac{m^2}{8} \sum_{p,q=1}^{N} [(\dot{z}_p^2 - \dot{z}_q^2) - g^2(t_p^2 - t_q^2) + 2gv(t_p - t_q)]^2$$

but g does not from $\partial T/\partial g = 0$ (it still does from $\partial T/\partial v = 0$). While the situation is once again analytically more complex it is not bleak;

$$\partial T/\partial g = m \sum_{p,q=1}^{N} (E_p - E_q)[-g(t_p + t_q) + v](t_p - t_q) = 0$$

and

$$\partial T/\partial v = mg \sum_{p,q=1}^{N} (E_p - E_q)(t_p - t_q) = 0.$$

Note that the last equation is linear in the product gv. Solve for it and substitute first of the pair—the result is a quadratic equation for g^2.

Let us now generalize to a $1/r^2$ gravitational force and three spatial dimensions. The Keplerian two-body problem apparently has seven constants of the motion; the scalar energy per unit mass E

$$E = \dot{\mathbf{r}} \cdot \dot{\mathbf{r}}/2 - \mu/r,$$

the vector angular momentum per unit mass $\mathbf{L}$,

$$\mathbf{L} = \mathbf{r} \times \dot{\mathbf{r}},$$

and the Laplace (-Runge-Lenz) vector

$$\mathbf{e} = \dot{\mathbf{r}} \times \mathbf{L}/\mu - \mathbf{r}/r.$$

These are not all independent,

$$\mathbf{e} \cdot \mathbf{L} = 0 \qquad \text{and} \qquad 2EL^2 = -\mu^2(1 - e^2).$$

I have used the standard symbolism for the relative location $\mathbf{r}$ of the secondary with respect to the primary, their masses m and $M(M \gg m)$, the relative velocity $\dot{\mathbf{r}}$, and the universal constant of gravitation, G; $\mu = G(M + m) \simeq GM$. The underlying equations of relative motion are

$$m\ddot{\mathbf{r}} = -\mu m\mathbf{r}/r^3.$$

The initial conditions, say $\mathbf{r}$ and $\dot{\mathbf{r}}$ at the time $t = t_o$, $\mathbf{r}$ at times t_1 and t_2, or the orbital element set, are the unknowns of the problem. The force center locations and velocities are related to topocentric values $\mathbf{R}$ and $\dot{\mathbf{R}}$ by

$$\mathbf{r} = \mathbf{R} + \rho, \qquad \dot{\mathbf{r}} = \dot{\mathbf{R}} + \dot{\rho}$$

where $\rho(\dot{\rho})$ is the location (velocity) of the observer with respect to the center of force.

Some subset of the kinematic variables $\mathbf{R}$ and $\dot{\mathbf{R}}$ are measured. For the remainder of this sextuplet we have available the solution to the equations of motion. Thus, if we write the topocentric location $\mathbf{R}$ as the product of the topocentric distance R and a unit vector of topocentric direction cosines ℓ

$$\mathbf{R} = R\ell,$$

then in an asteroid problem we observe the components of ℓ and have formulas for R, $\dot{R}$, and ℓ. When an artificial satellite is observed by a radar, then R, $\dot{R}$, and ℓ are observed whereas ℓ is known from the solution to the equations of motion. In this fashion, at each time of observation $t = t_1, t_2, \ldots, t_N$, we have two mutually exclusive and exhaustive subsets comprising the $\mathbf{R}$, $\dot{\mathbf{R}}$ sextuplet.

Quantities such as E, $\mathbf{L}$, and $\mathbf{e}$ are mixed; part of their value is measured and the remainder depends on the initial conditions. This was easy to see in the one-dimensional ball problem but the generalization is straightfoward. Therein when we measured z we used it in E while the unmeasured value of $\dot{z}$ was expressed in terms of the initial conditions, force model parameters, and so forth. The same occurs now. Whenever a kinematic variable is measured, its numerical value is used to help construct E, $\mathbf{L}$, or $\mathbf{e}$. For the remainder of the unmeasured kinematic variables the formulas are used to complete the calculation of E, $\mathbf{L}$, and $\mathbf{e}$. These formulas necessarily incorporate the initial conditions. This is how they enter the problem and why T depends on them.

Understanding this, following the precepts of the new principle of statistical estimation is straightforward. We construct

$$T = \frac{1}{2} \sum_{p,q=1}^{N} (E_p - E_q)^2 + \frac{w_L}{2} \sum_{p,q=1}^{N} |\mathbf{L}_p - \mathbf{L}_q|^2$$

$$+ \frac{w_e}{2} \sum_{p,q=1}^{N} |\mathbf{e}_p - \mathbf{e}_q|^2$$

$$- \sum_{p=1}^{N} \lambda_p \mathbf{e}_p \cdot \mathbf{L}_p - \sum_{p=1}^{N} \Lambda_p [2E_p L_p^2 + \mu^2 (1 - e_p^2)]$$

and minimize T with respect to the initial conditions and the Lagrange multipliers $\{\lambda_p, \Lambda_p\}$. An alternative procedure would be to analytically reduce the seven dependent invariants E, $\mathbf{L}$, $\mathbf{e}$ to five independent ones by explicitly incorporating their interrelationships. Finally, the quantities w_L and w_e are weights to adjust the differing physical dimensions. In natural units these would be unity. Rather than deal with them in an artificially simple context, let us see how they naturally disappear in a more realistic scenario.

Suppose that instead of the pure Kepler problem we also incorporated the oblateness of the Earth. This conservative force would add another term to the potential energy and destroy the spherical symmetry of the overall problem. Therefore, the angular momentum vector $\mathbf{L}$ would no longer be a constant of the motion, only the component along the geopotential's axis of symmetry would still be conserved (say L_z). We could handle the statistical adjustment of data in one of two ways. We could just use E and L_z in T or we could employ a concept known as adiabatic invariants.

Adiabatic invariants are quantities lacking first-order secular changes when periodic systems are perturbed (see Goldstein, 1980). Also, they all have the same dimensions, that of an action. Hence, if the time span of the data is not too long, then the definition of T can be extended to include adiabatic invariants along with rigorously conserved quantities. The principle advantage of this concept becomes apparent when dissipative forces must be considered in addition to symmetry-breaking ones. Once again, because of very slow rates of change, adiabatic invariants may exist where true constants of the motion can no longer. By the mechanism of adiabatic invariants, and limits on the duration over which the data reduction is performed consistent with the proof of adiabatic invariance, many more realistic problems can be treated by the formalism discussed herein.

6 A PREDICITION OF PHYSICAL STATISTICS

The relationship of physical statistics to statistical mechanics is straightforward; physical statistics is a logical precedent to statistical mechanics. The statement known as the fundamental postulate of statistical mechanics can be *derived* from physical statistics. Moreover, while the existence and uniqueness properties of the estimates following from the principle of physical statistics have so far been assumed, in this case a simple proof of these attributes is available. To summarize, we now no longer need to separately posit the statement known as the fundamental postulate of statistical mechanics. Just as rules for curve-fitting and the differential correction of orbits fall out of the principle of estimation of physical statistics, so too does the fundamental postulate of statistical mechanics.

Mechanical systems can be described in terms of generalized coordinates q_1, q_2, $\ldots, q_f$ and their corresponding momenta p_1, p_2, $\ldots, p_f$. The 2f-dimensional pseudo-Cartesian product space is known as phase space. The fundamental postulate of statistical mechanics asserts that the probability density of systems in phase space is uniform.

In general the density in phase space, $\rho(q, p, t)$, provides the probability that the system has coordinates and momenta within dq of q and dp of p. Naturally ρ is a real, non-negative function and it is normalized,

$$\int_\Gamma \rho(q, p, t) \ dq \ dp = 1.$$

The volume of phase space $V(\Gamma)$ accessible to the system is finite.

Liouville's theorem states that $d\rho/dt$ vanishes. Statistical physics instructs us to concentrate on invariants. For this problem ρ is clearly it. By analogy with our other estimation problems we want to minimize the functional $T[\rho]$,

$$T[\rho] = \int_\Gamma \int_\Gamma [\rho(q, p, t) - \rho(q', p', t)]^2 dq dp dq' dp'$$

subject to the normalization constraint. While this might become a very complex problem in functional differentiation, the form of T, the attributes of ρ, and the fact that $V(\Gamma)$ is finite make the problem trivial;

$$\rho = 1/V(\Gamma).$$

That is, the density distribution in phase space is uniform: equal regions in phase space have equal probabilities. As just discussed this result is known as the fundamental postulate of statistical mechanics. A formal uniqueness and existence proof of this result can be constructed under the weak assumption that a generalized Fourier-Bessel expansion for $\rho(q, p, t)$ exists.

7 CONCLUSION

An attempt has been made to secure the underpinnings of the statistical estimation of parameters when the data arises from a scientific experiment. This attempt has been successful because the unknown and unknowable errors of observation and their distribution functions have been de-emphasized. Rather, known quantities, such as constants of the motion or abstract geometrical notions, form the basis of Physical Statistics. The breadth of this new principle of estimation is wide enough to encompass curve-fitting problems, parameter estimation in experimental situations, and provide a basis for statistical mechanics.

REFERENCES

GOLDSTEIN, M. (1980) *Classical Mechanics*, Addison-Wesley, Reading, MA.

MORTON, B. J., and TAFF, L. G. (1987) Target localization from simultaneous bearings. Submitted to *J. Oper. Res.*

TAFF, L. G. (1981) *Computational Spherical Astronomy*, Wiley, New York.

TAFF, L. G. (1985) *Celestial Mechanics,* Wiley, New York.

TAFF, L. G. (1988a) New principle of statistical estimation. *Phys. Rev.* **37A**, 4943–4949.

TAFF, L. G. (1988b) Gamma-ray burst astrometry. *Astrophys. J.* **326**, 1032–1035.

TAFF, L. G. (1988c) The plate overlap technique: A reformulation. *Astron. J.* **95**, 409–411.

Postal Address

Space Telescope Science Institute
3700 San Martin Drive
Baltimore, MD 21218, USA

The Median and the Morphological Filters as Robust Tools

A. BIJAOUI

Observatoire de la Côte d'Azur, Département A. Fresnel
Nice, France

Abstract

The data smoothing is a current operation in signal and image processing. Many tools were built in order to reduce the noise. They use convolutions with given masks or with the Fourier transform. These linear operations spread the signal according to a given law; consequently they are very sensitive to data the values of which are far from the ones which can be predicted from local regular variations. These abnormal values can result from many non-Gaussian errors, like scratches on a photographic plate, or from the signal structure. It is important to use tools which eliminate the defects with a minimum spread. The median and more generally the morphological filters were developed for such a purpose.

The median filter is intensively used in astronomy field for spectrograms and image processing. The general idea lies in the determination of the median of all the values in a given window around each sample. The median is a more robust variable than the mean. An aberrant value does not modify the local median. By consequence, the median filter is often used for the background mapping and for the defects removal.

The morphological filters are more recent tools. They were developed in the framework of the *Mathematical Morphology*. In a first step, we introduce the two basic operators on binary fields, the erosion E and the dilatation D. The combinations of these topological operators are very useful for binary image analysis. For the morphological filters we define the umbra as the binary image obtained with a signal plot. The morphological filters are the results of the morphological operators on the umbra.

We discuss the main filters resulting from the erosion, the dilation, the opening $O = DE$, the closure $C = ED$. The operator $S = CO$ deletes the peaks and fills the holes. We present a morphological filter algorithm based on a pyramidal analysis.

1   The linear data filtering.

The data smoothing is a current operation in signal or image processing. Many tools were built in order to reduce the noise. They use convolutions with given masks or the Fourier transform. The linear filtering results from a set of given hypothesis on the nature of the noise to be reduced. It must be additive, stationnary with a normal law. The data themselves are linearly correlated on a large area.

In case of no causal signals the Wiener's filter is the optimal one [1]. It can be performed with the help of the Fourier's transform or by a local convolution with a specific mask.

In all kinds of signal processing the classical hypothesis on the nature of the noise are affected by the existence of some artefacts. This can be scratches or dusts on the photographic plate, cosmic impacts or scanning defects for a CCD frame.

The linear filtering spreads the defect in order to decrease its effect, but it does not remove it. No peculiar linear convolution can achieve any signal cleaning from non-Gaussian noise. We need to reject in a local window the aberrant values.

That may be done by a local detection followed by the exchange with an interpolated value. A classical solution lies in the use of order-statistic filters, and especially the median filter [2].

2   The median filter and the *pepper and salt* noise.

The principle of the median filter is very simple. Around each pixel we define a window of a given size and we determine the median value of the intensity distribution. Many algorithms are available, using a stack, or a quick sort technique. The computing time is similar to the one of a convolution with the same window.

As the median is defined by a rank in the distribution, it is little sensitive to the abnormal values. The median filter removes them, if they are really well localized. If the defect is too large compared to the window, the median filter considers it as a real object. Only a knowledge on the pattern distribution on the signal allows one to separate large size defects.

The median filter leads also to a smoothing of the Gaussian noise. The quality of this smoothing is always worse than the one which can be obtained with a linear filtering.

Of course, the defect removal is the classical application of this filter in signal processing. Only small size defects are removed. For large ones, which can be easily localized, their contours are interactively determined, and the internal regions are filled with a suitable interpolation.

The main application of the median filter is the background mapping [3]. The window size is larger than four times the stellar one. The subtraction shows the residual peaks corresponding to the stars. It is easy to show [4] that a bias exists between the median and the real background. A better background can be computed with an histogram model [5]. The use of the median filter leads to systematic errors, which can be very large for deep fields.

3   The mathematical morphology and the morphological filters.

Another way to remove peaks and holes is based on the morphological filters [6]. These new tools were recently developed from the mathematical morphology [7].

Let us consider a binary set A. The erosion E of A with a structuring element B is the set $\{x = a - b \quad a, x \in A \quad b \in B\}$. The dilation D is the set $\{x = a + b \quad a \in A \quad b \in B\}$. The Opening O is the product of E with D. A small difference exists generally between the original set and the resulting one: the capes of 1 pixel size are removed . The Closing C is the symmetrical operator. It fills the bays.

If we iterate O or F we obtain the same result: they are idempotent operators.

Let us now consider a digital signal $f(x)$. The plot of this signal delimitates a binary set, the umbra. The morphological filters are the application of the morphological operator on the umbra. Whith a specific structuring element, the erosion and the dilation operators are:

$$E(f(x_i)) = \inf(f(x_{i+k})) \qquad D(f(x_i)) = \sup(f(x_{i+k}))$$

where $k \in (-K, +K)$.

The Opening (or the Closing) is always the product E by D (or D by E). Their effects are simple: O removes all the peaks of a size little than the window one, and F fills the corresponding holes. The products FO or OF remove the peaks and fills the holes, but they are not idempotent operators.

In case of the median filter, we modify the signal in order to remove the peaks and the holes: with the morphological filters this is done in a more defined manner. We can iterate the process with larger and larger windows, in order to remove the larger peaks and to fill the larger holes. We can used for that purpose a dyadic technique.

The pyramidal image representation [8] is now a classical tool. Recently this concept was applied to the mathematical morphology [9]. A reduction of the size by a factor 2×2 is obtained at each step.

A similar filtering, without data reducing, is obtained with an *algorithme à trous* like the ones developed for the wavelet analysis [10]. The comparizons are taken in a window having always three elements, with a increasing distance at each step by a factor 2. We eliminate first the 1 pixel size peaks, then the 2 pixel ones, 3 − 4 pixels ones, and so one.

4   Conclusion: the correlated or uncorrelated events.

When we smooth a signal, we want to reduce the statistical fluctuations using the correlation between close data. That may be disturbed by uncorrelated events. The morphological filters are the adequate tools to detect and to remove them, but they do not smooth the data. They must only be used for the signal cleaning. As they are non linear operators they may lead to biases. A careful data examination must be done before the signal analysis.

The data processing must combined the morphological filtering and the linear one.

References:

[1] Pratt W.K. (1978) in *Digital Image Processing* p.427 J.Wiley and sons New York.

[2] Pratt W.K. (1978) in *Digital Image Processing* p.330 J.Wiley and sons New York.

[3] Sulentic J.W. and Lorre J.J. (1984) Sky and Telescope **407**.

[4] Irwin M.J. (1985) *Automatic analysis of crowded fields* Mont. Not. R.A.S. **214** 575-614.

[5] Bijaoui A. (1980) *Sky background determination and applications* Astron. Astrophys. **84** 81-85.

[6] Maragos P. (1987) *Tutorial on advances in morphological image processing and analysis* Opt. Eng. **26** 623-632.

[7] Serra J. Image Analysis and Mathematical Morphology Academic Press New York 1982.

[8] Burt P.J and Adelson E.H. (1983) *The Laplacian pyramid as a compact image code* IEEE Trans. Com. **31** 532-540.

[9] Toet A. (1989) *A Morphological pyramidal image decomposition* Pattern Rec. Let. **9** 255-261.

[10] Holdschneider M., Konland-Martinet R., Morlet J., Tchamitchan P. (1989) "The "Algorithme à trous" Centre de Physique Théorique CNRS-Luminy Marseille.

WFPC: Options for overcoming undersampling[1]

H.-M. Adorf

Space Telescope – European Coordinating Facility
European Southern Observatory
Karl-Schwarzschild-Str. 2
D-8046 Garching bei München, F.R. Germany

Abstract: The Planetary Camera (PC) in the blue and the Wide Field Camera (WFC) in all wavelength ranges will undersample the effective point spread function (PSF). Since countermeasures against insufficient sampling can be taken only at the time of observation, it is important to incorporate an appropriate sampling strategy into the observation design. This contribution reports on a study, underway at the ST-ECF, which investigates the WFPC undersampling problem in order to generate advice on the optimal design of WFPC imaging campaigns. Various sampling options are presented which may also be helpful in similar situations with ground-based equipment.

The Wide Field and the Planetary Camera both undersample the effective point spread function (PSF) in much of the covered wavelength range. This problem, though known for a long time, has not been subject to a detailed study so far. Recognizing the importance of the WFPC among the HST-instruments, we have earlier this year embarked on a study of the associated undersampling problem (Adorf 1989a, b) that encompasses (i) an assessment of the severity of the problem for both cameras in various wavelength ranges, (ii) a specification of a conceptual 'stochastic' model describing the 'clean' imaging process, (iii) a derivation of strict limits to the precision of photometric and astrometric measurements, given by the so-called "Cramer-Rao-bound" (Adorf 1987), and if possible (iv) the specification of algorithms tackling the problems of comparing and combining image frames that are relatively displaced by fractions of the detector pixel separation.

With the help of a conceptual 'stochastic' model of the imaging process the data gathering process is described in an idealized fashion by considering four different types of image degradations: (i) image *blurring* caused by diffraction in the camera optics and the detection of photons by pixels with finite apertures, (ii) *aliasing errors* arising from insufficient image sampling by the detector pixel-grid, (iii) *random noise* and (iv) *quantiziation errors* occurring at detector read-out. Most of the various 'dirty' effects occurring in WFPC imaging (Lauer 1988) are neglected for the sake of simplicity.

Several parameters enter into the problem description in an essential way, among these are the pixel (center) separation, the width of the optical PSF and the width of the detector pixels (the latter two being combinable into a width of an 'effective PSF'). Whereas pixel size and separation are of course well known, there are still some uncertainties remaining with respect to the width of the optical PSF. We are using the optical simulation software available from the STScI (Burrows & Hasan 1989) to estimate this important parameter for various wavelengths.

Space constraints do not permit an extensive discussion of sampling theory, which in two dimensions is considerably involved (for a review see e.g. Jerri 1977). Instead a picture gallery of some options shall be presented which may help to decide on how to best combat insufficient sampling. A few implications due to the peculiarities of HST operations are also mentioned.

A simple measure to improve sampling in a particular direction — taking advantage of the effective PSF being fairly circular symmetric — is to align one of the detector diagonals with a direction on

---

[1] *ST-ECF Newsletter* **12**, 1989, (in press)

the sky (Fig. 1a, b) in which spatial resolution shall preferentially be optimized. This option requires the specification of a special roll angle in the HST Phase II proposal forms.

In order to facilitate cosmic ray detection, HST operations foresee to routinely take a pair of co-frames for each WFPC target. Whereas by default these frames will be taken at identical positions, the observer may specify a 'fractional pixel offset' (Fig. 2) between the two exposures and thereby partly fill in 'holes' in the sampling pattern of the detector grid. The pointing accuracy of the HST Fine Guidance Sensor system should allow such sub-pixel offsets, when the exposures are taken in immediate succession, presumably even when taken in different subsequent orbits.

If even higher spatial resolution is required than what can be obtained with two interleaved co-frames, fourfold sampling schemes may be employed, where one may choose between the simpler 'square sampling' (Fig. 3) or the more involved 'hexagonal sampling' option (Fig. 4). The latter provides the highest resolution attainable with four frames — at the price of a more complex data analysis procedure.

The intrinsically complicated question of how two or more frames can be combined when their relative displacement and position angle is essentially random (a situation occuring when the target is revisited after some period) will also be addressed in due time.

The simple 'diagonal' and twofold interleaved sampling options merit consideration by every WFPC observer. Inhowfar the more sophisticated sampling schemes can be realized in practice will be revealed only during HST Orbital and Science Verification. While it is too early to draw definite conclusions, the need was felt to raise the awareness within the community with respect to the WFPC undersampling problem. The various possible options presented above should help the HST user to design a WFPC observing strategy which, at least to some extent, overcomes the adverse effects of undersampling.

References

Adorf, H.-M.: 1987, "Theoretical Limits to the Precision of Joint Flux- and Position-Measurements with CCD-Detectors", (unpublished)

Adorf, H.-M.: 1989a, "WFPC Undersampling. I. Problem Assessment", ST-ECF Technical Report

Adorf, H.-M.: 1989b, "On the HST Wide Field and Planetary Camera Undersampling Problem", in: *Proc. 1st ESO/ST-ECF Data Analysis Workshop, European Southern Observatory*, Garching bei München, Apr. 1989, P. Grosbøl, F. Murtagh, R.H. Warmels (eds.), (in press)

Burrows, C., Hasan, H.: 1989, "Optical Modelling Software User Manual, Version 6a", March 1989, Space Telescope Science Institute, Baltimore, MD 21 218

Jerri, A.J.: 1977, "The Shannon Sampling Theorem — Its Various Extensions and Applications: A Tutorial Review", *Proc. IEEE* **65**, No. 11, 1565–1596

Lauer, T.R.: 1988, "The Reduction of Wide Field / Planetary Camera Images", Princeton Observatory Preprints

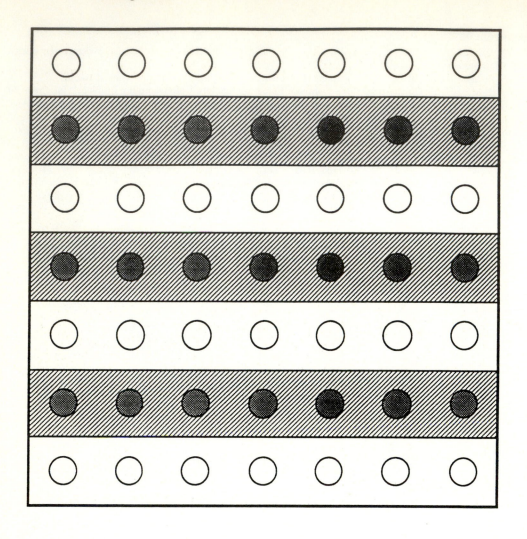

Fig. 1a: Spatial resolution along principal detector directions. The finest spatial detail that can be resolved is 1/2 cycle per pixel-center separation ("Nyquist frequency").

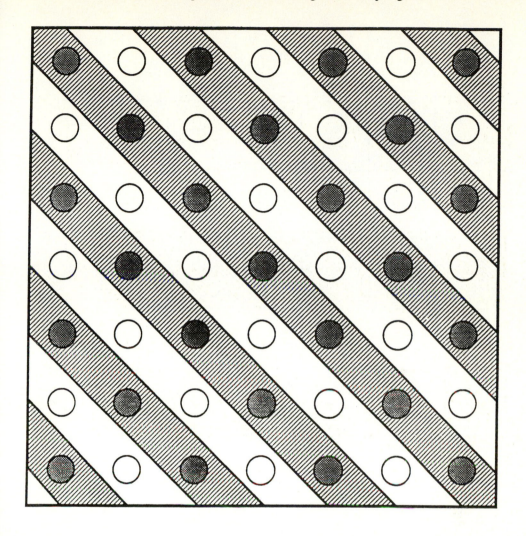

Fig. 1b: Spatial resolution along detector diagonals. Due to denser sampling, spatial resolution is improved along both detector diagonals by a factor $\sqrt{2}$, compared with the principal (horizontal and vertical) detector directions.

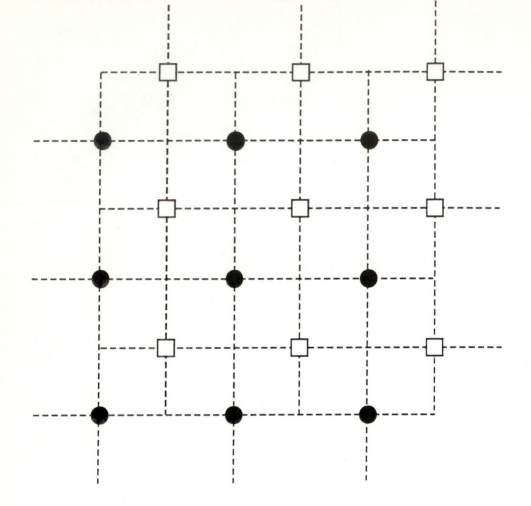

Fig. 2: Twofold interleaved sampling. Two image frames taken with a relative displacement of half a (diagonal) pixel separation can be combined into a single simple frame with its axes along the diagonals of the co-frames it is composed of. Spatial resolution of the combined frame is increased by a factor of two along the principal axes of the co- frames, but not at all along their diagonals. The shaded areas represent pixel shapes.

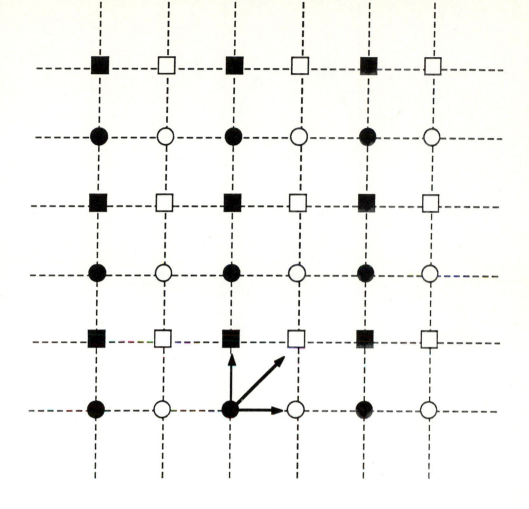

Fig. 3: Fourfold square sampling. Spatial resolution can be doubled in all directions by taking four interleaved frames, fractionally displaced by half a pixel separation as shown. As in the previous case a simple frame can be composed out of the four co-frames.

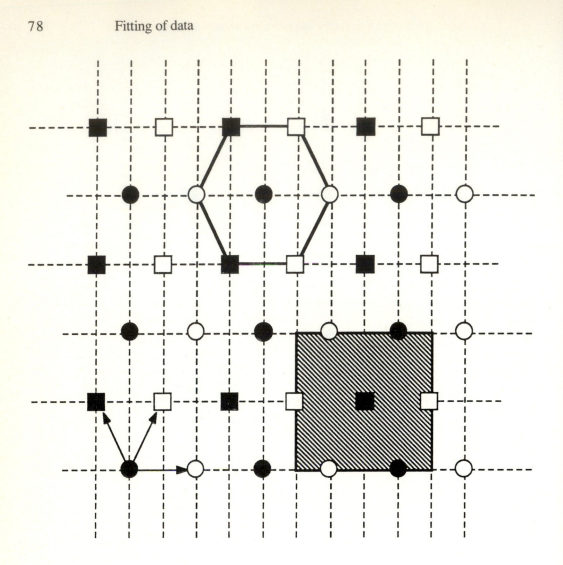

Fig. 4: Fourfold hexagonal sampling. For a circular symmetric point spread function optimal spatial resolution can be achieved by a 'hexagonal' arrangement of the four co-frames. However, this co-set of frames is more difficult to analyze.

WATREDAO: An Automated Statistical System for the Determination of Reliable Distances to Stars and Clusters

M.P. FitzGerald, G.L.H. Harris, E.J. Jylänne*

Guelph–Waterloo Program for Graduate Work in Physics, Waterloo Campus, University of Waterloo, Ontario N2L 3G1, Canada; and
Dominion Astrophysical Observatory, Herzberg Institute for Astrophysics, Victoria, British Columbia V8X 4M6, Canada.[a]

1.   INTRODUCTION

For some time we have been concerned by inconsistencies in the derived properties of open clusters; poorly derived properties include distance and calibration of the intrinsic properties of most luminous stars. Open cluster distances depend on the modulus of the Hyades and thence on fitting to increasingly more luminous portions of the ZAMS.[b] The fitting procedure depends on: (a) the inter-relation of the ZAMS M_V, $(B - V)_0$, and $(U - B)_0$, (b) the interstellar extinction relations, (c) the eye of the investigator, and (d) the propagated errors in individual members. In practice investigators use different intrinsic relations for (a) and (b), the 'fit by eye' is highly subjective, and the eye is quite poor at allowing for moderate and/or variable errors.[c]

Because of the above problems we have devised a statistically based approach for finding cluster (and individual stellar) moduli requiring V, $(B - V)$, at least one of another colour (usually $(U - B)$), MK spectral and luminosity class, *as well as the observational errors of these quantities.* If no errors are available WATREDAO can assign its own, or other, default errors of [0.022, 0.022, 0.027, 1.0, 0.70] to [V, $(B - V)$, $(U - B)$, Sp, LC] respectively. The method meets the following criteria:
(a) It mimics the good qualities of the eye,   (b) it is reproducible by any user,
(c) it gives reliable estimates of $\langle V_0 - M_V \rangle$, $\langle E_{B-V} \rangle$, differential extinction and errors therein,
(d) it can use MK as well as UBV data, (e) it calculates *photometric mk* classes,
(f) it assigns cluster membership and classifies stars as ZAMS, evolved or binary,

---

* Elizabet Jylänne died in October 1988. Her programming ability and knowledge of this area of Astrophysics has made this project feasible. She is sorely missed, both in her ability and in her person.

[a] The title refers to **W**aterloo **A**stronomers **T**echnique for **R**eduction of **E**xtinction and **D**istances to **A**stronomical **O**bjects. It is in recognition of the great support received from the University of **WAT**erloo and the **D**ominion **A**strophyscial **O**bservatory. For reader relief we did not incorporate our other strong supporter the National Science and Engineering Research Council of Canada.

[b] Zero Age Main Sequence.

[c] Variable errors are always present around A2 and F8 because of the 'humps' in the two colour relation; the Q–method is unreliable because the intrinsic ZAMS relation is not quite smooth; mean colour excesses are frequently wrong because the intrinsic relation is a function of Luminosity Class.

(g) it determines probability (and its reliability) of cluster and background group membership,   (h) it determines the cluster age,   (i) it is interactive,

(j) it can use any tabulated intrinsic relations, but currently adopts those from FitzGerald (1970) and Schmidt-Kaler (1983).[d]

Here we consider only the UBV system and fitting to the ZAMS. Intrinsic properties are interpolated with quadratic relations between values tabulated for 67 spectral sub-classes and 10 luminosity classes.[d]

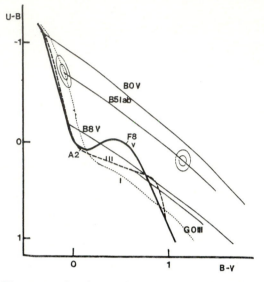

Mean, variance and weight

We define mean, variance, variance in the mean, and weight conventionally. In particular for observations of equal weight

$$w_X = 1/\sigma_X^2 , \qquad (1)$$

$$w_{\langle X \rangle} = 1/\sigma_{\langle X \rangle}^2 . \qquad (2)$$

For observations with variable weight

$$\langle X \rangle = \Sigma(w_i \, X_i)/\Sigma w_i , \qquad (3)$$

$$W_{\langle X \rangle} = \Sigma w_i , \qquad (4)$$

$$\sigma_{\langle X \rangle}^2 = 1/W . \qquad (5)$$

Figure 1. Intrinsic two-colour relations and extinction lines. Also shown are the observed and dereddened one and two sigma error ellipses for a B5 Iab star. If the star were class V the resultant intrinsic colours would be quite different.

2.   CORRECTION FOR EXTINCTION

WATREDAO corrects individual stars for extinction by one of three methods, primarily: (a) dereddening in the two-colour diagram, or (b) mean cluster extinction if (U – B) is unavailable or the error from (a) is much greater than the non-uniform extinction and includes the former in its bounds, or (c) MK class if (U – B) is unavailable. The extinction lines are curved and depend on intrinsic colour and E_{B-V} after Schmidt-Kaler (1983).

It propagates observed errors at the one and two sigma level allowing the reddening line to intersect the appropriate intrinsic relation, which we assume to have a *natural dispersion* of variable breadth ranging for [M_V; (B – V); (U – B)] from [0.05 to 0.1; 0.008 to 0.09; 0.017 to 0.1]. Figure 1 shows intrinsic extinction relations for various stars. Note the A2 and F8 humps which give rise to non-unique solutions. Figure 2 shows several graphical examples of how WATREDAO corrects for extinction and estimates one and two sigma errors in the intrinsic values. It calculates the visual extinction from the observed colour excess and intrinsic colour, and forms V_0 and (B – V)$_0$ *with their propagated errors* for each star.

---

[d] *Spectra*   1-66: O0-M5;   67: M5.5;   68-74: M6-M12;   75-81: C3-C9.
   Luminosity   1: VII = WD;   2: VI = ZAMS;   3–6: V–II = d–bg;   7-10: Ib-Ia0 = c.

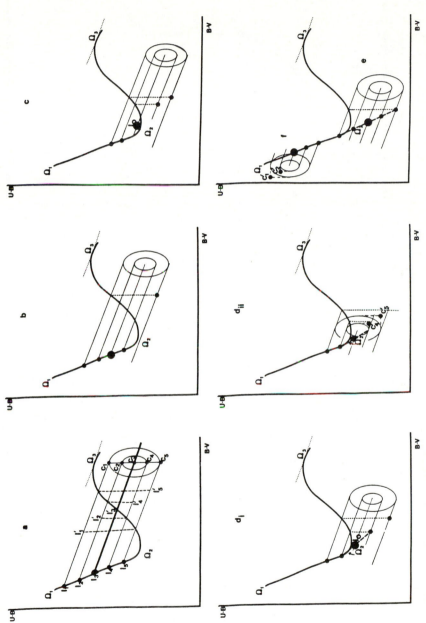

Figure 2. Seven dereddening cases.

The large filled circle represents the adopted intrinsic colour. Not all points are lettered in each diagram.
(a) Two independent solutions: (i) Solution is I_3 with one and two sigma errors of half the separation between I_4-I_2 and I_5-I_1 respectively; (ii) An equivalent solution about I_3'.
(b) The two sigma bound runs from I_1 to the dot shown. The dashed line is vertical.
(c) I_0 lies mid-way between the two intersections of the C_3 extinction line. The error bounds are defined as in (b).
(d$_i$); (d$_{ii}$) The C_3 extinction line does not cross an intrinsic line. I_3 is at the mid-point of the dashed line and error bounds are defined as in (b).
(e) As in (d$_{ii}$), but more extinction.
(f) The case for a star bluer than the intrinsic line. (U–B) is ignored if the two sigma ellipse fails to intersect it.

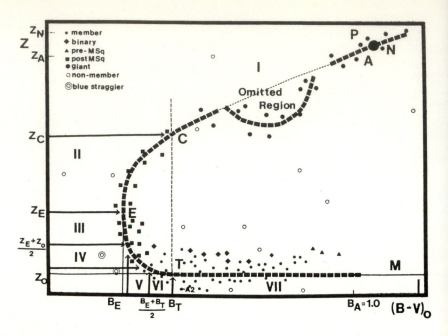

Figure 3. Z versus $(B - V)_0$.
The true distance modulus for the cluster is at Z_0, a horizontal line. Certain specific points in the diagram are labelled; each has a corresponding value of Z, $(B - V)$ and M_V^Z; these are used as age indicators. Initial fitting is done by fitting a 'lower' horizontal envelope at Z_0. Final fitting is accomplished by a complex maximum likelihood approach, requiring the region between M and T be horizontal, between T and E an ellipse, and between E and N a hyperbola with a cubic term. Fitting in regimes I to VII is described in the text (the large dashed portion of region I is omitted; P is a low weight point, determined from B_E, used to stabilize the cubic when no stars lie in the giant part of region I).

Since intrinsic relations are luminosity class dependent WATREDAO iterates to find spectral-luminosity class and intrinsic colours consistent with the distance modulus adopted. It then finds a new modulus (see below) and repeats the process, reaching convergence within 2 to 4 iterations provided membership assignment is not too difficult. Initial moduli within $Z_0 = (V_0 - M_V^Z)_{-5}^{+2}$ are satisfactory;

3.   ZAMS LOWER ENVELOPE FITTING

3.1   Preliminary Fitting

To find a cluster distance modulus WATREDAO fits a horizontal lower envelope moving it upwards (see Figure 3). It starts by forming a modulus and variance,[e] assuming each star is a ZAMS object:

$$Z = V_0 - M_V^Z ; \tag{6}$$

$$\sigma_Z^2 = \sigma_{V_0}^2 + \left(\frac{dM_V^Z}{d(B - V)_0} \right)^2 \sigma_{(B-V)_0}^2 . \tag{7}$$

The resultant σ_Z^2 values vary substantially depending on the slope of the ZAMS

---

[e] We use equations similar to (7) throughout to form variances allowing for the slope of a line.

and on the error in intrinsic colour; the latter are often very large ($\sim 0.^{m}1$) near A2 and F8 in the two-colour diagram. Figure 3 shows a schematic appearance for the diagram of Z versus $(B - V)_0$. Initially WATREDAO fits a lower horizontal envelope such that the distribution of points *below* Z_0 in the figure is *that expected from their propagated observational errors*, σ_Z. In particular, defining

$$R = Z - Z_0 \qquad (8)$$

and

$$\Upsilon = \left[\sum_{R>0} w(R/\sigma)^2 \right] \bigg/ \left[\sum_{R>0} w \right] , \qquad (9)$$

we hypothesize that Υ has value 1.0 for the correct value of Z_0. In practice Υ is affected by the discrete addition of individual stars as Z_0 is decreased. Consequently we fit a cubic (nearly linear) curve of log Υ versus Z_0 where Υ is evaluated: (a) mid-way between a newly added observation and that previous, and (b) at the average Υ immediately before and after a newly added observation.

The resultant variance in the fitted Z_0, $\sigma_{Z_0}^2$, is assumed to be the sum of: (a) $1/W_{\langle Z_0 \rangle; R>0}$, (b) the mean variance of the fitted curve above at Z_0, and (c) the square of the difference between the Z_0 above and with the faintest two points removed. Typical fitting accuracies are 0.04 to 0.3 depending on numerous factors, such as the number of ZAMS stars. The process is iterative, after each new Z_0 WATREDAO obtains new intrinsic parameters and extinctions for each star, it reviews membership (but leaves the decision to the researcher), and repeats the process if required.

3.2 Final Fitting

Once an acceptable excess and modulus are derived, and the membership decisions are satisfactory, a final fit may be done. If we consult Figure 3 we see that it is divided into regions I-VII to which we fit three sections: a horizontal line, an ellipse, and a hyperbola with cubic term. These are, respectively

$$Z - Z_0 = 0 , \quad (10)$$

$$\left(\frac{B - B_T}{B_E - B_T} \right)^2 + \left(\frac{(Z - Z_0) - Z_E}{Z_E} \right)^2 = 1 , \quad (11)$$

$$\left(\frac{B + B_T - 2B_E}{B_E - B_T} \right)^2 - \left(\frac{(Z - Z_0) - Z_E}{Z_C} \right)^2 + \left(\frac{(Z - Z_0) - Z_E}{Z_\chi} \right)^3 = 1 . \quad (12)$$

Region VII may be subdivided into 'spectral sub-groups'. WATREDAO fits these 7 (or more) regions to the three sections with a modified SIMPLEX version of maximum likelihood, requiring each region to have a weighted Υ (equation (9)) as close to 1.0 as possible: In region I the weighted deviations are minimized;[f] II-IV are fitted by bluemost envelope, and V-VII by lower envelope. The critical

---

[f] Weights for I are $w_{\rm I} = 1/(\sigma_Z^2 + 0.50)$ to allow for natural scatter of evolved stars.

points TECA give five independent parameters B_T or M_T, B_E, M_E, M_C and M_A which are age cubic functions of age. Based on 53 clusters evaluated by eye we find $\sigma_{\langle \log \tau \rangle} = 0.10$.

4.   ASSIGNMENT OF MEMBERSHIP

WATREDAO assigns membership by calculating membership probabilities, and their reliabilities, based on a variety of tests that can be performed, including information on star counts, proper motions etc. Each individual test has a calculated probability p_i and reliability r_i between 0.0 and 1.0 defined (with modifications for individual cases) as:

$$p_i = e^{-0.5(\Delta/\tau)^2} \tag{13}$$

$$r_i = \left[\frac{\varsigma^2}{\varsigma^2 + \tau^2} \right] \tag{14}$$

These are combined to form an overall probability and reliability:

$$P = \left[\prod_i p_i^{r_i} \right]^{1/\Re} \tag{15}$$

$$\Re = \sum r_i \tag{16}$$

where Δ is the difference between two predictions of the same quantity, τ^2 the sum of their variances, and ς^2 the minimum variance expected.[g] WATREDAO assigns membership based on the value of $[\Delta/\tau]$, the boundaries between [m, pm, pn, n] being at [0.8, 1.3, 2.15]. It also classifies stars as Main Sequence (just below and above ZAMS; possible, probable, and certain binary), post and pre-Main Sequence, blue straggler, giant or white dwarf.

5.   PRELIMINARY RESULTS

WATREDAO meets all the criteria set out above. It handles photographic and photoelectric data equally well. Typical results for clusters with about 20 to 4 ZAMS stars give errors of 0.116 to 0.316 *including an uncertainty error of 0.10, combined as variances.* In the future we intend to do Monte Carlo testing on the program, re-calibrate the ZAMS and intrinsic stellar properties as functions of MK class, and determine all open cluster distances, properties, and errors on a uniform basis. We are currently reworking the various FORTRAN subroutines to ensure consistency of nomenclature and the maximum transparency prior to releasing the program for general use.

6.   REFERENCES

FitzGerald, M.P. 1970, *Astronomy and Astrophysics*, **4**, 234.
Schmidt-Kaler, Th. 1983, *in Voigt, H.H. 1981, Landolt-Bornstein: Numerical Data and Functional Relationships in Science and Technology, Group* **VI**, *Vol.* **1** *and* **2**. **4.1.1**, 10-24.

---

[g] These have values close to those given as default errors in § I.

DISCUSSION

J.F.L. Simmons to L.G. Taff: Einstein would not have dismissed Newtonian mechanics, or Pauli the notion of distinguishable particles, in such a partisan way as you seem to dismiss mathematical statistics. Classical theories have their domain of validity, and in that domain often provide the only viable way of approaching a problem. This also applies to the results of various branches of mathematical statistics. There is often a degree of arbitrariness in using least-squares, maximum likelihood or any other optimising technique, and when physical principles can be used to decide the best suited optimisation procedure, then this should be done, but in many cases such principles are not apparent.

I did not address dynamics but a purely mathematical model for physical space-time. In the same way, I am questioning the appropriateness of an arbitrary branch of mathematics for physical experiments. Classical mechanics is numerically appropriate in a certain realm (low speeds) but Euclidean geometry is *conceptually* inappropriate for physical space-time. Its numerical success is accidental. Euclidean geometry does not possess the desirable invariance properties of physical theories – neither does any branch of mathematical statistics. Indeed, that there are branch*es* of geometry or of statistics should tell you of their inadequacy.

J. Nousek to L.G. Taff: When you dispense with the assumptions of "mathematical" statistics, you also loose the ability to perform certain tests of models, such as estimation of parameter limits, and goodness of fit. In particular, if the model is wrong there is no gain, or even reason to use "physical" statistics.

If the model is wrong, then why are you using it? All the benefits of mathematical statistics follow because of the extraordinarily strong assumptions you make a priori. Confidence limits, goodness of fit tests, and so on cannot be performed unless you assume a *specific* probability model for the unknown errors of observation. This is a false power and is solely a consequence of assuming you know the answer. Finally, if I did have a poor or inadequate model, then I would get a big value of T and this would be my clue. A big sum of squares of the residuals in mathematical statistics can merely say that you were unlucky.

S.P. Bhavsar to L.G. Taff: The very question bothering me in this context has been asked: one must clearly distinguish between the physical laws we are *assuming* and those we are *testing* when we use physical statistics.

Your point is well taken. I cannot use physical statistics to analyze the height/weight relationship of the men in this room. But that is because collecting height and weight measurements is not the same as measuring the charge-to-mass ratio of the electron. The latter is a scientific experiment, the former is merely data acquisition.

S.P. Bhavsar to L.G. Taff: What you say is okay when the physics is simple, like energy conservation. When you have a very complicated physical model, e.g. the light curve of a supernova, you *have* to use mathematical statistics – to fit a curve to the data.

Agreed. When the problem does not have a clear physical model, then physical statistics is inapplicable in practice, not in theory.

C.S. Cole to L.G. Taff: It seems you are asking us to give up assumptions about the data in favor of assumptions about the model. I often have more faith in the data than in the model.

Your already made assumptions about the model to build any framework in which to analyze the data. If your model demands energy conservation, then this is nothing additional, it is just a consequence of differential equations you have already adopted. Nothing new or extra is required. Mathematical statistics always requires on the order of N^2 additional, untestable assumptions regarding the N data points.

H. Eichhorn to L.G. Taff: I wonder if you are not putting yourself into a disadvantage by ignoring information about the covariance matrix of the measurements, which quite often you have. When you use, for example, two data sets of vastly different precision, your technique would produce "nests" of systematic residuals in the more precise data.

If (*if!*) you really do know something about the covariance matrix or the skewness or the kurtosis, then by all means use it. I created Physical Statistics to rest precisely on what one knows instead of the plethora of things one doesn't know (but must assume) to use mathematical statistics. Moreover, I agree that in certain very simple situations one can know such things, i.e. a meter stick graduated in cm. versus one in mm. However, in general, one has no such knowledge.

H. Eichhorn to A. Bijaoui: Could you please tell us what a Wiener filter is, and how it differs from least squares?

The Wiener filter is the one derived from least squares between the expected signal and the linearly predicted one. It is given by $W(\nu) = S(\nu)/(S(\nu) + N(\nu))$ where $S(\nu)$ is the density of the signal, and $N(\nu)$ the density of the noise.

H. Morrison to H.-M. Adorf: Your stochastic model for CCD imaging assumed that values in adjacent pixels will be uncorrelated, didn't it? Is this applicable if a stellar image covers more than one pixel?

Yes, since it is only the noise that I assume to be uncorrelated.

H.L. Marshall to H.-M. Adorf: Can you estimate the sensitivity loss between pixel center and edge now that the uncertainty of the measured flux is shown to be dependent on position within the pixel?

The sensitivity is not different between pixel center and edge, only the signal-to-noise ratio. But maybe there is some misunderstanding prevailing and we should discuss your question privately later.

M. Kurtz to H.-M. Adorf: Could you suggest a case where you would recommend that an observer uses a four-observation scheme, given the lower observing time and increased read-out noise?

Originally I had thought that HST WFC imagery in the UV would in fact *require* fourfold ("square") sampling, but due to problems totally unrelated to undersampling the WFC will presumably not be used much in that wavelength regime.

E. Feigelson to H.-M. Adorf: Will actual WF/PC STSDAS algorithms approach the Cramér-Rao accuracies [precisions!] you have calculated?

I fear they won't, since the current version of the STSDAS software has not been designed with undersampling problems in mind.

H.L. Marshall to M.P. Fitzgerald: Have you performed Monte Carlo simulations of color magnitude diagrams so that you may verify the estimators and their variances?

Not yet, we certainly intend to. We have compared our results with distances obtained by "old fitting methods" and the same intrinsic data and extinction lines. WATREDAO has always produced results within the old "error" bars, though usually with smaller variances than we assigned to the "old" method. We have also tested photographic data, and we find the same results as for photoelectric, namely with large final variances.

S.J. Hogeveen to M.P. Fitzgerald: (1) Do you provide WATREDAO to other astronomers as a tool? (2) Does your group (intend to) recalibrate the cluster distances themselves?

(1) Yes, once it is fully tested, and we are satisfied that the system is operating properly, and quite transparent to the user. (2) We intend to recalibrate the intrinsic properties of stars, M_V, $(B - V)_O$, $(U - B)_O$ as functions of MK class; then we intend to redetermine cluster distances on what we hope is now a uniform system with uniform methods.

SMALL SAMPLES

Le Jackknife et le Bootstrap

PH. P. NOBELIS

Département de Mathématiques
7, Rue René Descartes
67084 Strasbourg CEDEX
U43103 @ FRCCSC21

ABSTRACT

The Jackknife and the Bootstrap are the most known resampling methods which allow to avoid assumptions, as normality, on the observations. Each procedure is here precisly described and some examples are given. Then follow general view of resampling methods and applications in estimation problems as bias correction, nonparametric confidence intervals, etc. As it shown, Jackknifing and Bootstraping are closely related. Multidimensional cases are mentionned through linear regression and principal components analyses. This exposure is concluded with complete references on other applications of these methods.

1. INTRODUCTION

L'hypothèse du caractère gaussien, noté $\mathbb{N}$, des observations a longtemps été une contrainte dans l'utilisation concrète des méthodes statistiques. Ceci a constitué la principale motivation pour la mise en place de méthodes dites non paramètriques. Parmi celles-ci, le rééchantillonnage permet de considérer des échantillons observés de taille relativement petite. Les deux méthodes les plus utilisées sont le JACKKNIFE introduit par M. Quenouille (1949,1956) et développé, entre autres, par J. Tukey (1958) , et le BOOTSTRAP proposé par B. Efron (1979).

Ces démarches sont essentiellement utilisées dans des problèmes d'estimation : réduction de biais, évaluation de la variance d'un estimateur, construction d'intervalles de confiance non paramètriques et robustes, régression linéaire, étude des valeurs et vecteurs propres en Analyse en Composantes Principales, etc.

Dans toute la suite $X_1, X_2, \ldots, X_n$, désigne un n-échatillon de variables aléatoires (v. a.) mutuellement indépendantes issues d'une v. a. X, dont la loi est notée $\mathbb{L} = \mathbb{L}(X)$. On suppose que cette loi dépend d'un paramètre $\theta \in \mathbb{R}^k$ inconnu et que l'on dispose également d'une statistique

$$T = T_n = T(X_1, X_2, \ldots, X_n; \mathbb{L}),$$

estimateur de θ. Les réalisations des différentes v. a. seront notées par des miniscules.

2. DEFINITIONS

Considèrons les n statistiques suivantes

$$T^{(i)} = T_{n-1}(X_1, \ldots, X_{i-1}, X_{i+1}, \ldots, X_n), \quad i = 1, \ldots, n.$$

C'est la même statistique T calculée sur l'échantillon où, à tour de rôle, une observation est chaque fois omise.

Définition. _On appelle JACKKNIFE de la statistique_ T _la variable aléatoire_

$$JT = n^{-1}\Sigma_i (n T_n - (n-1) T^{(i)}).$$

La quantité $n T_n - (n-1) T^{(i)}$ est la _pseudo-valeur_ d'ordre i de T. Si T est, par exemple, la moyenne de l'échantillon, son Jackknife est encore la moyenne. Notons $\mathbb{L}_n$ la loi empirique associée à l'échantillon, c'est-à-dire celle qui affecte une probabilité n^{-1} à chaque valeur X_i.

Définition. _Soit_ $X_1^*, X_2^*, \ldots, X_n^*$ _un n-échantillon extrait de_ $X_1, X_2, \ldots, X_n$ _avec la loi_ $\mathbb{L}_n$. _On appelle BOOTSTRAP de la statistique_ T _la variable aléatoire_

$$BT = T(X_1^*, X_2^*, \ldots, X_n^*).$$

Comme la loi $\mathbb{L}_n$ converge vers $\mathbb{L}$ quand $n \to \infty$ (Théorème de Kolmogorov), celle de BT nous fournit une bonne évaluation de la loi de T. Il est posible de l'évaluer de trois manières différentes :

- par calcul direct,

- par développement de Taylor,

- par simulation : plusieurs échantillons $x_1^*, x_2^*, \ldots, x_n^*$ sont extraits sur lesquels des réalisations de BT sont obtenues; un histogramme peut ensuite être construit sur celles-ci.

Les deux méthodes peuvent être généralisées. Considérons le simplexe de dimension n :

$$\mathbb{S}_n = \{ \, {}^t p = (\, p_1, \ldots, p_n \,), \, p_i \geq 0, \, \Sigma_i \, p_i = 1\}.$$

Chaque $p \in \mathbb{S}_n$ est vecteur de réechantillonnage. Soit l'application

$$T : p \rightarrow T(p) = T(\, x_1^p, \ldots, x_n^p \,),$$

où $x_1^p, \ldots, x_n^p$ désigne un n-échantillon extrait de $x_1, \ldots, x_n$ au hasard et où chaque x_i apparait un nombre np_i de fois. Ainsi :

- $T(\, p^0 \,)$, avec ${}^t p^0 = (\, 1/n, \ldots, 1/n \,)$, est la valeur de T sur l'échantillon observé dans sa totalité;

- $T(\, p^i \,)$, avec ${}^t p^i = (\, 1/(n-1), \ldots, 1/(n-1), 0, 1/(n-1), \ldots, 1/(n-1) \,)$, et le 0 en $i^{ème}$ position, est la valeur de $T^{(i)}$;

- $T(\, p^* \,)$, avec ${}^t p^* = (\, n_1/n, \ldots, n_n/n \,)$ et $\Sigma_i \, n_i = n$, est une des réalisations possibles de T^*.

3. ESTIMATION DE LA VARIANCE DE T

Très souvent la quantité Var(T), caractéristique importante pour l'évaluation de l'efficacité de T, est inconnue. Les deux méthodes ci-dessus permettent de l'évaluer. La démarche est la suivante. Posons :

$$S_J^2 (T) = (\, (n-1)/n \,) \, \Sigma_i \, (\, T^{(i)} - \overline{T}_J \,)^2 \text{ avec } \overline{T}_J = (1/n) \, \Sigma_i \, T^{(i)},$$

$$S_B^2 (T) = \text{Var}(\, T^* \,).$$

Cette dernière quantité ne peut être, en général, calculée de manière exacte; on est ainsi amené à procéder par simulation : soit b échantillons "bootstrappés" $(\, X_1^*, X_2^*, \ldots, X_n^* \,)_k$ et T_k^*, $1 \leq k \leq b$, les réalisations de T^* pour chacun d'eux; on construit un estimateur de Var($\, T^* \,$) par

$$S_B^2(T) = (\ 1/(b-1)\ ) \sum_k (\ T_k^* - \overline{T}^*\ )^2 \quad \text{avec} \quad \overline{T}^* = (1/b) \sum_k T_k^*.$$

Il est aisé de voir que, lorsque $T = \overline{X}$, l'on obtient

$$S_J^2(\overline{X}) = S^2 = (\ 1/(n-1)\ ) \sum_i (\ X^{(i)} - \overline{X}\ )^2,$$

$$S_B^2(\overline{X}) = (\ (n-1)/n\ )S^2,$$

Remarquons que dans ce cas particulier la variance de T^* a pu être calculée directement. Les deux estimations, ci-dessus, de la variance d'une statistique sont liées. En effet, si on désigne par $T_{LIN}(p)$ la statistique de rééchantillonnage, linéaire en p, passant par $\overline{T}_J$ pour p^0 et associée à T, c'est-à-dire :

$$T_{LIN}(p) = \overline{T}_J + {}^t(\ p - p^0\ )\ u,$$

alors nous avons le résultat ci-dessous.

Proposition. L'égalité suivante est satisfaite.

$$S_J^2(T) = (\ n/(n-1)\ )\ S_B^2(\ T_{LIN}\ ).$$

Ainsi, de point de vue de la variance, le Jackknife de T peut être considéré comme un Boostrap sur T_{LIN}. En général l'évaluation par Jackknife est moins bonne que celle réalisée par Boostrap; elle dépend en fait de la qualité de l'approximation de T par T_{LIN}. Pour $T = med(\ X_1, X_2, \ldots, X_n\ )$, la médiane de l'échantillon, par exemple, si $\mathbb{L}(X)$ admet θ pour unique médiane et si sa fonction de répartition admet une dérivée f (la densité) positive et continue sur un voisinage de θ, alors on peut montrer que

$$\text{Lim}_{n \to \infty} \mathbb{L}(\ nS_J^2(T)\ ) = \mathbb{L}(\ Y/(\ 4f^2(\theta))\ ),$$

$$\text{Lim}_{n \to \infty} \mathbb{L}(\ \sqrt{n}\ (\ T - \theta\ )\ ) = \text{Lim}_{n \to \infty} \mathbb{L}(\ \sqrt{n}\ (\ T^* - T\ )\ ) = \mathbb{N}(\ 0;\ 1/(\ 4f^2(\theta))\ ),$$

où Y est une variable aléatoire d'espérance 2 et de variance 20.

4.ESTIMATION ET REDUCTION D'UN BIAIS

Dans ce paragraphe nous donnons une technique générale de réduction du biais d'un estimateur par Jacknife ou Bootstrap.

Définition. Si T *est un estimateur d'un paramètre* $\theta \in \mathbb{R}$, *on appelle BIAIS de* T *la quantité*

$$B(T) = E[T] - \theta .$$

Proposition. Le Jacknife permet d'éliminer le biais d'ordre $1/n$.

$$B(T) = O(1/n) \quad \Rightarrow \quad B(JT) = O(1/n^2).$$

Ce résultat peut se généraliser de deux manières différentes. Pour la première on procède de la façon suivante.

Proposition. Si $B(T) = O(1/n^a)$, *alors le Jacknife défini par*

$$JT_a = (n^a T - (n-1)^a \overline{T}_J)/(n^a - (n-1)^a),$$

est tel que $B(JT_a) = O(1/n^{(a+1)})$.

Pour la seconde nous supposons qu'il existe $L+1$ estimateurs T_j de $\theta \in \mathbb{R}$, tels que

$$E[T_j - \theta] = \sum_l h_{jl}(n) b_l(\theta).$$

Proposition. Si l'estimateur

$$JT^{(L)} = \mathrm{Det} \begin{vmatrix} T_1 & \cdots & T_{L+1} \\ h_{1,1} & \cdots & h_{L+1,1} \\ \cdots & \cdots & \cdots \\ h_{1,L} & \cdots & h_{L+1,L} \end{vmatrix} \Big/ \mathrm{Det} \begin{vmatrix} 1 & \cdots & 1 \\ h_{1,1} & \cdots & h_{L+1,1} \\ \cdots & \cdots & \cdots \\ h_{1,L} & \cdots & h_{L+1,L} \end{vmatrix} ,$$

appelé JACKKNIFE GENERALISE D'ÒRDRE L, *existe alors il est sans biais.*

$$E[JT^{(L)}] = \theta .$$

A condition de connaître les expressions exactes des estimateurs, nous pouvons ainsi réduire leur biais. Lorsque cette quantité est inconnue, on peut utiliser

$$B_J(T) = (n-1) \, (\,\overline{T}_J - T\,);$$

$$B_B(T) = E[\, T^* - T\,].$$

L'espérance ci-dessus ne peut pas être, en général, calculée. On procède alors à l'extraction de b échantillons "bootstrappés" $(X_1^*, X_2^*, ..., X_n^*)_k$ avec T_k^* , $1 \le k \le$ b, les valeurs de T associées. $B_B(T)$ est ensuite estimé par

$$B_B(T) = (\,1/b\,) \, \Sigma_k \, (\,T_k^* - T\,).$$

Comme pour la variance de T les deux évaluations précédentes du biais sont liées entre elles. En effet, si on désigne par $T_{QUA}(p)$ la statistique de rééchantillonnage, quadratique en p, passant par T pour p^0 et par $T^{(i)}$ pour p^i, c'est-à-dire :

$$T_{QUA}(p) = T + \,^t(\, p - p^0\,) \, v \, + \,^t(\, p - p^0\,) \, w \, (\, p - p^0\,) \, /2,$$

où le vecteur v et la matrice w sont convenablement choisis, alors on a le résultat ci-dessous.

Proposition. *L'égalité suivante est satisfaite.*

$$B_J(T) = (\, n/(n-1)\,) \; B_B(\,T_{QUA}\,) \, .$$

Ainsi chaque fois qu'un estimateur T est proche d'une fonction quadratique sur le simplexe $\mathbb{S}_n$, le biais du Jackknife , qui est plus facile à mettre en œuvre, se comporte, à une constante près, comme celui du Bootstrap.

5. INTERVALLE DE CONFIANCE D'UN PARAMETRE

Dans ce paragraphe nous présentons des procédés de construction d'intervalles de confiance. C'est l'aspect qui a le plus été développé dans les diverses études des propriétés du Jackknife. J. Tukey a émis la conjecture selon laquelle la statistique :

$$\sqrt{n} \, (\, JT - \theta\,) \, / \; \sqrt{ (1/(n-1)) \; \Sigma_i \; (nT_n - (n-1)\,T^{(i)}_{n-1} - JT\,)^2 }$$

est approximativement distribuée selon une loi de Student à n - 1 degrés de liberté. Mais plusieurs contre-exemples, où ceci n'est pas vérifié, ont été mis en évidence. Cependant cette affirmation a été prouvée pour une très large classe de statistiques dont celles appelées les U-statistiques. On peut alors, dans ces cas, en utilisant la démarche classique, construire des intervalles de confiance de paramètres. Pour ce qui est du Bootstrap, B. Efron a proposé les procédés suivants.

i) <u>Méthode des Quantiles</u>. Un grand nombre b de simulations sont effectuées sur l'échantillon observé. Pour chacune d'elles, $(X_1^*, X_2^*, ..., X_n^*)_k$, on évalue la réalisation de T, notée T_k^* , et ceci pour tous les k, $1 \leq k \leq b$. Ces réalisations permettent alors de construire l'histogramme de la loi $\mathbb{L}(T^*)$. A chaque ω donné, $0 < \omega < 1$, on associe le quantile empirique d'ordre $t = t(\omega)$, par la relation

$$\omega = \#(T_k^* \leq t) / b.$$

On définit l'intervalle de confiance du paramètre étudié au seuil α , par

$$[t(\alpha /2) ; t(1-(\alpha /2))].$$

ii) <u>Méthode des Quantiles avec biais corrigé</u>. Notons $\Phi(t)$ la fonction de répartition d'une loi gaussienne centrée réduite. Pour un ω donné, $0 < \omega < 1$, on considère le nombre z_ω , ainsi que le nombre z_0, définis par

$$\Phi(z_\omega) = 1- (\omega/2) \text{ et } \Phi(z_0) = \#(T_k^* \leq T) / b.$$

Alors un intervalle de confiance du paramètre étudié au seuil α avec biais corrigé est donné par

$$[t(\Phi(z_\alpha - z_0)) ; t(\Phi(z_\alpha + z_0))].$$

Si la moitié exactement des T^* sont inférieurs à T, alors $z_0 = 0$ et on retrouve l'intervalle construit par la méthode simple des quantiles. La démarche utilisée tient compte de la position de T dans l'ensemble des échantillons "bootstrappés" et donne en général des intervalles de confiance asymétriques.

6. REGRESSION LINEAIRE

Nous indiquons dans ce paragraphe l'application des deux méthodes de rééchantillonnage dans la situation la plus élémentaire. Considérons le modèle linéaire multidimensionnel

$$Y = X \, \Theta + \varepsilon.$$

avec $Y = {}^t(y_1, ..., y_n)$ la variable dépendante, $\Theta = {}^t(\theta_1, ..., \theta_k)$ le vecteur des paramètres, X la matrice des données dont la $i^{\text{ème}}$ ligne est ${}^t x_i = (x_{i1}, ..., x_{ik})$, variable indépendante, et enfin le vecteur des erreurs $\varepsilon = {}^t(\varepsilon_1, ..., \varepsilon_n)$ supposées indépendamment identiquement distribuées selon une loi gaussienne centrée de même variance σ^2. On suppose que le rang de la matrice X est $rg(X) = k$ ou, ce qui est équivalent, la matrice ${}^t XX$ est inversible. Il est bien connu que la méthode des moindres carrés permet alors d'obtenir l'estimateur

$$T = ({}^t XX)^{-1} \, {}^t X \, Y.$$

L'estimation de la variance-covariance de T est donnée par

$$(\Sigma_i (y_i - {}^t x_i \, T_i)^2 \, / \, (n\text{-}k)) \, ({}^t XX)^{-1}.$$

Avec les mêmes notations que précédemment, $T^{(i)}$ désigne la statistique T sans sa $i^{\text{ème}}$ composante. On a

$$T^{(i)} = T - \{ \ (({}^t XX)^{-1} \, {}^t x_i \, (y_i - {}^t x_i \, T_i)) \, / \, (1 - {}^t x_i \, ({}^t XX)^{-1} \, x_i) \ \}.$$

La version multivariée de l'estimation de la variance par Jackknife s'écrit dans le cas présent

$$S_J^2 \, (T) = ((n\text{-}1)/n) \, \Sigma_i \ (T^{(i)} - \overline{T}_J) \, {}^t(T^{(i)} - \overline{T}_J),$$

$$= ((n\text{-}1)/n) \, ({}^t XX)^{-1} \, \{ \ \Sigma_i \ x_i \, {}^t x_i \, (y_i - {}^t x_i \, T_i)^2 \ \} \, ({}^t XX)^{-1},$$

avec $\overline{T}_J = (1/n) \, \Sigma_i \, T^{(i)}$. Le Bootstrap quant à lui est appliqué par rééchantillonnage des résidus et on note

$$y_i^* = {}^tx_i\ T_i + \varepsilon_i^*, \quad \text{pour } i = 1,\dots, n.$$

Ensuite la méthode des moindres carrés induit

$$T^* = ({}^tXX)^{-1}\ {}^tX\ Y^*.$$

La variance des résidus est ici évaluée par $\quad (1/n)\ \Sigma_i\ (y_i - {}^tx_i\ T_i)^2$, ce qui permet d'écrire l'estimation de de la covariance

$$(1/n)\ \Sigma_i\ (y_i - {}^tx_i\ T_i)^2\ ({}^tXX)^{-1},$$

expresion à comparer avec celle obtenue dans la situation classique et avec le Jackknife ci-dessus. Plusieurs résultats de convergence des estimateurs de $f(\Theta)$, pour des fonctions f, ont été établis.

7. ANALYSE EN COMPOSANTES PRINCIPALES

Schématiquement, le Bootstrap appliqué à l'A. C. P., permet d'obtenir
- des intervalles de confiance des valeurs propres de la matrice des variances-covariances ou de celle des corrélations. Ces intervalles sont construits simultanément;
- des cônes de confiance des vecteurs propres associés, permettant, le cas échéant, une interprétation plus rigoureuse des résultats;
- un intervalle de confiance de la qualité globale de représentation par projection de l'ensemble des points sur un nombre donné d'axes factoriels.
Pour plus de précision il convient d'examiner les références correspondantes ci-dessous.

REFERENCES

Nous présentons les références les plus importantes sur l'ensemble des domaines qui ont été étudiés par le Jackknife et le Bootstrap.

ABRAMOVITCH L. , SINGH K. (1985) : " Edgeworth corrected pivotal statistics and the Bootstrap."
Ann. Statist., 13, pp 116- 132.
ARVESEN J. N. (1969) : " Jackknifing U - Statistics. "

Ann. Math. Statist., 40, pp 2076 - 2100.

ARVESEN J. N. , SALSBURG D. S. (1973) : " Approximate tests and confidence intervals using the Jackknife. "

Perspectives in Biometrics, Ed. Elashoff, New York pp 123 -149.

ATHREYA K., GHOSH M., LOW L. , SEN P. (1984) : " Laws of large number for Bootstrapped means and U - statistics. "

J. Statist. Plan.Inf., 9, pp 185 - 194.

BABU G. J. , SINGH K. (1983) : " Inference on means using the Bootstrap."

Ann. Statist., 11, pp 999 - 1003.

BENASSEN J. , JUNCA-HOLMES S. (1982) : " Synthèse bibliographique sur les méthodes du Jackknife et du Bootstrap. "

Rapport Technique n° 8202, Unité de Biométrie, ENSAM-INRA-USTL, Montpellier.

BERAN R. , SRIVASTAVA M. (1985) : " Bootstrap tests and confidence regions for functions of a covariance matrix."

Ann. Statist., 13, pp 95 - 115.

BICKEL P.J., FREEDMAN D.A. (1981) : "Some asymptotic theory for the Bootstrap. "

Ann. Statitst., 9, pp 1196 - 1217.

BICKEL P.J. , FREEDMAN D.A. (1984) : " Asymptotic normality and the Bootstrap in stratified sampling. "

Ann. Statitst., 12, pp 470 - 482.

BISSEL A. F. (1975) : " The Jackknife. "

Bull. in Appl. Statist., 4, n° 1, pp 55 - 64.

BOOS D.D. , MONAHAN J. F. (1986) : " Bootstrap methods using prior information."

Biometrika, 73, pp 77 - 83.

BRILLINGER D. R. (1964) : " The asymptotic behaviour of Tukey's general method setting approximate confidence limits (the Jackknife) when applied to maximum likelihood estimates. "

Rev. Inst. Int. Statist.,32, pp 202 - 206.

CHAKRABARTY R. P. , RAO R. N. K. (1968) : " The bias and stability of the Jackknife variance estimator in ratio estimation. "

J. Am. Statist. Assoc., 63, pp 746 - 749.

CHENG K. F. (1982) : " Jackknifing L - estimates. "

Can. J. Statist., 10, pp 49 - 58.

CRESSIE N. (1981) : " Transformations and the Jackknife."

J. R. S. S. (B), 43, n° 2, pp 177 - 182.

DAVISON A. C. , HINKLEY D. V. , SCHECHTMAN E. (1986) : " Efficient Bootstrap simulation."
Biometrika, 73, pp 555 - 566.
DICICCIO T. , ROMANO J.P. (1988) : " A review of Bootstrap confidence intervals."
J. R. S. S., 50, pp 338 - 354.
DUNCAN G. (1978) : " An empirical study of Jackknifed constructed confidence regions in non linear regression."
Technometrics, 20, pp 123 - 129.
EFRON B. (1979) : " Computers and the theory of Statistics : thinking the unthinkable. "
S.I.A.M., 21, pp 460 - 480.
EFRON B. (1981) : " Censored data and the Bootstrap."
J. Am. Statist. Assoc., 76, pp 312 - 319.
EFRON B. (1982) : " The Jackknife, the Bootstrap, and other resampling plans. "
S.I.A.M. CBMS - NSF Monog., 38.
EFRON B. (1985) : " Bootstrap confidence intervals for a class of parametric problems."
Biometrika, 72, pp 45 - 58.
EFRON B. (1987) : " Better Bootstrap confidence intervals."
J. Am. Statist. Assoc., 82, pp 171 - 185.
EFRON B. , DIACONIS P. (1984) : " Computer - intensive methods in statistics."
Scientific American, pp 96 - 108.
EFRON B. , GONG G. (1983) : " A leisurely look at the Bootstrap, the Jackknife, and Cross-Validation. "
Am. Statitstician, 37, pp 36 - 48.
EFRON B. , STEIN C. (1981) : " The Jackknife estimate of variance. "
Ann. Statist., 9, pp 586 - 596.
EFRON B. , TIBSHIRANI R. (1986) : " Bootstrap methods for standard errors, confidence intervals, and other measures of statistical accuracy. "
Statist. Science, 1, pp 54 - 77.
FOX T., HINKLEY D. , LARNTZ K. (1980) : " Jackknifing in non-linear regression."
Technometrics, 22, n° 1, pp 29 - 33.
FREEDMAN D.A.(1981) : " Bootstrapping regression models. "
Ann. Statist., 9, pp 1218 - 1228.
GLEASON J.R. (1988) : " Algorithms for balanced Bootstrap simulations."
Am. Statistician, 42, pp 263 - 266.
GRAY H. , SCHUCANY W. (1972) : " The generalized jackknife statistic."

Marcel Dekker, New York.

HALL P. (1986) : " On the number of Bootstrap simulations required to construct a confidence interval. "

Ann. Statist., 14, pp 1453 -1462.

HOLMES S. (1989) : " La méthode à la Cyrano (Bootstrap). Quelques raffinements et leur application à l'Analyse Multidimensionnelle. "

A paraître Compt. Rendus Congrés INRIA, Conf. Inv., Nice.

LECOUTRE J.-P. , TASSI Ph. (1987) : " Statistique non paramétrique et robustesse. "

Economica, Paris, pp 275- 319.

MILLER R.G. (1974) : " The Jackknife - a review. "

Biometrika,, 61, pp 1 - 17.

PARR W. , SCHUCANY W. (1980) : " The Jackknife : a bibliography."

Int. Statist. Rev., 48, pp 73 - 78.

QUENOUILLE M.(1949) : " Approximate tests of correlation in time series."

J. R. S. S. (B), 11, pp 64 - 84.

QUENOUILLE M.(1956) : " Notes on bias in estimation. "

Biometrika, 43, pp 353 - 360.

RUBIN D. B. (1981) :" The bayesian Bootstrap. "

Ann. Math. Statist., 9, n° 1, pp 130 - 134.

SINGH K. (1981) :" On the asymptotic accuracy of Efron's Bootstrap. "

Ann. Statist., 9, n° 6, pp1187 - 1195.

TUKEY J. (1958) : " Bias and confidence in not quite large samples, abstract. "

Ann. Math. Stat., 29, p 614.

WAINER H., THISSEN D. (1975) : " When Jackknife fails. "

Psychometrica, 40, pp 113 - 114.

WATKINS T. A. (1971) : " Jackknifing stochastic processes. "

Ph. D. Dissertation, Texas Tech. Univ.

WICHMANN B.A. , HILL I.D. (1982) : " An efficient and portable pseudo-random number generator."

Appl. Statist., 31, pp 188 - 190.

WU C. F. J. (1986) : " Jackknife, Bootstrap and other resampling methods in regression analysis (with discussion)."

Ann. Statist., 14, pp 1261 - 1294.

YOUNG G. A. (1988) : " A note on Bootstrapping the correlation coefficient."

Biometrika, 75, pp 370 - 373.

Krigeage in Astronomy: Application to the Study of the Galactic Plane.

J. CABRERA-CAÑO (1,2), E. J. ALFARO (1), A. POLVORINOS (2).

(1) Instituto de Astrofisica de Andalucia. Granada. Spain

(2) Universidad de Sevilla. Sevilla. Spain.

ABSTRACT.

In Astronomy, we are very often confronted with two-dimensional estimation or contouring problems. We try to illustrate in this communication how *Kriging*, an statistical tool developed by geostaticians, may also result a very powerful and valuable technique in our field.

Only a very light introduction of the mathematical support is presented, together with a basic bibliography. An example of application to the structure of galactic plane shows, at least, that its *cartographic* abilities should not be disregarded.

1 INTRODUCTION.

Kriging is a well known statistical tool in fields like Geology and Mine Planning. Developed initially by Matheron in Fontainebleau, its name honours the work of D. G. Krige who was probably the first to use this type of approach to solving problems related to the estimation of natural resources and in particular to grade ore calculation. Outside these fields, *Kriging* is virtually unknown, and yet could be very useful when tackling many problems in Astronomy.

The aim of this report is two-fold: a) to set out the fundamental mathematical ideas on which this technique is based, b) to give an example of how this technique could be used in the field of Astronomy.

The paper has been divided into three major sections: A geological problem. Mathematical notes. A problem of Astronomy.

2 A GEOLOGICAL PROBLEM.

Let us consider the typical case of an aquifer and let us suppose that we need to know the depth of its water level at one particular point, by estimating this value from a discrete set of explorations undertaken in an area close to the site in question.

The mathematical problem inferred from this is one of two-dimensional interpolation on a non-regular grid, or *automatic cartography*, as it is more simply called. There are various algorithms capable of carrying out this task, *Kriging* being one of them. The major advantage of *Kriging* is that not only does it afford an estimate of the values of the function at each point, but moreover it provides a measurement of the error involved in this estimate. To undertake this task, *Kriging* adopts a stochastic model for the variable under study. This is without doubt the corner stone of the method and in it lies its main advantage.

3 MATHEMATICAL NOTES.

The theory of *Kriging* is extensive and complex and clearly it cannot be explored here. Reference to the primary bibliography is preceptive. Nevertheless, we shall try to give some insights into the main features in order to justify the astronomical application presented below. Let us return to the problem of the water level of the aquifer. Let $z = z(x)$ be the depth of the water level at point x $(z \in \Re, x \in \Re^2)$. The stochastic model for the *regionalized variable* z is based on the supposition that it correspond to one realization:

$$z(x) = Z(x, \omega) \; \omega \in \Omega \tag{1}$$

Given a set of values $z_i = z(x_i)$ of such realization on points $x_1, ..., x_n$, *Kriging* is the best linear unbiased estimator *(BLUE)* of $z(x)$ at generic point x. If the *kriged* value is expressed as $z^*(x)$, it can be written as:

$$z^* = \sum_{\alpha=1}^{n} \lambda^\alpha(x) z_\alpha \tag{2}$$

in which the weights λ^α are the solution of the *Kriging equations*.

The formulation of *Kriging equations* requires a knowledge of the statistical properties of the stochastic process. However these are not known *a priori* and have to be estimated from the available data. If we consider the simplest case, a stationary process, the following conditions are fullfilled:

i) The mean $m = E(z(x))$ is independent of x, and

ii) The covariance function $C(x, y) = E((z(x) - m)(z(y) - m)) = C(x - y)$ depends only of the relative position of the two points x and y.

If we simplify even more and consider the problem to be isotropic, the covariance function can be expressed as a function of the distance h between the points x and y,

$$C(x, y) = C(h) \; ; \; h = d(x, y) \tag{3}$$

Thus, if the process is stationary and isotropic m and $C(h)$ can easily be obtained from the set of data $z(x_i)$. In this particular case, the *Kriging equations* can easily be formulated from the definition of *BLUE*.

Since *Kriging* is by definition unbiased, the following equation must be verified

$$E(z^*(x)) = \sum_{\alpha=1}^{n} \lambda^\alpha E(z(x_\alpha)) = (\sum_{\alpha=1}^{n} \lambda^\alpha)m = m \tag{4}$$

which lead to the first condition for the parameters λ

$$\sum_{\alpha=1}^{n} \lambda^\alpha = 1 \tag{5}$$

On the other hand it is easily shown that the estimation variance is:

$$Var(z - z^*) = K(\lambda^1, ..., \lambda^n) = C_0 - 2\sum_{\alpha=1}^{n} \lambda^\alpha C_\alpha + \sum_{\alpha=1}^{n}\sum_{\beta=1}^{n} \lambda^\alpha \lambda^\beta C_{\alpha\beta} \tag{6}$$

where
$$C_0 = C(0) , \ \ C_\alpha = C(x, x_\alpha) \text{ and } C_{\alpha\beta} = C(x_\alpha, x_\beta)$$

then, we obtain the estimated value z^* by minimizing this variance under the restriction of $\sum_{\alpha=1}^{n} \lambda^\alpha$; which lead, introducing the Lagrange multiplier μ, to the *Kriging equations*:

$$\begin{cases} \sum_{\beta=1}^{n} \lambda^\beta C_{\alpha\beta} - \mu = C_\alpha \ (\alpha = 1, ..., n) \\ \sum_{\alpha=1}^{n} \lambda^\alpha = 1 \end{cases} \tag{7}$$

their solution provide us the n parameters λ^α, from which the estimated value z^* is obtained, and the parameter μ which allows us to determine the error involved in the estimate, from the relationship

$$\sigma_k = min[Var(z - z^*)] = C_0 - \sum_{\alpha=1}^{n} \lambda^\alpha C_\alpha + \mu \tag{8}$$

Thus, *Kriging* is no more than a weighted average of experimental points where the weights depend on the correlation properties of the stochastic process involved.

We have discussed here the basic principles of this technique, considering only the simplest case. Undoubtedly, the simplest is at the same time the most unusual and

Table 1.-Basic Bibliography.

GENERAL REFERENCES

Matheron, G. 1971. The theory of regionalized variables and its applications: Cah. Cen. Morphol. Math. n.5 ENSMP, Paris, 211 p.

Journel, A.G. and Huijbregts, C.T. 1978. Mining geostatistics. New York, Academic Press, 600 p.

David, M. 1977. Geostatistical Ore Reserve Estimation. Elsevier Scientific Publishing Company. New York. 364 p.

UNIVERSAL KRIGING

Matheron, G. 1969. Le Krigeage Universel. Cah. Cen. Morphol. Math. n. 1 ENSMP, 82 p.

Huyjbregts, C.T. and Matheron, G. 1971. Universal Kriging (an optimal method for estimating and contouring in trend surface analysis). Cah. Ins. Min. Metall. Spec. v-12 p 159-169.

Delhomme, J.P. 1976. Application de la theorie des variables regionalises dans les sciences de l'eau: Unpublished Docteur-Ingenieur thesis. Universite Pierre et Marie Curie, Paris. 130 p.

IRF-k AND GENERALIZED COVARIANCE ESTIMATION

Matheron, G. 1973. The intrinsic ramdom functions and their applications. Adv. Appl. Prob. v-5 p 439-468.

Starks, T.H. and Fang, J.H. 1982. On the estimation of the generalized covariance functions. Jour. Math. Geol. v-14 p 57-64.

Delfiner, P. 1976. Linear estimation of nonstationary spatial phenomena. in Guarascio (Ed.), Advanced Geostatistics in the Mining Industry. D. Reidel Dordrecht, p 49-68.

Marshall, R.J. and Mardia, K.W. 1985. Minimum norm quadratic estimation of components of spatial covariance. Jour. Math. Geol. v-17 p 517-526.

Kitanidis, P.K. 1985. Minimum variance unbiased quadratic estimation of covariance of regionalized variables. Jour. Math. Geol. v-17 p 195-208.

Starks, T.H. and Sparks, A.R. 1987. Estimation of the genarilized covariance function: II. A response surface approach. Jour. Math. Geol. v-19 p 769-783.

to continue we shall note the principal lines of development of *Kriging* which are necessary when dealing with more realistic problems.

Leaving aside the mathematical difficulties, the *Kriging* equations are often formulated in terms of the *variogram* $\gamma(h) = Var(z(x) - z(y))$ instead of using $C(h)$. This allows for the fact that in certain problems $C(h)$ does not exist, whereas $\gamma(h)$ does. When this is the case, this is accomplished by simply substituting $C(h)$ by $-\gamma(h)$ in the *Kriging equations*.

When the process is not stationary in mean, the *Kriging equations* have to be generalized, taking into account the trend of data. This trend is modelled by a two-dimensional polinomial of order k. The development of this idea gives rise to what is known as *Universal Kriging*.

Once the trend is eliminated, it is necessary to estimate the covariance function or the variogram of the transformed data. This problem presents a fundamental difficulty; these estimates results necessarily biased as it may be shown.

To overcome this difficulty, G. Matheron et al. have developed the theory of *Intrinsic Random Functions of order* k and *Generalized Covariance Functions* for which, unbiased estimator may be implemented. We shall finish these mathematical notes with a brief and, in our opinion, fundamental bibliography on the subject.

4 A PROBLEM OF ASTRONOMY.

Let us consider that we wish to obtain the height pattern of the youngest component of the galactic disc, in a radius of 4 Kpc around the Sun. To do this we select a sample of object younger than 10^7 years, whose coordinates are determined with sufficient precision. For this particular example we have chosen a sample of open clusters from Lyngå catalogue (1981) which comprises 82 items. The distribution of the sample on the X, Y plane cannot be considered to be uniform and as it is well known delimits the local spiral structure. Thus we are dealing with a typical *Kriging* problem which has been solved using *Universal Kriging* and *Generalized Covariance functions*.

The results are shown in figure 1. The dashed lines represent levels below the galactic plane. The most outstanding result of this work is, in our opinion, the appearance of an ellipsoidal depression centered on $X = -1500$pc $Y = 500$pc, with a major semi-axis of near 3 Kpc parallel to the X axis.

REFERENCES.

Lyngå, G., 1981. Catalogue of Open Cluster Data. CDS, Strasbourg.

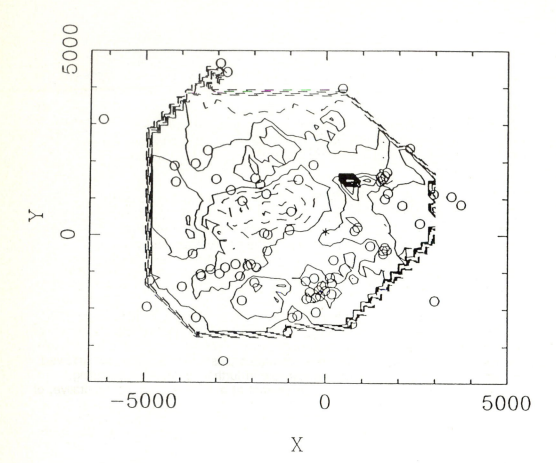

Figure 1.- Estimated Z. Dashed lines correspond to Z below the galactic plane. Interval between two consecutive isolines is 50 pc.

Bootstrap, Data Permuting and Extreme Value Distributions – Getting the most out of Small Samples

BHAVSAR S.P.

University of Kentucky, U.S.A.

SUMMARY

The use of statistical methods on samples of astronomical data are discussed. We discuss the application of bootstrap to estimate the errors in measuring the two-point correlation function for galaxies and the topology of the large-scale structure. We discuss a technique to estimate the statistical significance of filamentary structures in the distribution of galaxies. Extreme value statistical theory is applied to the problem of the nature of the brightest galaxies in rich clusters.

INTRODUCTION

What can a statistician, feet firmly planted on the ground, offer an astronomer whose head is up in the stars? The answer of course is: "quite a lot", which is why we are at this meeting. I shall talk about my own experience, giving an account of how statistics has enriched my astronomical research. Besides formalizing errors and making estimates of uncertainties more rigorous, it has actually opened up new ways to do analysis, providing new insights and approaches to problems which previously had proved elusive.

Small samples are a common occurrence in the field of Astronomy. Data has to be gathered from the laboratory of the universe, over which one has no control. Astronomers have to take what is given, and very often that is very little. In this context, I shall discuss three separate topics which I have been involved in, where statistics has played a major role in furthering our understanding. Please note that this in no way is an attempt at a review, or is representative, of the many uses of statistics in astronomy. It is a description of my personal involvement and excitement at the realization of how much the rich field of statistics has to offer to astronomy. The three applications that I shall talk about here are:

1) The application of bootstrap to estimate the standard errors in measuring the galaxy-galaxy correlation function, and other measures of galaxy clustering.

2) A technique developed to estimate whether large-scale filamentary structures in the universe are statistically significant.

3) An application of extreme value statistical theory to the understanding of the nature of the brightest galaxies in rich clusters.

BOOTSTRAP

The bootstrap was invented by Efron in 1977 (eg see Efron 1979). The method uses the enhanced computing power available in this age to explore statistical properties that are beyond the reach of mathematical analysis. At the heart of this method is the procedure of creating from the available data set, hundreds or thousands of other "pseudo-data" sets. In older methods one assumed or derived some parametric distribution for the data and derived statistical measures. With bootstrap, any property of the original data set can be calculated for all the pseudo-data sets and the resulting variance of a measured quantity calculated over the entire ensemble. We have applied the bootstrap re-sampling technique (to my knowledge for the first time in astronomy) to illustrate its use in estimating the uncertainties involved in determining the galaxy-galaxy correlation function (Barrow, Bhavsar and Sonoda 1984).

The two-point correlation for galaxies is the factor by which the number of observed galaxy-galaxy pairs exceed the number expected in a homogeneous random distribution, as a function of their separation. It is an average taken over the entire sample and measures the clumping of galaxies in the universe. Usually a galaxy's coordinates in the sky are relatively easy to measure and known accurately. The same cannot be said for a galaxy's distance from us which is a difficult and time consuming measurement. Thus the most common catalogs available are sky surveys, for which the two-point correlation function has been extensively determined; though in the last few years 3-D spatial catalogs of limited extent have become available.

With a sky catalog one measures the angular two-point correlation function. This was pioneered by Totsuji and Kihara (1969), and Peebles (1973) and followed up extensively by the work of Peebles (1980) and his collaborators. They found that the angular correlation function for galaxies is a power law. If the projected distribution has a power law correlation then the spatial two-point correlation is also a power law with a slope that is steeper by -1 than the angular correlation (Limber 1953). The angular two-point correlation for galaxies to magnitude limit 14.0 in the Zwicky catalog (Zwicky et $al.$ 1961-68) is given by

$$\omega(\theta) = \left(\frac{0.06}{\theta} \right)^{0.77} \tag{1}$$

Let me describe in a little more detail the actual determination of the above measurement. The sky positions (galactic longitude and latitude) of all galaxies brighter than apparent magnitude 14.0 and in the region $\delta>0°$ and $b^{II}>40°$ forms our sample. This consists of 1091 galaxies in a region of 1.83 steradian. The angular separation between all pairs of observed galaxies is found and binned to give $N_{oo}(\theta)$, the number of pairs observed to be separated by θ. This is compared to the expectation for the number of pairs that would be obtained if the distribution was homogeneous, by generating a Poisson sample of 1090 galaxies in the region and determining the pair-wise separations between each of the observed 1091 real galaxies and the 1090 Poisson galaxies. This gives us $N_{op}(\theta)$, the number of pairs between observed and Poisson galaxies separated by θ. The angular two-point correlation function is

$$\omega(\theta) = \frac{N_{oo}(\theta) - N_{op}(\theta)}{N_{op}(\theta)} \qquad (2)$$

The binning procedure is decided upon before performing the analysis. The data is binned so that there are roughly an equal number of observed-Poisson pairs in each bin. A plot of *log* $\omega(\theta)$ versus *log* θ shows the correlation function to be a power law. The slope of the best fitting straight line determined by some well defined statistical measure gives the value of the exponent in equation (1). Our independent determination of this exponent gives a value of 0.77, consistent with earlier determinations. The question is: from the limited data available, how accurately does this describe the correlation function for galaxies in the universe? It is assumed that our sample is a "fair sample" of the universe (peebles 1980), implying our faith in the assumption that the region of the universe that we sample is not perverse in some way. The *formal errors* on the power law fit to the data do <u>not</u> give us the true *variability* in the slope of the correlation function that may be expected. These formal errors indicate to us a goodness of fit for the best power law that we have found to fit the data; but the question of the statistical accuracy of the value 0.77 representing the slope of the power law for the two-point correlation function for galaxies in the universe, remains uncertain.

The bootstrap method is a means of estimating the statistical accuracy of the two-point correlation from the single sample of data. A pseudo data set is generated as follows. Each of the 1091 galaxies in the original data is given a label from 1 to 1091. A random number generator picks an integer from 1 to 1091, and includes that galaxy in the pseudo data sample. A galaxy is picked from the original data in this manner through 1091 loops. The pseudo data sample will contain duplicates of some galaxies and will not contain all the galaxies of the original data set. The angular correlation function can be calculated for this new data set in exactly the same manner as the original by fitting it to a power law of the form

$$\omega(\theta) = \left(\frac{\theta_o}{\theta}\right)^{\gamma} . \qquad (3)$$

This entire procedure of generating a new data set and determining its angular correlation can be repeated hundreds of times using different sets of random numbers. The samples generated in this way are called bootstrap samples. The frequency distribution for the values of the slope, γ, and the correlation length, θ_o, can be plotted for the ensemble of bootstrap samples to estimate the variance of γ, and θ_o respectively.

We generated 180 bootstrap samples for the 1091 galaxies in the Zwicky data, determined the two-point correlation function for each as described by equation (2), and fit the power law described by equation (3) to each correlation function. Figure 1 shows the distribution of γ obtained for these 180 samples. Figure 2 shows the distribution of θ_o.

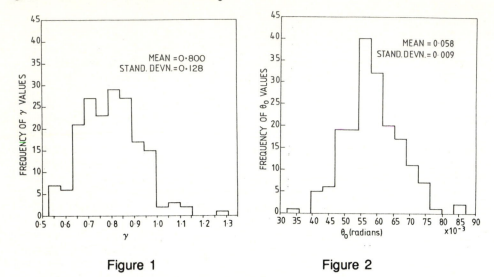

Figure 1   Figure 2

We can determine the standard deviation in the values of γ and θ_o. These are found to be

$$\sigma(\gamma) = 0.13 \tag{4}$$

$$\sigma(\theta_o) = 0.01 \tag{5}$$

This means that 68 percent of the values of γ lie in an interval whose width is 0.26, or in other words, the bootstrap samples show that 68% of the γ lie in an interval 0.67 to 0.93, and the rest outside this interval, on either side. Similarly 68% of the values of θ_o lie in an interval 0.05 to 0.07. Both these intervals are much larger than the formal errors that have been assigned to the quantities γ and θ_o in the literature. It is worth noting that this method does not provide a best estimate for the value of γ or θ_o, but provides a statistical accuracy for these values. The *average* of the bootstrap samples is not an indication of the *true* value any more than the value obtained from the particular set of data is.

In 1915 Sir Ronald Fisher calculated the variance of statistically determined quantities (such as the slope γ here) assuming that the data points on the graph for the two variables were drawn at random from a normal probability distribution. In his time it would have been unthinkable to bootstrap because computation was millions of times slower and expensive. Today's computing power enables us to glean information from the available data without making assumptions about its distribution.

THE STATISTICAL SIGNIFICANCE OF LARGE-SCALE STRUCTURE

a) *filaments*

How can we quantify the presence of some visually prominent feature in the universe? In particular how "real" are the linear or "filamentary" distributions of galaxies suggested by recent surveys? Their implications for evolution of structure in the universe remains controversial because it is extremely difficult to assess their statistical significance. Though the visual descriptions have been rich in describing form, they are subjective, leading to attributes too vague to model or compare with numerical simulations (Barrow and Bhavsar 1987).

Recently my collaborators and I pioneered the use of a pattern recognition technique - the minimal spanning tree; MST - to identify the existence of filaments (Barrow, Bhavsar and Sonoda 1985 [BBS]) and invented a statistical technique to establish their significance (Bhavsar and Ling 1988a [BL I]; 1988b [BL II]). Though apparent to the eye, this has been the first objective, statistical and quantitative demonstration of their presence (see Nature **334**:647).

The famous Lick Observatory galaxy counts (Shane and Wirtanen 1967) are a good example of the problem. The wispy appearance of this sky map (Groth et al. 1977) has evoked a strong image of the universe permeated by a galactic lace-work of knots and filaments. Yet doubts about these large-scale filamentary features remain, and for good reason. The human propensity toward finding patterns in data sets at very low levels has been responsible for some embarrassing astronomical blunders: the most notable example being the saga of the Martian canals[1] (Evans and Maunder 1903; Lowell 1906; Hoyt 1976). To test the validity of controversial visual features an understanding of human visual bias and methods to overcome it are necessary.

There have been several quantitative descriptions of galaxy clustering. Most notably the correlation function approach (Peebles 1980) has been path-breaking in the early studies of structure in the universe. The practical difficulty of measuring high order correlations, however, makes this approach unable to provide information on the profusion of specific shapes of various sizes suggested by recent data. Clustering algorithms and percolation methods have not been very successful at discriminating visual features because they each use only one parameter (respectively, the density enhancement and the percolation radius). This leads to an over simplification of the problem of recognizing patterns. In particular, the difference in visual appearance between the CfA survey and earlier simulations, that were evident to the eye, were only weakly reflected in their percolation properties (Bhavsar and Barrow 1983; 1984). Two essential requirements needed to quantify any particular feature that may be noticed in the data are; first, an objective, well defined, and repeatable method to identify this feature of interest, and second, a means of evaluating the statistical significance of this candidate object.

---

[1] A more recent controversy about a visual feature on Mars, photographed in July 1976 by the Viking orbiter, is that of a mile-long rock resembling a humanoid face (Carlotto 1988)!

The Minimal Spanning Tree or MST (Zahn 1971) is a remarkably successful filament tracing algorithm that identifies filaments in galaxy maps (BBS; Ling 1987; BL I). The technique, derived from graph theory, constructs N-1 straight lines (edges) to connect the N points of a distribution so as to minimize the sum total length of the edges. This network of edges uniquely connects all (spans) N data points, in a minimal way, without forming any closed circuits (a tree). In order to distil the dominant features of the MST from the noise, the operation of "pruning" is performed. A tree is pruned to level p when all branches with k galaxies, where $k \leq p$, have been removed. Pruning effectively removes noise and keeps prominent features. A further operation, "separating" removes edges whose length exceeds some cut-off, l_c. This gets rid of the unphysical connections in the network, just as the eye distinguishes well separated features. A pruned and separated MST is called a "reduced MST". Figure 3 illustrates the construction of the MST and the processes of pruning and separating on a simple point set. The reader should refer to BBS for more details.

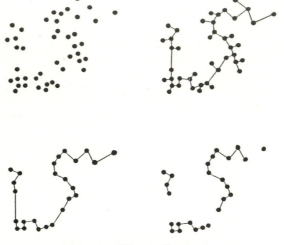

Figure 3

The reduced MST identifies basic linear patterns in galaxy surveys and can measure them in a quantitative fashion. It is quite a good representation of how the eye-brain forms a global impression of most data sets of random or clustered point distributions. For example the MST does a fairly good job of picking out the constellations from a catalog of bright stars (Hillenbrand 1989; Hillenbrand, Bhavsar and Gott 1989).

How can we determine if the filaments that are identified are statistically significant or not? Are they unique to the distribution of galaxies or does strong small-scale clumping coupled with occasional chance alignments produce the false impression of large-scale linear structure? For this we need to know what kind of filamentary structures occur just by random chance in similarly clumped distributions. One way to do this is to use bootstrap to generate pseudo-samples of data and test them for features. We shall describe the determination of the robustness of another measure, the topology of large-scale clustering, using bootstrap samples later. For the problem of filaments, a nagging doubt is that the small scale clumping is producing the strong *visual* impression of filamentary structure. We can test this by applying the MST to samples that have the same small scale clumping as the data, but no large scale correlations. The data permuting technique (BL I, BL II) described below achieves this.

Suppose that our eyes (and the eyes of our colleagues) detect features that are of the order of 30 Mpc across. We choose $\underline{p}$ and $\underline{l}_c$ such that the reduced MST also identifies these features. If small-scale clustering merely conspires to form large-scale features like these in a statistical way, then rearranging the galaxy distribution on the larger scales (say greater than 10 Mpc for instance), but leaving it intact on the smaller scales (in this case 10 Mpc and smaller), should have no overall effect. The original 30 Mpc features may disassociate but other similar features will come into prominence, and be identified by the reduced MST using exactly the same criteria as before. On the other hand the continued absence of such features after repeated rearrangements of the data would suggest that the original 30 Mpc features were unique to the data and not due to chance illusions.

In practice, this randomizing operation is easily performed by taking a cube of data, dividing it into smaller cubes (cubettes) and randomly shuffling the cubettes within the original volume. We have written a computer program which, given an integer n, divides a data cube of side L Mpc into n^3 cubettes of side L/n Mpc, randomly translates and rotates each cubette and reassembles them into a new cube of side L Mpc (analogous to manipulating a Rubik's cube, including the inside). Many different random realizations are obtained for each value of n, and n is varied from n=2 to n=n_{max}, where n_{max} is determined by the correlation length or some analogous measure. Note that the clustering within any cubette, of length L/n, is unchanged. All the MST techniques can now be used on this "fake" data cube to identify filaments and compare them to the original. The same procedure can also be applied in 2-D to an appropriately chosen shape extracted from 2-D data. The use of the MST ensures that exactly the same criteria will be applied to identify filamentary structures every time. If filamentary features are identified at the same level in these permuted versions of the data as they were in the original data, then we conclude that the original features are spurious, the result of statistical chance. Otherwise, the expected number of spurious filaments and their variance can be obtained for these permuted versions, at different length scales, giving us the statistical significance of the original features.

One might ask at this point: What determines a filament? Does it depend arbitrarily on the choice of $\underline{l}_c$ and $\underline{p}$? Actually our procedure provides a working definition of a filament. Filaments can be identified at all levels, depending on the values of the parameters we choose to reduce the MST. If a tree is not pruned enough the small filaments persist even after the permutations. For some minimum choice of the pruning level the features are unique to the original distribution, showing the statistical presence of linear structures, and providing a measure of what to call a filament. Our experiments show that for some distributions this point is never reached because there are no filaments present. Remember, our motivation is to objectively define a visual feature, then check for its statistical significance. That a linear feature identified at a particular $\underline{p}$ and $\underline{l}_c$ is unique to the data, in fact <u>defines</u> for us "a filament."

Samples to be analyzed have to be chosen with care. For the above procedure to be of value it is imperative that the data cube be a <u>complete</u> sample and free from obscuration. Since this is stressed and elaborated on in detail in BL I we shall not say any more here. We used the CfA red-shift catalog (Huchra <u>et al</u> 1983), in the region $\delta > 0°$ and $b'' > 40°$, with complete red-shifts down to apparent magnitude 14.5 and corrected for Virgo in-fall. We volume-limited this truncated cone by applying an absolute magnitude cutoff at -18.5 mag. Inside this volume-limited truncated cone, which contains 489 galaxies, we inscribed the largest possible cube, measuring 41.6 Mpc (H_o=50 Km s^{-1}/Mpc) on each side. The data in this cube is complete and practically free of galactic obscuration. For details of the positioning of the cube and its dimensions the reader should refer to BL I. Figure 4a shows the data cube described above,its reduced MST and the end-to-end lengths and total lengths of the identified filaments. This cube was permuted/reassembled many times for each value of n, from n=2 to n=10. *We did not find filaments at the level of the original data in <u>any</u> of these realizations.* In fact they are present at a significantly lower level. Figure 4b shows one of these realizations for n=5. Again, for more details and figures, refer to BL I.

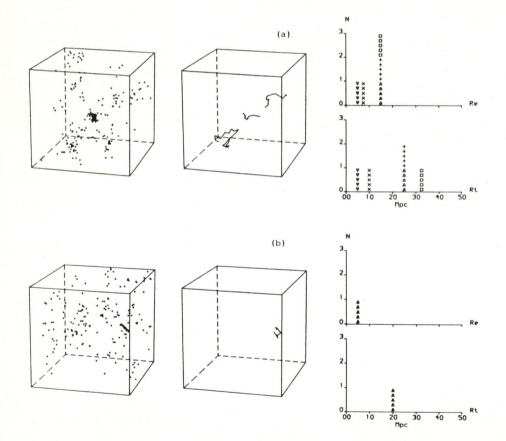

Figure 4a and 4b

We have also shown (Bhavsar and Ling 1988b, hereafter referred to as BL II) how easily our eye-brain is fooled into seeing large-scale filamentary features where none exist. The MST on the other hand identifies real filaments in toy models (see figure 1a and 1b in BL I) but is <u>not</u> fooled by spurious ones. This is shown using a construction now called Glass patterns (Glass 1969), produced by taking a random distribution of dots (figure 5a) which is then superimposed on itself with a slight rotation (figure 5b); figure 5c is the pruned and separated MST of the pattern in figure 5b.

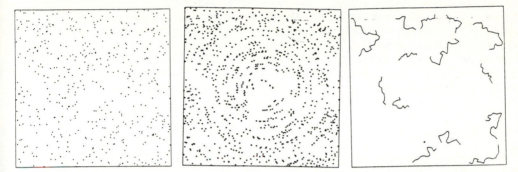

Figure 5a   Figure 5b   Figure 5c

The result is astonishing, the reader is urged to make transparencies of figure 5a and experiment. The visual impression is that of broken circular concentric rings of dots. Actually the dots have no auto-correlations, beyond the local pairings, but the eye-brain is fooled because it misinterprets a low level of global coherence (a result of the rotation) for large-scale linear features.

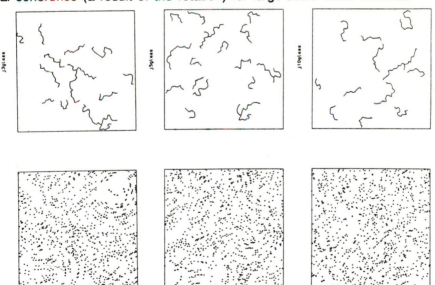

Figure 6

The MST is <u>not</u> fooled into identifying the perceived filaments. When the data permuting technique described above is used it shows that the filaments that the MST does identify are not statistically significant (see BL II for details.) Figure 6 shows the results of using the data permuting technique on figure 5b for n=3,5 and 10. Not only is the MST not fooled by the spurious filaments in figure 5b, but the ones it *does* identify are shown to be not statistically significant.

Questions about the type of initial conditions and mechanisms in the early universe that make filamentary structures remain unanswered. Can they result simply from gravitational instability on initial random fluctuations in matter, or do they require specific initial conditions, or even non-gravitational mechanisms? We propose to use the demonstrated success of the MST techniques to answer this question by analyzing filamentary structure among numerical simulations and comparing them with data.

b) *topology*

An analysis of the present day topology of the large-scale structure in the universe (Weinberg, Gott and Melott 1987) is directly related to the topology of the initial density fluctuations. The series of papers by Gott and his collaborators detail the way a measure of the topology - the genus of contours of a smoothed density distribution of galaxies - can be studied to determine the type of initial density fluctuations that existed in the early epochs of the universe.

The available data is sparse, and the relation between the genus and the threshold density has to be made from this available sample. The process involved in obtaining this relation involves smoothing the data, drawing the contour and then determining the genus per unit volume.

Figure 7, taken from Weinberg, Gott and Melott (1987) shows the contours for the CfA data (Huchra *et al.* 1983), drawn at various thresholds, ν, of volume fractions, containing respectively 7%, 16%, 24% and 50% of the low density regions, and the same fractions of the high density regions.

The plot of genus versus the threshold volume, ν, is shown in figure 8 taken from Gott *et al.*(1989). The dark circles are the values determined from the data. The dark line has been drawn to pass through these points. The question of the reliability of these points has been determined by creating bootstrap samples of the original data and determining the genus for these samples for each value of threshold. The procedure is tedious, and the error bars shown correspond to the variation obtained from four realizations, though this is by no means adequate, and further bootstrap samples are needed to determine the variance reliably.
The variations are shown around each data point. The open squares show the mean values obtained for the bootstrap models. As mentioned the bootstrap is a procedure for determining the variance and not the mean values. No significance should be attached to the position of the squares.

7% low   16% low   24% low   50% low

 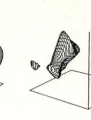  

7% high   16% high   24% high   50% high

 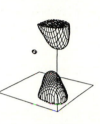  

CfA survey data
Figure 7

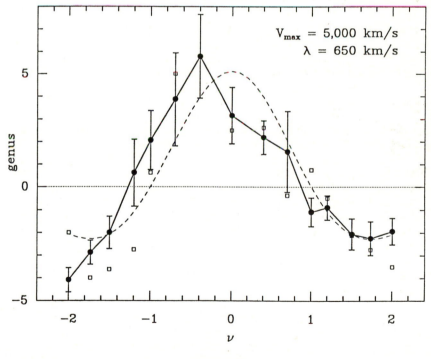

CfA

$V_{max} = 5,000$ km/s
$\lambda = 650$ km/s

genus

ν

Figure 8

EXTREME VALUE THEORY AND BRIGHTEST GALAXIES

This is an application of the statistical theory of extreme values to astronomy. Once again the name of the illustrious statistician Sir Ronald Fisher comes up, for it was he who developed the theory of extreme values. The problem has to do with the remarkably small dispersion (≈ 0.35 mag) in the magnitudes, M_1, of the first brightest galaxies in clusters. This has made them indispensable as "standard candles", and they have, for this reason, become the most powerful yardsticks in observational cosmology. There has existed, however disagreement as to the nature of these objects. The two opposing viewpoints have been: that they are a class of "special" objects (Peach 1969; Sandage 1976; Tremaine and Richstone 1977); and at the other extreme, that they are "statistical", representing the tail end of the luminosity function for galaxies (Peebles 1968, Geller and Peebles 1976).

In 1928, in a classic paper, R.A. Fisher and L.H.C. Tippett had derived the general asymptotic form that a distribution of extreme sample values should take - independent of the parent distribution from which they are drawn! This work was later expanded upon by Gumbel (1958).

From extreme value theory, for clusters of similar size, the probability density distribution of M_1 for the "statistical" galaxies is given (Bhavsar and Barrow 1985) by the distribution in equation 6, which is often referred to as the first Fisher-Tippett asymptote, or Gumbel distribution.

$$p_{Gum} = a \exp [a(M_1 - M_o) - e^{a(M_1 - M_o)}], \tag{6}$$

where a is the parameter which measures the steepness of the luminosity function at the tail end. M_o is the mode of the distribution and is a measure of the cluster mass, but with only a logarithmical dependence on the cluster mass.

We can compare the above distribution with the data to answer the question: "Are the first-ranked galaxies statistical?" Figure 9 shows the maximum likelihood fit of equation (6) to the magnitudes of the 93 first-ranked galaxies from a homogeneous sample of richness 0 and 1 Abell clusters. This excellent data for M_1 was obtained by Hoessel, Gunn and Thuan (1980,[HGT]). The fit is very bad. A Kolmogorov-Smirnov test also rejects the statistical hypothesis with 99% confidence. If all the galaxies are special, it is not possible to have a theoretical expression for their distribution in magnitude. Though we can make a simple argument that if they are formed from normal galaxies as a standard "mold", we expect a Gaussian distribution for their magnitudes or luminosities. This is in fact what is assumed for their distribution by most observers, as seen from the available literature. This possibility is explored also, and figure 9 shows a Gaussian with the same mean and variance as the data compared with the data for both cases where the magnitudes or the luminosities have a Gaussian distribution. The case where the luminosities are distributed as a Gaussian is called expnormal (in analogy to lognormal) for the magnitudes. Neither, it can be seen from figure 9 is acceptable.

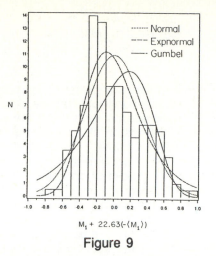

$M_1 + 22.63(-\langle M_1 \rangle)$

Figure 9

This result does not necessarily imply that all brightest galaxies are special. It does demand though, that not <u>all</u> first ranked galaxies in clusters are statistical. The question as to their nature, implied by their distribution in magnitudes, was recently addressed by this author (Bhavsar 1989).

It was shown that a "two population model" (Bhavsar 1989) in which the first brightest galaxies are drawn from two distinct populations of objects - a class of special galaxies <u>and</u> a class of extremes of a statistical distribution - is needed. Such a model can explain the details of the distribution of M_1 very well. Parameters determined purely on the basis of a statistical fit of this model with the observed distribution of the magnitudes of the brightest galaxies, are exceptionally consistent with their physically determined and observed values from other sources.

The probability density distribution of the magnitudes of the special galaxies is assumed to be a Gaussian. This is the most general expression for the distribution of either the magnitudes or luminosities of these galaxies, if they arise from some process which creates a standard mold with a small scatter, arising because of intrinsic variations as well as experimental uncertainty in measurement. The distribution of M_1 for the special galaxies is given by

$$p_{sp}(M_1) = \frac{1}{\sigma_{sp}\sqrt{2\pi}} \exp\left\{\frac{(M_1 - M_{sp})^2}{2\sigma_{sp}^2}\right\}. \qquad (7)$$

If a fraction, d, of the rich clusters have a special galaxy which competes with the normal brightest galaxy for first-rank, then the distribution that results is derived in Bhavsar (1989). This expression, $f(M_1)$, which describes the distribution of M_1 for the first-ranked galaxies in rich clusters, for the two population model is given by equation (8)

$$f(M_1) = d[p_{sp}(M_1)\int_{M_1}^{\infty} p_{Gum}(M')dM' + p_{Gum}(M_1)\int_{M_1}^{\infty} p_{sp}(M')dM'] + (1-d)p_{Gum}(M_1). \qquad (8)$$

The first term in the above expression gives the probability density distribution of special galaxies with the condition that the brightest normal galaxy in that cluster is always fainter. The second term gives the probability density distribution of first-ranked normal galaxies, in clusters containing a special galaxy, but the special galaxy is always fainter. The last term gives the probability density of normal galaxies in clusters that do not have a special galaxy. Equation (8) is our model's predicted distribution of M_1 for the brightest galaxies in rich clusters. The parameters in this model are: i) σ_{sp} - the standard deviation in the magnitude distribution of the special galaxies; ii) M_{sp} - the mean of the absolute magnitude of the special galaxies; iii) a - the measure of the steepness of the luminosity function of galaxies at the tail end; iv) M_{ex} - the mean of the absolute magnitude of the statistical extremes given by $M_{ex}=M_o-.577/a$, we shall instead use the parameter $b = M_{sp}-M_{ex}$, the difference in the means of the magnitudes of special galaxies and statistical extremes; and v) d - the fraction of clusters that have a special galaxy.

We have chosen the maximum-likelihood method, being the most bias free, and therefore best suited, to determine the values of the parameters. There are five independent parameters and 93 values of data. We maximize the likelihood function, defined by

$$\pounds = \prod_{i=1}^{93} f(M_1[i]), \qquad (9)$$

where the function $f(M_1[i])$ is the value of $f(M_1)$ defined in equation (8), evaluated at each of the 93 values of M_1 respectively for i = 1 to 93. The values of the parameters that maximize $\pounds$ give the maximum-likelihood fit of the model to the data. The parameters, thus determined, have the following values:

$$\sigma_{sp} = 0.21 \text{ mag} \qquad (10)$$

$$M_{sp} = -22.80 \text{ mag} \qquad (11)$$

$$b = M_{sp} - M_{ex} = -0.53 \text{ mag} \qquad (12)$$

$$a = 4.33 \qquad (13)$$

$$d = 0.65 \qquad (14)$$

Figure 10 compares the data to the model [equation (8)] evaluated for the parameter values determined above. The fit is very good. Note that the fit is

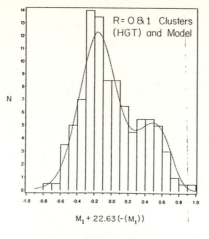

$$M_1 + 22.63 (-\langle M_1 \rangle)$$

Figure 10

calculated using all 93 independent observations, and not tailored to fit this particular histogram.

A further detailed statistical analysis by Bhavsar and Cohen (1989) of alternatives to the assumed Gaussian distribution of the magnitudes of special galaxies, along with a study of the confidence limits of the parameters determined by maximum-likelihood and other statistical methods has determined a "best" model. It turns out that the models in which the luminosities have a Gaussian distribution work marginally better. This may have been expected, luminosity being the physical quantity. This model requires 73% of the richness 0 and 1 clusters to have a special galaxy which is on average half a magnitude brighter than the average brightest normal galaxy. As a result, about 66% of all first-ranked galaxies in richness 0 and 1 clusters are special. This is because in 7% of the clusters, though a special galaxy is present, it is not the brightest.

Although it is generally appreciated that some of the brightest galaxies in rich clusters are a morphologically distinct class of objects (eg cD galaxies); we have approached the problem from the viewpoint of the statistics of their distribution in M_1, and conclude that indeed some of the brightest galaxies in rich clusters are a special class of objects, distinct from the brightest normal galaxies. Further we have been able to model the distribution of these galaxies. We have presented statistical evidence that the magnitudes of first-ranked galaxies in rich clusters are best explained if they consist of two distinct populations of objects; a population of special galaxies having a Gaussian distribution of magnitudes with a small dispersion (0.21 mag), and a population of extremes of a statistical luminosity function. The best fit model requires that 73% of the clusters have a special galaxy that is on average 0.5 magnitudes brighter than the brightest normal galaxy. The model also requires the luminosity function of galaxies in clusters to be much steeper at the very tail end, than conventionally described.

REFERENCES

Barrow, J. D., and Bhavsar S. P. 1987, Quart.J.R.A.S., **28**, 109 (BB).
Barrow, J. D., Bhavsar, S. P., and Sonoda, D. H. 1984, M.N.R.A.S., **210**, 19p
-------- 1985, M.N.R.A.S., **216**, 17 (BBS).
Bhavsar, S. P. 1989, Ap. J. **338**, 718.
Bhavsar, S. P., and Barrow, J. D. 1985, M.N.R.A.S. **213**, 857 (BB).
Bhavsar, S. P., and Barrow, J. D. 1983, M.N.R.A.S., **205**, 61p.
-------- 1984, in Clusters and Groups of Galaxies eds. F. Mardirossian, M.
  Giuricin, and M. Mezzetti (Dordrecht: Reidel) p. 415.
Bhavsar, S. P., and Ling, E. N. 1988, Ap. J. (Letters), **331**, L63 (BL I).
-------- 1988, Pub. Ast. Soc. Pac. **100**, 1314 (BL II).
Carlotto, M. J. 1988, Appl. Optics, **27**, 1926.
Efron, B. 1979, Ann. Stat., **7**, 1.
Evans, J. E., and Maunder, E. W. 1903, M.N.R.A.S., **63**, 488.
Fisher, R. A., and Tippett, L. H. C. 1928, Proc. Cambridge Phil. Soc. **24**, 180.
Geller, M. J., and Peebles, P. J. E. 1976, Ap. J. **206**, 939.
Glass, L. 1969, Nature, **223**, 578.
Gott, J. R. *et al.* 1989, ap. J. **340**, 625.
Groth, E. J., Peebles, P. J. E., Seldner, M., and Soneira, R. M. 1977, Sci. Am.,
Gumbel, E. J. 1966, Statistics of Extremes, (New York: Columbia Univ. Press).
Hillenbrand, L. 1989, B.A. thesis, (Princeton University).
Hillenbrand, L., Bhavsar S. P., Gott, J. R. *in preparation*.
Hoessel, J. G., Gunn, J. E., and Thuan, T. X. 1980, Ap.J. **241**, 486 (HGT).
  237, 76 (May).
Hoyt, W. G. 1976, Lowell and Mars, (Tucson: University of Arizona Press).
Huchra, J., Davis, M., Lantham, D.W., and Tonry, J. 1983, Ap. J. Suppl., **52,** 89.
Ling, E. N. 1987, Ph. D. thesis, (University of Sussex).
Limber, D. N. 1953, Ap.J., **117**, 134
Lowell, P. 1906, Mars and Its Canals, (New York: The Macmillan Company).
Peach, J. V. 1969, Nature, **223**, 1140.
Peebles, P. J. E. 1968, Ap.J. **153**, 13.
-------- 1973, Ap.J. **185**, 413.
Peebles, P. J. E. 1980, The Large Scale Structure of the Universe (Princeton,
  New Jersey: Princeton University Press)
Peebles, P. J. E.,and Groth, E. J. 1975, Ap. J., **196**, 1.
Sandage, A. 1976, Ap.J. **205**, 6.
Shane, C. D., and Wirtanen, C. A. 1967, Pub. Lick Obs., Vol. **22**, Part 1.
Totsuji, H., and Kihara, T. 1969, Pub. Ast. Soc. Japan, **21**, 221.
Tremaine, S. D., and Richstone, D. O. 1977, Ap. J. **212**, 311.
Weinberg, D. H., Gott, J. R., and Melott, A. L. 1987, Ap. J. **321**, 2.
Zahn, C. T. 1971, IEEE Trans. Comp., **C20**, 68.
Zwicky, F., Herzog, E., Wild, P., Karpowicz, M., and Kowal, C. T. 1961-
68,Catalogue of Galaxies and Clusters of Galaxies, (Cal. Inst. of Tech,
Pasadena)

Suketu P. Bhavsar, Department of Physics and Astronomy,
University of Kentucky, Lexington KY 40506-0055, USA.

Uncertainties in Measuring Geometrical Parameters of Elliptical Galaxies

G. FASANO and C. BONOLI

Padova Astronomical Observatory

SUMMARY

We present a statistical comparison of the data available in the literature concerning the geometry of the isophotes in elliptical galaxies. The discrepancies found in position angles and ellipticities are well beyond the errors quoted in the various photometries, both in the inner and in the outer regions, suggesting the influence of systematic errors to be largely underestimated. In particular, probably because of the possible effect of bad flat fielding and sky removal, the mean disagreement in the outer isophotes is surprisigly high. To interpret quantitatively this result, we have analyzed several synthetic galaxy images for which an incorrect flat fielding or sky subtraction has been simulated. Simple expressions for the expected errors, mainly depending on the estimated background uncertainty and on the ellipticity of the isophote, are given in some particular cases.

1. COMPARISON OF DATA IN THE LITERATURE

In the two dimensional surface photometry of galaxies it is well known that the outer luminosity distribution of an elliptical galaxy can be drastically changed because of poor sky removal, whereas its inner luminosity distribution results from the convolution of the true light distribution by the combined atmospheric and instrumental point spread function (PSF). Many authors have examined these problems as far as their effect on the observed luminosity profile is concerned (see Capaccioli 1988 for a recent review). On the contrary, their influence on the geometry of elliptical isophotes is not well explored.

The errors quoted in the various photometric surveys for position angle (θ) and ellipticity (ϵ) are in general small. For instance, from internal consistency and from the analysis of the noise observed in the profiles, Djorgovski (1986) obtained: $\Delta\theta = 5°$ and $\Delta\epsilon = 0.02$. On the other side, the errors expected to arise from applying some fitting algorithm to noisy isophotes, are even smaller (Bender and Möllenhoff [1987] give $\Delta\theta \leq 2°$ and $\Delta\epsilon \leq 0.005$ in the worst case).

To check the reliability of such estimates, we have tested the mutual agreement of the various photometries available in the literature. To this purpose, we collected ellipticity and position data for all the objects found in the literature and studied by at least two authors. For each object and for several values of the isophotal semi–major axis a_i, the data from each couple of authors have been compared, getting a set of absolute differences $\Delta\theta_i$ and $\Delta\epsilon_i$. We have then plotted in figures 1a-d all the $\Delta\theta_i$ and $\Delta\epsilon_i$ versus the quantities $r_i^{(25)}$ and $< \epsilon >_i$, which represent the major axes normalized to the diameters D_{25} as given in the RC2 (de Vaucouleurs *et al.* 1976) and the average ellipticity respectively. In the same figures are reported the standard deviations from zero obtained binning the distributions.

Some interesting conclusions can be drawn from these figures:
The mechanisms which produce systematic errors in position angles and ellipticities mostly operate in the inner and outer regions of the profiles.
The observed discrepancies are quite larger than the errors quoted in the various surveys (they also are almost randomly distributed among them).

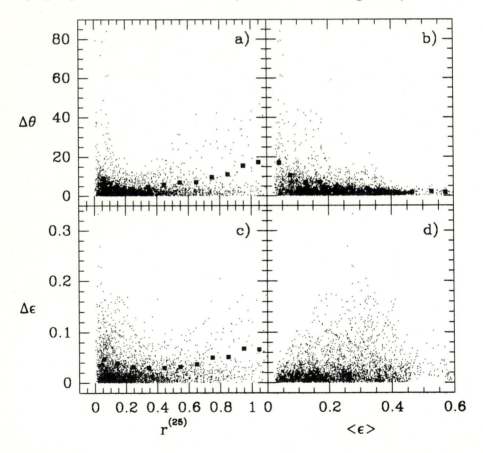

Figure 1

- The lower the ellipticity of the isophote, the higher the error on its position angle estimate. Although this statement is an obvious one, we note that large errors are possible also in the ellipticity range 0.2÷0.4, whereas errors of 20° are quite normal for ϵ <0.2.
- The errors in ellipticity probably do not depend on the ellipticity itself (the correlation showed in the figure 1d is an apparent one, caused by the very definition of $<\epsilon>$).

The large discrepancies found in the inner regions are obviously due to the convolution of the true light distributions by different PSF's. Franx (1989) analyzed the effects of a noncircularly symmetric PSF on position angles and ellipticities of the inner isophotes, getting analytical expressions for $\Delta\theta$ and $\Delta\epsilon$, whose amplitude is consistent with figure 1a. On the contrary, the disagreement observed in the outer profiles appears surprisingly high, suggesting that possible residual trend in the flat fielding and sky removal, distorts the outer isophotes more than commonly accepted.

3. THE OUTER REGIONS

To get an empirical idea of the errors which would be expected in the ellipticity and position angle outer profiles, we have analyzed several synthetic galaxy images for which an incorrect flat fielding or sky subtraction has been simulated. In both cases, assuming the mean value of the background to be unchanged, the excess of surface brightness can be expressed, in the first order approximation, as:

$$\Delta\mu = -2.5 Log\{1 + \xi \times (1 + 10^{0.4\delta\mu})\} \qquad (1)$$

where the quantities ξ (relative error of the adopted flat field or sky level) and $\delta\mu$ (surface brightness difference between galaxy and sky) depend in general on the position inside the frame. If we assume the galaxy surface brightness profile to follow the $r^{1/4}$ law, we can easily demonstrate that the excess of surface brightness $\Delta\mu$ corresponds in each point to an excess of radius (normalized to the radius itself):

$$|\Delta r|/r = 0.4805 \times (r/r_e)^{-1/4}|\Delta\mu(r)| \qquad (2)$$

where r_e is the effective radius of the profile in the given direction. We shall see in the following that the quantity $(|\Delta r|/r)$, averaged along any given isophote, turns out to be useful to parameterize the errors $\Delta\theta$ and $\Delta\epsilon$ in the outer geometrical profiles. This is quite reasonable since such quantity should express in one sense the degree of distortion of the isophote.

For the sake of simplicity, among the possible background distortions which give rise to systematic deviations in the geometrical outer profiles, we confined ourselves to consider only those producing an apparent tilt of the sky intensity plane around the y axis.

Besides the $r^{1/4}$ law luminosity profile, the model galaxies have been assumed to have constant ellipticity (ϵ_o) and position angle (θ_o). In particular we have set, once and for all, $\theta_o = 0°$ and $\theta_o = 60°$ for the model galaxies employed in the analysis of the ellipticity and position angle errors respectively. Although quite arbitrary, the latter assumptions are useful since they produce values of $\Delta\theta$ and $\Delta\epsilon$ close to the respective upper limits, without giving rise to any discontinuity in the profiles.

Several noiseless galaxy models have been generated covering the ellipticity range $0.1 \div 0.6$ (step 0.1). For each model three different values $(0.01, 0.025, 0.04)$ of the parameter ξ_{max} (relative background difference between the borders of the frame) have been used to simulate bad background removals. Then, the *quasi elliptical* outer isophotes of each image have been analysed by means of an ellipse–fitting procedure, in order to estimate the differences $\Delta\theta = |\theta - \theta_o|$ and $\Delta\epsilon = |\epsilon - \epsilon_o|$.

In figures 2a,b we have plotted the quantities $log(\Delta\theta)$ and $log(\Delta\epsilon)$ which result from our set of simulations, versus the corresponding values of $log(|\Delta r|/r)$ computed,

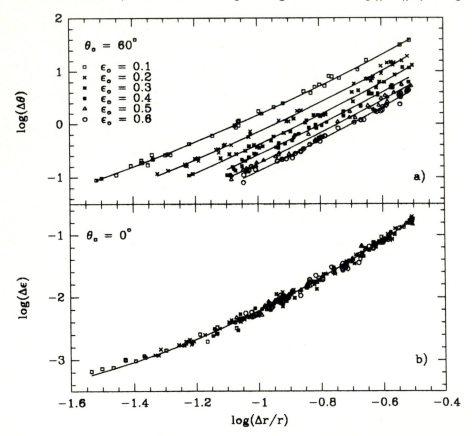

Figure 2

for each isophote, from equations (1) and (2). The isophotes having different ellipticities are represented in figure 2a with different symbols and the full lines reproduce a second-order polynomial fits to the data.

Note that the correlation in figure 2a between $log(\Delta\theta)$ and $log(|\Delta r|/r)$ shows a marked dependency on the ellipticity of the isophotes, whereas the one in figure 2b, involving $log(\Delta\epsilon)$, is quite independent on ϵ. This result confirms and quantifies in a completely independent way, the general behaviour of the errors discussed in connection with the figure 1. Note also that the values of $\Delta\theta$ and $\Delta\epsilon$ in figure 2 are fairly consistent with the respective ones in figure 1.

REFERENCES

Bender R. and Möllenhoff C. : 1987, Astron. Astrophys., **177**, 71

Capaccioli M.: 1989, In Proc. of the *Second Extragalactic Astronomy Regional Meeting*, Cordoba, Argentina

de Vaucouleurs G., de Vaucouleurs A. and Corwin H.G.: 1976, *Second Reference Catalogue of Bright Galaxies*, Austin, Univ. of Texas press (RC2).

Djorgovski S.: 1986, *Ph. D. Dissertation, University of California, Berkeley*

Franx M., Illingworth G. and Heckman T.: 1989, Astron. J., **98**, 538

Address of the authors:
Osservatorio Astronomico di Padova
Vicolo dell'Osservatorio 5
35122 – Padova (Italy)

DISCUSSION

G. Fasano to P. Nobelis: How can we introduce weighting of the data in the bootstrap and jacknife theories?

One way of weighting data is to extract samples from them using the distribution defined by the weights in place of the usual empirical one. I don't know whether a study in that direction exists.

H. Eichhorn to P. Nobelis: I take it that the main usefulness of the methods is when one has no information on the covariance matrix of the **y** because the estimator for Σ you quote is not the best available if you have information on the covariance matrix of **y**. Is this so?

This is true for standard linear regression. But the general situation is unclear and in particular for overparametrized models the estimation of the covariance will be falsely optimistic (the number of parameters, k, is close to the sample size, n).

J. Cuypers to P. Nobelis: Are there any rules for the number of simulations you have to do to use the bootstrap?

Some studies like Hall, P. (1986) were made in that direction.

E. Feigelson to E.J. Alfaro: Can kriging be applied to 3-dimensional data without assuming isotropy? If so, it may provide a powerful tool in studying "voids and filaments" in the spatial distribution of galaxies.

J. Pfleiderer: I don't think it should be applied. Kriging is a method of interpolating or extrapolating known data showing some tendencies in the distributions, but not following underlying physics. Applying it to voids and filaments would amount to this: given a filament and void (existing for some physical reason), does it mean that you expect more filaments and voids for reasons of mathematically simple distribution extrapolations?

E.J. Alfaro: Kriging may, of course, be applied to three-dimensional data, but isotropy is perceptive. If this condition is not initially accomplished, it may be approached by transforming of spatial coordinates before kriging. Concerning the distribution of galaxies, it is worth noting that kriging deals with the values of an $R^n \to R$ function and not with the topological properties of the experimental network used in estimating it, although this network evidently influences the estimation.

C. Jaschek to E.J. Alfaro: Do you have a physical explanation for the depression in the galactic plane of the cluster depression?

We have not yet analysed the astronomical results of this paper. However we wonder if these phenomena concern a spiral arm mechanism or on the contrary are they a consequence of the global dynamics of the disc.

A. Bijaoui to E.J. Alfaro: How does the resolution of the grid depend on the correlation function?

The grid of estimated points is not dependent on the correlation function and its size can be freely chosen by the user depending only on the computing time available. The estimated error of the "kriged" values is, on the contrary, dependent on the correlation function and on the distribution of neighbouring experimental points.

F. Murtagh to S.P. Bhavsar: One thing worries me in connection with the use of the MST for testing filamentary structure. You begin with a well-defined mathematical structure. Then you use ad-hoc rules – pruning, splitting – to pick out filaments. This is specific to the assumed structure. Your are essentially rigging the rules, so it does not surprise me that permuting your "cubettes" leads to the loss of this structure.

Yes, the MST is a well-defined, objective procedure; on the other hand pruning and separating require parameters. In the written version you see an answer to this. In fact the procedure can be used to *define* filaments. We explore the parameter space available and find what we should call filaments – the linear structures that are unique to the data but disappear.

H.L. Marshall to S.P. Bhavsar: You looked for filaments because the data suggest this kind of structure but why not look for a priori defined structures? For example, some have suggested the counts are "frothy" in appearance, thus why not look for spherical shapes?

It would be great if we had a general pattern recognition algorithm for non-specific structure. Visual impressions suggest filaments, and so we have investigated these structures.

L.G. Taff to S.P. Bhavsar: Here you have a *complete* sample of all galaxies to some limiting magnitude. When you fit a power law to this data you get one exponent. This data can yield no other number and because it is a complete sample there are no errors or uncertainties in it. What can the bootstrap method add?

It gives us an answer to the following question: if this data sample is a small but characteristic sample of all the galaxies in the universe, what is the variance in some quantity obtained from this sample as a measure of the true value for the universe.

S.J. Hogenveen to S.P. Bhavsar (comment): The 1091 galaxies are one sample as we see it in our corner of the universe. However it is just one outcome of an "ensemble" of statistically possible outcomes. "Bootstrap" in a way simulates the ensemble and allows you to determine the reliability of the data you extract from the original sample.

E. Feigelson to S.P. Bhavsar (concerning the debate between L.G. Taff and S.P. Bhavsar on whether $\gamma = -.77 \pm 0.03$ or ± 0.13): This discussion shows that the same data can give different statistical results depending on the question being asked. This also occurs in choosing what regression line one might fit to bivariate data (see T. Isobe's talk yesterday). Astronomers should be explicit regarding their goals, and choose statistical method appropriate for them.

J. Pfleiderer to S.P. Bhavsar: The data set gives a well-defined distribution with small error. It can be expected that this is somehow representative of what would happen elsewhere but this extrapolation is more uncertain, hence the larger uncertainty. Extreme example: two galaxies define a slope with error zero but any extrapolation by bootstrap would result in an infinite uncertainty.

One has to distinguish between the formal errors obtained by fitting complexly correlated data points by some regression line, which is not valid in every case. Bootstrap gives us a measure of the variance in some quantity obtained from the data when formal statistical errors are hard to define.

Mircea Pfleiderer to S.P. Bhavsar: The ability to perceive a large part of our environment at once – which cannot be done by a computer – is called, by biologists – "complex comprehension".

Yes our eye-brain is by far the best hardware/software we have for pattern recognition. Thank you for your comment on the label for what it does.

MODELLING AND INFERENCE

Choosing the Correct Model for the Adjustment of Observations

H. EICHHORN
Department of Astronomy,
University of Florida, Gainesville, U.S.A.

ABSTRACT

It is pointed out that quantities to be estimated (target quantities) can rarely be measured directly but are computed from directly measurable quantities by means of a model. A negative excess in the distribution function of the adjustment residuals may indicate a realistic model. The simplest case of a least squares adjustment, known as OLS (ordinary least squares) in statistics, is rarely adequate for the state-of-the-art treatment of most contemporary problems. The rich literature ought to be consulted on how to handle the more general problems. The paper points out some of the traps into which investigators frequently fall and emphasizes the need to utilize **all** available conditions and all available information. The incorporation of additional observations and the enforcement of additional legitimate constraints can only increase the accuracy of the target parameter estimates. The concept of "efficiency" is explained and it is pointed out that the adoption of "natural" (i.e., orthogonal) adjustment parameters helps in choosing an efficient adjustment model.

1   THE DISTRIBUTION FUNCTION OF ADJUSTMENT RESIDUALS

In the terminology of modern statistics, the process of measuring is a process of estimation. It is intuitively obvious that there exist true values of the target quantities (the quantities to be measured); it is also obvious that the process of measurement can never produce these true values. This is so, because even the most precise measuring device necessarily produces readings with a limited number of digits (the "least count") while an infinite number of significant figures would be required to reproduce the true values. Likewise, assuming the existence of meaningful true values of quantities to be measured may generally no longer be correct once the precision of the measurements exceeds a certain value; remember, for example, Heisenberg's uncertainty relationship.

The estimation of any target quantities (parameters), which involves measurements, requires modeling. Only rarely can a target quantity be measured directly. The temperature of a gas, for example, is defined in terms of the average kinetic energy of the molecules which constitute it, but it would be quite impossible to ascertain the temperature of a gas by measuring the kinetic energies of even a sample of the molecules in the gas. We **calculate** the gas temperature from what we have actually measured, that is the length of a column of liquid.

Let the vector of target quantities be $\underline{l}$ and that of the directly measurable quantities be $\underline{k}$. The model (or more specifically, the calibration model) is then a mathematical relationship of the form $\mathbf{g}(\underline{k}, \underline{l}; \underline{p}_1, ..., \underline{p}_n) = \mathbf{0}$. Assume that it can be solved with respect to $\underline{k}$. The components of the vector of quantities $\underline{p}^T = (\underline{p}_1, ..., \underline{p}_n)$ are parameters which must be known to calculate $\underline{k}$ from $\underline{l}$. If they are not known a priori, they must be regarded as additional target parameters and also estimated in the course of the adjustment, together with $\underline{l}$. Note that $\underline{l}$ and $\underline{p}$ are of the same character as far as the mathematics of the adjustment process is concerned, even though it so happens that we are interested only in an estimate of $\underline{l}$.

The adjustment residuals computed after estimating $\underline{l}$ and $\underline{p}$ on the basis of the model will show a certain frequency distribution, according to the experience of countless investigators most often a normal one. This indicates that in every measuring situation there are circumstances not considered in the calibration model because they are in practice beyond the control of the experimenter, which manifest themselves as scatter in the measurements. We can only speculate what these factors might be and how the calibration model might have to be modified to account for them. Not that this has not been tried, and quite frequently so with even astounding success when unsuspected effects were discovered, but the frustrating experience of all who have been engaged in making readings and deriving target quantities from them has taught us that **there is no perfect calibration model.** In view of this empirical fact (I wonder if it has ever been explictly pronounced) we act as if measurements were intrinsically and of necessity subject to some kind of an uncertainty principle, namely that precise measurements, even when made on presumed identical setups with perfect calibration models, will scatter around a mean and that the residuals against this mean will obey a certain distribution function. A Gaussian normal distribution function seems plausible because of the Central Limit Theorem (but there is no proof that this must be the actual distribution function in all situations) and is in fact quite often found to represent quite accurately the real distribution of the residuals, especially when there is a large number of measurements.

Consider now the idealized situation in which we have a perfectly accurate calibration model and a measuring device capable of making readings with a certain precision. Also assume that there will be no blunders. (Suppose we measure well–defined different lengths with one particular device that reads to a least count of 1mm, say.) Let the lengths be arbitrary and let them be uniformly distributed. The measuring process would then yield readings which are always precise to the nearest millimeter, and any residual between 0 and .5 millimeters would occur with equal frequency. **In this situation, the residuals would not be normally distributed.**

The frequency function of residuals is discrete and therefore cannot, in reality, be truly normal but more important, any measuring process is so precise that an excessive residual whose absolute value exceeds a certain limit will never be encountered. We should therefore expect that the distribution function of the residuals will be the flatter (with negative "excess"; cf. Hufnagel 1937), the better and more realistic a calibration model we have. From the above it would appear that a negative excess in the distribution function of the residuals could indicate that we are approaching a realistic (i.e., nearly completely parameterized) calibration model. Literature on the subject is hardly availabe, since the mathematical representation of the distribution function of adjustment residuals has not yet become a component of the standard procedure in adjusting observations, even though some authors (e.g., Branham 1986) have urged that it should.

2   ROBUSTNESS: HOW IMPORTANT?

We should expect that very large errors, even when their occurrence is predicted by the normal distribution function, will not appear when reasonably precise measurements are reduced based on reasonably accurate calibration models. Is a least squares adjustment then still the way to find the most accurate values of the target quantities derivable from the available data? The Gauss-Markoff theorem (cf. Mood and Graybill 1963) would suggest that this is so, at least as long as the unobservable errors (estimated by the residuals eventually obtained) "are uncorrelated, with mean 0 and variance σ^2". The Gauss-Markoff theorem states in addition that the least squares estimators are also unbiased. In practice, this means that (under certain conditions), least squares estimates are unbiased and have, by definition (as it were) the smallest possible variances. But this is no guarantee that they are also the most accurate!

If the distribution function of the residuals is expected to be uniform, one might perform an L_∞ (Chebychev) adjustment which "spreads" the residuals between its constant maximum absolute values. The estimates which such an adjustment yields depend very little on the specifics of the distribution function of the measuring errors.

Where does this leave L_1 adjustments as advocated by, e.g., Branham (1982)? As Branham suggests, they are excellent for detecting blunders, but what is the physical-geometric justification for postulating (implicitly, to be sure) exactly that class of error distribution functions which will yield the minimum of the absolute values of the residuals as the most accurate (in the technical sense) method for estimating adjustment parameters? Note that the L_1 method does not give unambiguous results for the parameter estimates, since the representation can (within limits) often be changed in such a way that it leaves the sum of the absolute values of the residuals unchanged.

It seems that much contemporary work is concerned not so much with finding the most accurate estimate of a target quantity (that estimate which has the highest probability to be closest to the true value) but rather one which has certain statistical properties such as robustness. The more "robust" an estimator is the less will a value which it yields depend on the peculiarities of the error (or residual) distribution function. The values yielded by the L_∞ (Chebyshev) method, for example, depend only on the largest residuals and not at all on the others. But is this a sound principle on which to base an adjustment? There is no evidence that robustness is in any way correlated with accuracy, and while it may be comforting to know that an estimate of a particular target quantity will not be very sensitive to the peculiarities of the error distribution function, it may – within the framework of our present knowledge – well be possible that an estimate computed from a less robust estimator has, in fact, a better chance to be nearer the true value.

Chebychev estimation has its place in curve fitting rather than in the adjustment of observations. If one seeks that polynomial (for example) whose absolute deviation from the sine oscillates several times between the same maximum values, one must estimate the coefficients of this polynomial by the L_∞ criterion. It will be interesting to contemplate the distribution of the residuals: δ = function minus approximation. Plotted against the argument, it is reasonable to expect the residuals to oscillate several times between the permitted accuracy limits. The curves $y = \varepsilon$ and $y = -\varepsilon$ will be tangent to $y = \delta(x)$ at several points (cf. Hastings 1955) and therefore linger around the extremes $\pm\varepsilon$. The distribution function can generally be expected to be concave toward the x-axis but to dip abruptly to zero beyond $\mp\varepsilon$.

Whenever the absolute values of the several extremes of the differences "reality minus model" are numerically (almost) equal when parameters were estimated on the basis of an almost, but not quite correct model, one can argue that the distribution function of the residuals "actual function minus modeled function" will be similar to that of the residuals: Chebychev approximation minus actual approximated function. When a Gaussian error distribution function is now superimposed upon the distribution function of the approximations, we may well also get a bell curve with negative excess, which can under these circumstances, however, not be interpreted as orginating in an almost perfect reduction model.

It will be wise to remember that not all of statistics is concerned with point parameter estimation and that certain statistical concepts have little or no meaning in the context of data adjustment. Consider for example, that certain statistics, such as the average (median, geometric mean &c.) height of the males over 21 years of age in a certain population are not in any sense the "true" values of the target quantity "height", deviations from which may be regarded as errors caused by an incomplete calibration model or imperfections in the measuring process. These are rather independently defined parameters which specify the shape of a distribution function. Unlike in the adjustment of observations, there cannot be any meaningful effort to reduce the dispersion of the deviations from a somehow defined standard height. While it may well be desirable, in general statistics, to find an estimator (a somehow defined standard) which depends as little as possible on the peculiarities of the distribution function of the quantities considered, I see no virtue in doing this in the context of adjusting observations when the aim is to find the most accurate estimate of the target quantities. It seems that the method of maximum likelihood (which leads necessarily to the method of least squares for distribution functions which allow the Gauss-Markoff theorem to be applied) accomplishes this by definition.

Following, e.g., Brown (1955), the normal equations, whose solutions are the sought estimates of the adjustment parameters, are obtained by maximizing (subject to certain constraints) the function $f(\xi) = \exp(-\tfrac{1}{2}\xi^T\sigma^{-1}\xi)$, with ξ being the vector of the adjustment residuals and σ their covariance matrix. This involves finding the derivatives of $f(\xi)$ with respect to each component of ξ, which leads to minimizing the quadratic form $\xi^T\sigma^{-1}\xi$. It follows therefore that there is a class of functions $F(f)$ for which the distribution function $F(\xi)$ will also be maximized when $f(\xi)$ is maximized. As we already know from the Gauss-Markoff theorem, a normal distribution is not a necessary condition for a least squares adjustment to be applicable and physically meaningful.

3   LEAST SQUARES AND SO–CALLED OLS

The regrettable lack of communication between astronomers and statisticians is responsible for a number of misunderstandings, one of which is the term "ordinary least squares", occasionally referred to as OLS in the statistical literature. By this, statisticians mean finding that vector $\hat{\underline{x}}$ which minimizes $|\underline{1}-M\hat{\underline{x}}|$ when $\underline{x}$ and $\underline{1}$ are subject to the "equations of condition" $M\underline{x} = \underline{1}$, with M a given matrix. Usually, the order $\underline{n}$ of $\underline{x}$ is smaller than the order $\underline{m}$ of $\underline{1}$ — the dimensions of M are obviously $\underline{m} \times \underline{n}$. The variants of this standard problem have been discussed by Lawson and Hanson (1974, pp. 5-22). Neither Lawson and Hanson nor statisticians in general include in what they consider least squares any of the generalizations that have been so ably summarized by Brown (1955) in his now classical paper (the only case known to me when an in-house technical report acquired any but local and temporary importance). Brown's algorithm is able to handle the case in which the equations of condition may be nonlinear, in which each equation of condition may contain any number of components of the vector $\underline{1}$ whose covariance matrix need not be an identity matrix and not even diagonal. Eichhorn and Standish (1981) showed that the generalizations described by Brown can be reduced to the case where each equation of condition contains exactly one (fictitious) observation, but the fictitious observations will have a nondiagonal covariance matrix. Unfortunately, even though Jefferys (1980, 1981) has given, in the Astronomical Journal, a compact and comprehensive restatement of Brown's algorithm and introduced and discussed some further generalizations and refinements, it seems that even the majority of astronomers understand by "least squares" not much more than the statisticians understand by OLS.

An experienced investigator will often, with little difficulty, be able to point out ways in which many published results of least squares adjustments could have been more accurate if the authors had taken proper advantage of all available techniques and used all the implicit and explicit constraints on the adjustment parameters which the physics and the geometry of the experimental situation impose.

Take the example of constructing a star catalogue from measuring the coordinates of the photographic images of stars on plates which were all exposed on the same camera-telescope. While experience has shown that in such a situation one cannot take it for granted that the effective focal length (equivalent to plate scale) of the optics will remain the same from one plate to the next, it is also obvious that the plate scales cannot vary beyond a certain limit; as first approximation one may postulate that they are normally distributed with a certain (known) dispersion about an (unknown) mean. This constrains the range of values which the indvidual scale values can assume, and it is therefore a waste of available information to set up the adjustment in such a way that the scales may assume any value for any plate, regardless of what they turn out to be on the other plates which were exposed on the same instrument. Loss (or nonutilization) of available information is, however, always accompanied by a loss of accuracy and could therefore be justified only if the loss of accuracy is totally insignificant and/or unnoticeable, or if the complete imposition of all available conditions would require an additional effort (in these days, of programming) that the gain in accuracy would not justify.

In making this statement, I am taking for granted that the investigator always aims at the most accurate parameters that can be obtained from the existing observations.   This is generally true for astronomers and geodesists, but not necessarily for physicists.   It is conceivable that it is sometimes easier to obtain more precise data in a physical laboratory than to introduce complicated additional sophistication into the adjustment algorithm.   In any case, Eichhorn (1978) has published an algorithm for the estimation of parameters which are (such as the plate scales in the above quoted examples) subject to stochastic constraints.   Similar algorithms were developed in the geodetic community and are there known by the name of "least-squares collocation".

Obvious constraints are likewise not utilized if one ignores the fact that the value of a particular parameter to be estimated must be near a known number. This is another procedure which causes considerable loss of accuracy.   Again taking an example from photographic astrometry, consider the various parameters which quantify the deviations of the projection from the geometry imposed by a strictly gnomonic projection.   One of these, e.g., is the coefficient of cubic distortion. Presuming that the photographic objective was properly designed and manufactured to give distortion free images, this coefficient will be reasonably close to zero; yet when the model for plate adjustment provides for its estimation, it is typically introduced as a free parameter and could therefore - in principle - assume any value.   This is in contrast with what we know, namely that its absolute value must be small. It is well known that the coefficients of scale and of cubic distortion are strongly correlated when they are unrestricted (cf. Eichhorn and Williams 1963, cases VI-IX) and therefore individually poorly determined. This correlation can be considerably loosened by taking advantage of knowing that cubic distortion must be small, or, what is mathematically equivalent, that the cubic distortion (or whatever parameter one considers) is near any known value. The principal difficulty here is to decide how near, that is to decide which variance to assign to the known approximation value. Eichhorn and Gatewood (1967) successfully applied this technique to the determination of the coma coefficient in the southern Hyderabad zone of the Astrographic Catalogue.

4   AGAINST METHOD
(not quite in the spirit of Paul Feyerabend)

Potential accuracy in the estimation of certain parameters will also be unneccessarily lost when available straightforward constraints are not exhaustively utilized. The determination of the cluster parallax of the Hyades is a parade example. To appreciate what is happening, one needs to remember that astronomical research did not start only after the wide availability of computers had reduced to a triviality the considerations concerning the amount of arithmetic necessary to solve a given problem. Previously, investigators had to consider carefully the amount of computing labor. This led to the developments of special "methods" which consisted of reducing the most general equations in a given problem to sets of equations which likewise permitted one to solve for the target parameters but whose solution required much less labor than the solution of those equations which govern the problem in its full generality. Circumstances such as these led to the development of the "convergence point method" and of "Hertzsprung's method" for estimating the distances of member stars of a star stream. Both these methods are based on a partial utilization of the available constraints; in its crudest form the condition that the velocities of all member stars have the same vertex. As a consequence, the proper motions of all stars are directed toward the vertex of the common velocity direction. The estimation procedure of the distances by the convergence point method then consists of finding the convergence point by investigating the position angles of the proper motions. The individual distances of each member star can be computed whenever the speed of the stream has been found from the measured radial velocities. The information contained in the individual radial velocities remains totally unutilized and is thus wasted.

On the other hand, the Hertzsprung method utilizes the gradients of the proper motion components across the cluster (cf. Upton 1970) and is thus, within certain conditions, insensitive to the values of the proper motions; note that the results from this "Hertzsprung method" would remain unchanged if the same constant bias were added to each proper motion component. There has been discussion in the literature as to which of these two methods is "better"; the truth is that they both waste available information and are therefore inferior to an adjustment which utilizes all observations to their full extent and enforces all existing constraints.

One might think that this is obvious now that we can use computers, but alas, this is but one of the many examples when ingenious but inferior restricted solutions ("methods"), which had been devised before computers became available, survive today. "This is the way it has always been done" is an argument one hears surprisingly often, even now when the reasons for "doing it this way" have long since lost their validity.

A more sophisticated justification for choosing a "method" is to point out that certain methods are insensitive (or at least not very sensitive) to systematic errors of certain types. We have noted that the "Hertzsprung method" for deriving cluster distances is insensitive to constant biases in the proper motions, whereas the convergence point (and with it, the derived distances) will change when the proper motion components are affected by constant systematic errors, (such as would be caused, e.g., in right ascension by a faulty value for the motion of the equinox). On the other hand, the adjustment residuals of the proper motions would show a systematic trend with the coordinates indicating that the reduction model itself was incomplete; a bias in the proper motions should have been provided for in the model. This would, no doubt, have increased the formal errors of the estimated coordinates of the convergence point considerably, but one cannot simply pretend that the material is more accurate than it actually is just so one gets pretty numbers as result.

5   PARAMETERS COMMON TO DIFFERENT SOLUTION SETS

Another manner of neglecting available constraints is this: Suppose there are a number of situations which lead to an adjustment problem, such that some but not all of the adjustment parameters are common to all situations. An example for this would be the calibration of a number of thermometers, all of which use mercury as the expanding liquid. The expansion coefficient of mercury (pretend it is not yet known) must be the same for all thermometers. This is often accomplished (after a fashion) by determining an individual value for this expansion coefficient from the measurements with each individual thermometer, and then regarding as the finally accepted value the (weighted) mean of all the individual values. This very common practice deserves a few remarks. Formulated mathematically, the situation is as follows: Let there be a number n of adjustment situations such that the system of normal equations for the estimation of the ν-th set of parameters is

$$\begin{pmatrix} \mathbf{A}_\nu & \mathbf{B}_\nu \\ \mathbf{B}_\nu^T & \mathbf{C}_\nu \end{pmatrix} \begin{pmatrix} \underline{\mathbf{a}}_\nu \\ \underline{\mathbf{b}} \end{pmatrix} = \begin{pmatrix} \underline{\mathbf{l}}_\nu \\ \underline{\mathbf{m}}_\nu \end{pmatrix} \qquad \nu = 1, \ldots, \underline{n} \qquad (1)$$

Note that there is one subset, $\underline{\mathbf{b}}$, of the adjustment parameters which occurs in each of the systems. It can be shown that the estimate of the vector $\underline{\mathbf{b}}$ is obtained by

$$\underline{\mathbf{b}} = [\sum_{\nu=1}^{n} (\mathbf{C}_\nu - \mathbf{B}_\nu^T \mathbf{A}_\nu^{-1} \mathbf{B}_\nu)]^{-1} \sum_{\nu=1}^{n} (\underline{\mathbf{m}}_\nu - \mathbf{B}_\nu^T \mathbf{A}_\nu^{-1} \underline{\mathbf{l}}_\nu) \qquad (2)$$

Obviously, this expression for $\underline{\mathbf{b}}$ is **not** formally identical with any of the $\underline{\mathbf{b}}$ one would have obtained if one of the equations (1) were formally solved for $\underline{\mathbf{b}}$. Van Hamme and Wilson (1984) gave an impressive numerical example of this by using the statistically correct way of finding estimates for the system parameters of binaries when photometric information as well as radial velocity measurements are available. These authors' example also shows that the best obtainable estimates for the system parameters are not simply the weighted averages of the estimates one obtains from reducing the photometric and radial velocity measurements independently of each other. If we remember that $\underline{\mathbf{a}}_\nu = \mathbf{A}_\nu^{-1} (\underline{\mathbf{l}}_\nu - \mathbf{B}_\nu^T \underline{\mathbf{b}})$, we also see that a less than perfect estimate for $\underline{\mathbf{b}}$ will also give less than perfect estimates for each of the $\underline{\mathbf{a}}_\nu$.

One should keep in mind that it is the purpose of an adjustment to find the "best" values of certain target parameters which can be obtained from the available observations under the constraints and conditions to which they are subject. Allowing of the $\underline{n}$ systems (1) to determine its own "best" $\underline{\mathbf{b}}_\nu$, as it were, produces the absurd situation that there are several different "best" estimates for $\underline{\mathbf{b}}$, whose (weighted) arithmetic mean is regarded as a still better one. What can be better than best, though? And furthermore, the (weighted) mean of the $\underline{\mathbf{b}}_\nu$ is not identical to the $\underline{\mathbf{b}}$ one gets from equation (2).

6   SYSTEMATIC ERRORS AND MODELING DECISIONS

In spite of a mathematically quite clear situation, some authors frequently argue for estimating a $\underline{\mathbf{b}}$ from only one or, in any case, not all of the systems (1). I consider their reasoning faulty even though I am - following the "Fricke imperativ" in this age of computers - in favor of solving all of the systems (1) and computing a separate preliminary estimate of $\underline{\mathbf{b}}$ from each of them before computing the definitive $\underline{\mathbf{b}}$ from equation (2). (the late Walter Fricke once said to me in a conversation: "Wenn ich eine Loesung veroeffentliche, dann habe ich hundert gemacht!")   If all the separate estimates of **b** thus obtained are compatible with each other within their tolerance, one may take this as an indication (but not as an empirical proof!) that the mathematical model relating the measured quantities to the target quantities is substantially complete, meaning that all significant aspects of the physics and the geometry of the situation are adequately accounted for. When one finds significant discrepancies between the various individual estimates of **b**, one can often explain these most plausibly by invoking so-called systematic errors. Systematic errors could, in principle, always be removed by including in the model the effect (or effects) which cause them. While no adjustment can remove systematic errors from the **results** (cf. Eichhorn and Williams 1963), one can always eliminate systematic trends from the adjustment **residuals** by the proper model; as a rule, this requires – ceteris paribus – the estimation of at least one additional parameter. Separate solutions should therefore be made to verify that the reduction models are appropriate, but after this has been accomplished, there is no longer any justification for computing, from separate systems, different "best" estimates for the same set of parameters.

Thirty five years ago, when I was working on a parallax program (taking plates on the hand-guided twenty-six inch McCormick refractor, measuring them on a primitive one-screw measuring engine and reducing the measurements with a mechanical calculator), I learned that one could save half the labor without losing much accuracy in the result by measuring the plates in right ascension only; the parallax factors in declination are so small that the declination results do not contribute much and besides, the measurements in declination might be subject to undetected systematic errors which could introduce a bias in the ultimately resulting parallaxes. It is quite true that under the circumstances, reasons of work economy force the adoption of those results as definitive which are believed to be least affected by errors. Everyone who has ever labored on a mechanical desk calculator will appreciate the many hours of arduous button punching (and cranking) that went into obtaining results which are simple by contemporary standards.

This is how things were without computers. Only the most careful planning, with an eye on labor economy, made it possible to get any results at all. But now, circumstances have changed: We have computers, and this should be reason enough to reexamine procedures which were established in the times, within vivid memory of many still very active researchers, when modern computers existed only in the imagination of some visionaries.   The inclusion of additional information (conditions, observations) obviously cannot cause a decrease in the accuracy of the estimates of the adjustment parameters. (Charity forbids me to quote some authors who have seriously proposed the opposite.) If one is certain that – to stay with the example – parallaxes found from investigating right ascension information only are free from systematic errors, one could use these estimates to calibrate the declination measurements by calculating what the declination residuals would be were they calculated assuming the right ascension result as valid.   This would reveal the character of the suspected systematic errors in the declination measurements, or rather show where the mathematical model used for calculating declinations from the directly measured quantities (e.g., relative coordinates of the somehow defined centers of blobs of silver grains in photographic emulsions carried by glass plates) was deficient and needs to be modified.   This involves judgment and at this time cannot be performed automatically by a computer.   With a proper calibration model also for the declination measures, the right ascension and declination information could now be combined to arrive at a yet more accurate estimate of the target parameter (or parameters).   In summation, a subset of the available material, known (or presumed) to be based on a faultless mathematical model, can be used to calibrate the rest of the information, that is, to guide the investigator toward constructing an appropriate model.

7   EMPIRICAL REDUCTION MODELS

Every problem of observation adjustment postulates a (vector of) relationships $\underline{F}(\underline{a},\underline{x}) = \underline{0}$ between the vector $\underline{a}$ of target quantities, and the vector $\underline{x}$ of observable quantities. The vector of the target quantities may (and usually will) include parameters one is not really interested in (the vector $\underline{p}$ from Section 1 above) but which one has to know in order to compute those components of $\underline{a}$ which are the target quantities proper of the investigation.   Let photographic astrometry again serve as source for an illustrative example: There, the target quantities proper are the coordinates of the stars, but it is impossible to estimate these without, at the same time, also estimating the plate parameters which are of no intrinsic astronomical interest. All adjustment algorithms presume that the analytical form of **F** is perfectly known. In all rigor, however, it never is: There are always questions as to the existence or significance of certain effects (e.g. is there or is there not, a magnitude equation in the positions? Does the objective produce noticeably comatic images, or may one model the images as coma free?) which prevent us from knowing the exact form of **F** and all the parameters in it that need to be estimated.

Some investigators tend to "try out" various models and eventually settle on the simplest which will still represent the observations properly. This is basically what needs to be done, as long as it is done correctly. An example for how **not** to simplify a model is occasionally provided by authors who calculate orbital elements of binaries. Once in a while, one encounters stars whose orbital planes are nearly normal to the line of sight. In these cases, the line of nodes and thus the position angle of the node as well as the argument of the periastron are so poorly defined that their nominal standard deviations one gets from the adjustment seem to indicate that the orbital elements cannot be properly determined, especially as the estimate for the inclination is in this case also very uncertain. Those who look in addition at the covariance matrix of the elements will notice very large correlations between the estimates of these three elements. It is wrong, however, to conclude that such poorly determined elements will also predict the future relative positions of the components only very poorly. This is not necessarily so. Therefore, as some investigators obtain estimates for the inclination which do not significantly differ from zero, they repeat the adjustment under the now a priori assumption that the orbital plane is perpendicular to the line of sight - the longitude of the node and the argument of the periastron are no longer asked for and are replaced by a new element, the position angle of the periastron. The adjustment on the basis of this (incomplete) model will then also give estimates for the other orbital parameters which are formally more precise than those originally obtained. It is easy to see, however, that this formally obtained result is fallacious; in reality, the estimates of <u>all</u> elements are correlated with that of the inclination, and an error in the estimated inclination is propagated, multiplied by the appropriate correlation, to the estimations of all other elements and thus increases the formal measures for their uncertainties. Assuming at the outset that the inclination is zero and therefore (incorrectly) suppressing its uncertainty is thus unrealistic and not only leads to artificially small formal standard deviations for the estimates of the parameters that were carried in the adjustment, but also to biases in the estimates themselves, for the reason that the consequences of the correlation with the suppressed parameters are simply left inconsidered. Firneis and Firneis (1975) showed this generally and in a specific example.

If one does not know exactly what the function $\underline{F}(\underline{a},\underline{x})$ looks like, the problem has more the character of a curve fitting problem than that of adjusting observations. We must realize that the curve fitting problem is intrinsically quite different from the problem of adjusting observations. It is presumed that there is a set of values x_ν, which are generated by a function $x = x(\underline{a})$ where $\underline{a}$ is a vector of parameters, which could themselves be variable (i.e., coordinates) or be constant. The points x_ν may also have superimposed upon them random quantities ξ_ν whose distribution function is $\phi(\xi)$. The function $x = x(\underline{a})$ may be known analytically (for instance, $x = \sin y$ may be given) or only in tabular form through a set of sample values. The problem then is to find some function $x^*(\underline{b})$ such that $|\underline{x}^* - \underline{x}| < \delta$ over the whole domain of x, where we may have $|x^* - x| = \delta$ any number of times. When $x(\underline{a})$ is known, we have the problem of finding approximation functions x^* that are easier to compute than x itself which may not be rigorously computable. In order to utilize, for example, the sine function (which is not directly computable) in a computer, it is necessary to find a rational (and therefore computable) approximation $\sin^* y$, whose value does not deviate, over a certain domain, from $\sin y$ by more than a pre-set amount. After the form of $\sin^* y$ has been chosen (from an infinity of possibilities), the parameters $\underline{b}$ can be determined in a number of ways; usually by subjecting $x - x^*$ to a Chebychev adjustment.

The situation encountered in practice (and treated as a problem in data adjustment) is that the form of $x(\underline{a})$ is known in its general outlines but not quite exactly, and that we do not have the values x_ν but only $x_\nu + \xi_\nu$. We shall therefore definitely have to estimate the ξ_ν, which leads to a genuine problem of data adjustment rather than curve fitting. However, this leaves open the problem of finding the accurate form of x(<u>a</u>). The physics and geometry of the situation might ideally lead to <u>x</u> = <u>ay</u> + <u>b</u> (where <u>y</u> would be a known quantity, not in need of adjustment, and <u>a</u> and <u>b</u> are the components of the adjustment parameter), but the model, that is, the idealized physics and geometry usually neglect some effect which cannot be neglected within the given framework of precision. If the adjustment (regardless which minimum principle is enforced) is performed on the basis of the deficient model (i.e., if in the example above, only <u>a</u> and <u>b</u> are estimated) meaning that only the parameters in the incomplete model are estimated, one will not only (see the example given by Firneis and Firneis) obtain unrealistically low formal error estimates for these parameters but the results will in addition be biased, that is they will not estimate the target parameters themselves, but rather quantities which – if the equations were set up not too unrealistically – will only be strongly correlated with the target quantities proper.

Consider this example. Let the coordinates of stellar images measured on a plate exposed with an objective that gives slightly radially distorted images be adjusted on the basis of a model which does not allow for estimation of a coefficient of the radial distortion. One way to describe radial distortion is to say that the scale value is not a constant but varies with the distance from the optical center. If radial distortion is not large (or conspicuous), the dependence of the scale factor on the distance from the optical center will be weak and one will not get conspicuously discordant results if one bases an adjustment on the assumption that there is no radial distortion. If one defines (properly) the scale factor as the scale factor at the optical center, the estimate one gets for it when one neglects to provide for weak distortion in the model will by systematically too large (or too small, depending on the sign of the coefficient of the distortion term).

Eichhorn and Williams (1963) have pointed out that a function of parameters will be unavoidably affected by a parameter variance whenever only estimates, but not the true values of the parameters, are available. In other words, all functions evaluated with estimated parameters will have systematic errors because the accidental errors of the parameters propagate themselves as systematic errors into all functions in which they themselves are the arguments. It can also be shown that each additional parameter which does not reduce the dispersion of the residuals increases the total parameter variance. Not infrequently (cf. the example of the orbital parameters of binaries quoted above), the structure of the equation systems whose solutions are the target parameter estimates is such that the estimates are highly correlated. (While high correlations frequently occur in tandem with large condition numbers κ, these do not measure the same property. Note that the condition number can be manipulated at will by, e.g., changing the units in which the unknowns are reckoned, while no change of scale alone will influence the correlations). This makes it especially difficult to decide if a particular parameter ought to be carried in, or left out from the adjustment model. Brosche (1966) pioneered the procedure to find a set of orthogonal parameters which may be added to or removed from the model without changing the estimates of the other parameters, or their formal errors.

8   EFFICIENCY AND NATURAL PARAMETERS

Eichhorn (1990) has defined a quantity called "efficiency." There is an orthogonal transformation of the original adjustment parameters which converts them to orthogonal and therefore mutually noncorrelated parameters. These are, of course, linear forms of the original ones. (We assume tacitly without restricting generality that the problem has been linearized.) It can be proved that the product of the standard deviations of these "natural" parameters is the minimum possible product of the standard deviations of all orthogonal transforms of the original parameters. This product is proportional to the reciprocal of the square root of the determinant of the matrix of the system of the original normal equations, because it is equal to the square root of the product of the eigenvalues of the inverse of this matrix. The "efficiency" e itself, which varies between zero and 1, is defined as the n-th root (n being the order of the vector of adjustment parameters) of the product of the variances of the above defined natural parameters divided by the product of the variances of the original parameters, which is the product of the main diagonal terms of the inverse of the matrix of the normal equations. The closer e is to 1, the less correlation there is between the original unknowns. The natural unknowns themselves always have an efficiency of 1. If one finds a model overloaded (that is, if it turns out that so many parameters are carried in the adjustment that their estimates are all highly correlated and consequently poor) it may become necessary to reduce the number of unknowns. The procedure to follow then is this: First calculate the efficiency of the originally sought adjustment parameters. This does not involve calculating the eigenvalues or eigenvectors of the matrix. If $\underline{e}$ turns out small (below 0.2, say), there are significant correlations between the estimates of some of the parameters. One would then calculate the natural parameters (this **does** require calculating the eigenvectors and eigenvalues of the system matrix of the normal equations) and their standard deviations. Those of the natural parameters which are apparently not significantly different from zero can be discarded without any of the remaining changing their values.

Let $\mathbf{M}\,\underline{a} = \underline{l}$

be the system of normal equations in their original form. We assume that the system is scaled so that $\mathbf{M}^{-1}$ is the covariance matrix of the adjustment parameters $\underline{a}$. Since

$$\mathbf{M} = \mathbf{P}\,\Lambda\,\mathbf{P}^T \tag{3}$$

with $\mathbf{P}$ being the matrix of the normalized eigenvectors of the (positive definite) matrix $\mathbf{M}$ and Λ the diagonal matrix of its eigenvalues, the natural parameters $\underline{b}$ are given by

$$\mathbf{P}^T\,\underline{a} = \underline{b} = \Lambda^{-1}\,\mathbf{P}^T\,\underline{l}.$$

Their covariance matrix is obviously Λ^{-1}, that is the diagonal matrix of the reciprocals of the eigenvalues of $\mathbf{M}$. If one now decides to drop some of the components of $\underline{b}$ from the adjustment because they are not significantly different from zero, one can still find the original parameters $\underline{a}$ from

$$\underline{a} = \mathbf{P}\,\underline{b}$$

even if some of the components of $\underline{b}$ are now equal to zero. Note, however, that according to the law of error propagation, the covariance matrix of $\underline{a}$ is still $\mathbf{P}\,\Lambda^{-1}\,\mathbf{P}^T = \mathbf{M}^{-1}$ as is obvious from equation (3). This shows that elimination of a parameter from the model cannot render those retained more accurate. These results, far from surprising, only underscore the statement that leaving out a number of parameters cannot improve the accuracy of those left in the model, and that the formal error estimates computed on the basis of an unrealistically truncated model are plainly meaningless.

If the original parameters themselves are changed to a different set but with the same number of components (by, e.g., introducing the probably less sensitive position angle of the periastron instead of the argument of the periastron), the argument above remains the same.

9   CONCLUSION

In conclusion - and this is perhaps somewhat gratuitous - we observe that many investigators still adjust data frequently as they would have fifty years ago. We would do well to heed Branham's (1986) admonition "A complete study of an over-determined linear system should include the least squares and L_1 solutions, their respective errors, the condition number of the linear system, and possibly a study of the singular values of the system and a graph of the distribution of the residuals". In addition, we should compute the natural adjustment parameters and their errors, in essence therefore the complete covariance matrix, together with its eigenvalues and eigenvectors, of the original adjustment parameters, as well as their efficiency, which seems to me to be a more fundamental number for characterizing the problem than the condition number of the linear system whose value can be changed by all sorts of manipulations.

ACKNOWLEDGEMENTS

My colleagues R.L. Branham, C.S. Cole, E.D. Feigelson, J.D. Scargle, H.C. Smith, L.G. Taff, A.R. Upgren and R.E. Wilson have read a draft of this paper and provided valuable critique.

LITERATURE

Branham, R.L. 1982. Astron. Journ. **87**, 928.

Branham, R.L. 1986. Celes. Mech. **39**, 239.

Brosche, P. 1966. Veroeff. Astron. Rech. Inst. Heidelberg No. 17.

Brown, D.C. 1955. Ball. Res. Labs. Rept. No. 937, Aberdeen, MD.

Eichhorn, H. 1978. MNRaS **182**, 355.

Eichhorn, H. 1990. Bull. of the Astron. Inst. of Czechoslovakia, to be published.

Eichhorn, H. and Gatewood, G.D. 1967. Astron. Journ. **72**, 1191.

Eichhorn, H. and Standish, M.E. 1981. Astron. Journ. **86**, 156.

Eichhorn, H. and Williams, C.A. 1963. Astron. Journ. **68**, 221.

Firneis, M.G. and Firneis, F.J. 1975. Astron. Nachr. **296**, 95.

Hastings, C. 1955. Approximations for Digital Computers. Princeton: Princeton U. Press.

Hufnagel, L. 1937. In: E. von der Pahlen, Lehrbuch der Stellarstatistik, p. 30 ff. Leipzig: Joh. Amb. Barth.

Jefferys, W.H. 1980. Astron. Journ. **85**, 177.

Jefferys, W.H. 1981. Astron. Journ. **86**, 149.

Lawson, C.L. and Hanson, J.R. 1974. Solving Least Squares Problems. Englewood Cliffs, N.J.: Prentice Hall.

Mood, A.M. and Greybill, J.A. 1963, Intro. to the Theory of Statistics, p. 348. New York, San Francisco, Toronto, London: McGraw Hill Book Co.

Upton, E.K.L. 1970. Astron. Journ. **75**, 1097.

Van Hamme, W. and Wilson, R.E. 1984. Astron. and Astrophys. **141**, 1.

Function minimization with several variables

G. Russo, F. Barone, L. Milano

Dipartimento di Scienze Fisiche, Università di Napoli, Italy

and

Istituto Nazionale di Fisica Nucleare, Sez. di Napoli, Italy

ABSTRACT

Function minimization is often used in astronomy, but when several variables are present, e.g. if they are parameters of a model to compare with observations, then it may happen that the function has many minima. In these cases, a "global" optimizer, rather then a "local" optimizer is needed. We explain here the difficulties one can find, and the methods presently available to solve a global optimization problem. The experience with two of these methods, the Controlled Random Search and the Simulated Annealing, is presented.

1. INTRODUCTION

In a science like astronomy, which is largely based on observational data, it is often necessary to use optimization techniques. With this term we refer to the problem of finding the minimum (or maximum) of a function in the region of interest. A practical example is the minimization of the sum of squared residuals of observed data from a theoretical function, a technique very common to determine the best parameters of a model that match the observations.

Every one who has faced optimization problem knows that, except the very simple cases, the problem is not solved by simply taking a routine from a numerical library. Very simple case is not a synonymous for analitical function, as one might suspect. Consider for example the function (Rastrigin's function):

$$f(x, y) = x^2 + y^2 - \cos 18x - \cos 18y \ (-1 < x < 1, -1 < y < 1).$$

which does not seem to have a complicated aspect: most of the routines in the commercially available libraries fail to find the global minimum, at $x = y = 0$, where $f = -2$. This because the function has about 50 local minima, arranged in a lattice configuration in the region of definition, and the algorithms usually contained in the libraries are local optimizer. What we should learn from this example is that, when facing an optimization problem, one has to first explicitly and in detail pose the problem. There are a number of questions to answer:

- is it a global optimization, or a local optimization problem?

- is it a constrained or an unconstrained problem?

- what is the nature of the constraints? (linear, non linear, simple bounds etc.)

2. THE OPTIMIZATION PROBLEM

An optimization problem begins with a set of independent variables or parameters, and often includes conditions or restrictions that define acceptable values of the parameters. Such restrictions are termed the constraints of the problem. The other essential component is a single measure of goodness, termed the objective function, which depends on some way on the variables. The solution of an optimization problem is a set of allowed values of the variables for which the objective function assumes an optional value. In mathematical terms, optimization usually involves maximizing or minimizing.

Let us assume that the ultimate objective of an optimization problem is to compute the solution. It is also helpful to assume that optimization problems can be posed in a standard form, such as:

$$\text{minimize} \quad f(x)$$
$$x \in R^n$$
$$\text{subject to} \qquad c_i(x) = 0 \ , \ i = 1, \ldots m'$$
$$c_i(x) \geq 0 \ , \ i = m' + 1, \ldots m$$

f (x) is the objective function, and $c_i(x)$ are the constraints. Any point $\hat{x}$ that satisfies the constraints is said to be feasible, and the set of all feasible points is termed the "feasible region". Only feasible points may be optimal.

It is also clear that different kind of minima exist: a simple classification makes the following distinction:

- strong local minima

- weak local minima

- global minimum.

2.1 LOCAL OPTIMIZATION

For local optimization methods we refer to Gill, Murray and Wright (1981) for a theoretical description, and to the NAG library for a comprehensive collection of subroutines which covers almost all practical cases.

2.2 GLOBAL OPTIMIZATION

Global optimization aims at determining the point(s) in the feasible region for which the objective function obtains its smallest value, the global minimum. In other words, global optimization means determination not just of a local minimum, but the smallest local minimum in the feasible region. The first important difference is that, while for local optimization one easily decides if the solution has been reached, by looking at the gradient, no such general rule exists for global optimization. This means that, in pratical applications, once a minimum is found, one has still to continue the search and the iterations, until a suitable stop criterion has been reached, but often without being certain of the result. A good summary of global optimization is contained in Törn and Zilinskas (1989).

3. CLASSIFICATION OF PROBLEMS

Global optimization problems may be classified in Constrained and Unconstrained problems; these latter are those for which the feasible region A is R^n, and may be further classified in "Solvable problems" and General Unconstrained problems, where solvable has a mathematical definition, but can be simply interpreted as: "there exists an algorithm which solves the problem in a finite number of steps". Constrained global optimization problems can, on the other hand, be classified as Special Form problem and General Constrained problems. A typical Special Form problem occurs when the function has a quadratic form; this means that the algorithm will take into account the particular form of the objective function. Constraints can be of several different forms; a possible classification is the following:

- Simple bounds

- Linear functions

- Sparse linear functions

- Smooth non-linear functions

- Sparse non-linear functions

- Non-smooth non-linear functions.

4. CLASSIFICATION OF METHODS

All methods can be divided in two classes, with respect to accuracy: those with guaranteed accuracy and those without. In order to obtain a guaranteed accuracy, it is necessary to perform an exaustive search of the feasible region, therefore with a large computing effort. These methods are called Covering Methods, while the other will be classified as follows:

Direct methods

- Random Search methods

- Clustering methods

- Generalized Descent methods

Indirect Methods

- methods approximating the level sets

- methods approximating the objective function.

4. EXAMPLES

We will now present in detail two methods, in order to give an idea of their simplicity, but also to explain why a high number of function evaluation is usually necessary.

4.1 THE CONTROLLED RANDOM SEARCH (CRS) PROCEDURE

This global optimization method is a direct, clustering-random method which only requires function evaluations, but no derivatives, and is applicable in the presence of constraints in a very easy way. To some extent, the CRS (due to Price, 1978) follows the idea of Becker and Lago (1970), but combines random search and clustering concepts in a single, continuous process.

The essential features of the algorithm are the following: an initial search domain, V, is defined by specifyng limits to the domain of each of the n variables and a predetermined number N of trial points are chosen at random over V, consistent with the constraints. The function is evaluated at each trial point and the position and function value corresponding to each point are stored in an array, A, named the search array. At each iteration a new trial point P is selected randomly from a set of possible trial points whose positions are related to the configuration of the N points currently held in memory, and always checked against the constraints; the function is evaluated in P, obtaining f_P, which is compared with f_M, where M is the point with the greatest function value among the N in memory. If $f_P < f_M$ them M is replaced, in A, by P, otherwise a new point P is chosen amoung the set of potentical trial points. As the algorithm proceds, the current set of N stored points tend to cluster around minima which are lower than the current value of f_M.

To make the procedure more efficient than pure random search, the probability of success ($f_P < f_M$) at each trial point must be sufficiently high: thi is accomplished, in the CRS, by defining the set of potential (next) trial poits in termus of the configuration of the

N points currently stored. At each iteration, $n + 1$ distinct points R_1, $R_2, \ldots, R_{n+1}$ are chosen at random from N ($N >> n$) in memory, and these constitute a simplex of points in n-space. The point R_{n+1} is taken (arbitrarily) as the pole (designed vertex) of the simplex and the next trial point, P , is defined as the image point of the pole with respect to the centroid, C , of the remaining n points:

$$\vec{P} = 2 \cdot \vec{C} - \vec{R}_{n+1}$$

where $\vec{P}, \vec{C}$ and $\vec{R}_{n+1}$ represent the position vectors, in n-space, of the corresponding points.

4.2 THE SIMULATED ANNEALING

This procedure is due to Metropolis et al. (1953) and has recently been used in a number of problems in physics.

A parameter T_0, in the same units of f, is chosen; ranges for each variable are assigned:

$$x_j^{min} < x_j < x_j^{max}; \; j = 1, n$$

A random point is chosen in this interval, and f is evaluated in this point:

$$f_i \; = \; f(\{x_j\}_i)$$

Then a random displacement is given to $\{x_j\}_i$:

$$\{x_j\}_{i+1} \; = \; \{x_j\}_i + \{\Delta x_j\}_i$$

taking into account the range for each x_j, and f is re-evaluated:

$$f_{i+1} \; = \; f(\{x_j\}_{i+1})$$

If $f_{i+1} \leq f_i$, then $\{x_j\}_{i+1}$ replaces $\{x_j\}_i$ and the process iterates with new random displacements; otherwise a probability function

$$P = e^{-f/T}$$

is computed, and a random point

$$R \in [0,1]$$

is chosen.

If $P \leq R$, then $\{x_j\}_{i+1}$ replaces $\{x_j\}_i$, and the process iterates with new random displacements; otherwise $\{x_j\}_i$ is unaltered but the process continues iterating with random displacements. This process continues until $\{x_j\}_i$ is not replaced for several iterations. At this point, the process is assumed to converge for that particular value of T, but a new lower value of T is chosen, and the process iterates again until convergence. The new value of T is chosen according to an "annealing" rule; one can choose, for example

$$T_m = K^m . T_0$$

with $K = 0.8$.

5. APPLICATIONS

We have successfully applied the CRS algorithm to an astronomical application: the determination of parameters of eclisping binaries based on comparison of observations with sinthetic light and radial velocity curves, determined with the method of Wilson (1979). The paper by Barone et al. (1988) gives full description of the application. Such application solves the problem of the uniquenes of the solution, since the global minimum is found, instead of the local one, as with the classical differential correction approach followed in the original method. However, the drawback is the big amount of CPU time necessary to solve a system, which is not always justified if different approaches can be followed.

REFERENCES

Barone, F., Maceroni, C., Milano, L., Russo, G.: 1988, Astron. Astrophys. 197, 347.

Becker, R.W., Lago, G.V.: 1970, "A Global Optimization Algorithm", in Proceedings of the 8th Allerton Conference on Circuits and Systems Theory, Monticello, Illinois, pag.3.

Gill, P.E., Murray, W., Wright, M.H.: 1981, "Practical Optimization", Academic Press, London.

Metropolis, N., Rosenbluth, A.W., Rosenbluth, M.N., Teller, A.H., Teller, E.: 1953, J. Chem. Physics 21, 1087.

Price, W.L.: 1976, Computer Journal 20, 367.

Törn, A., Zilinskas, A.: 1989, "Global Optimization", Lecture Notes in Computer Science 350, Springer-Verlag, Berlin.

Wilson, R.E.: 1979, Astrophys. J. 234, 1054.

The Third Catalogue of Nearby Stars.
Errors and Uncertainties

W. GLIESE – H. JAHREISS

Astronomisches Rechen–Institut, Heidelberg, F.R.G.

Summary   The forthcoming third edition of the *Catalogue of Nearby Stars* will originate from a data base which contains at present about 4700 stars. More than 1900 objects could be added within the last decade, and more than 75 per cent of these have still unknown trigonometric parallaxes and were selected due to their photometric or spectroscopic distances. This demonstrates how important it is to get reliable calibrations of colour-luminosity or spectral type-luminosity relations.

Recent observations proofed that more than one hundred stars from former editions are almost certainly much further away from the Sun than the 25 parsec we defined as our distance limit. In the past most of these stars were included on the basis of mis-classified spectra; they present a good example how important redundant information is to cope with erroneous data.

Introduction   Summing up all nearby objects contained in former publications up to [1], and about 1900 recently detected nearby stars or suspected objects we have altogether 4700 objects including the known components of binaries. Of course, the most important quantity of a star contained in our compilation is its distance. At Yale Observatory a new edition of the *General Catalogue of Trigonometric Stellar Parallaxes* (YPC) will be soon completed [2]. Thankfully we acknowledge that Prof. W. van Altena supplied us with an almost final 1989-version which we are using now.

During the last two decades trigonometric parallax work - often in combination with $UBVRI$ photometry - concentrated on faint nearby stars. Therefore, the number of nearby stars with available trigonometric parallaxes of high accuracy, i.e. $\sigma_\pi/\pi \leq 14\%$ corresponding to $\Delta M \leq 0^m.30$, could be almost tripled, suggesting that it is worthwile to repeat calibrations already carried out in the

Table 1: Standard errors of various spectral type-luminosity calibrations for dwarf K and M stars

system	dK	dM	source of calibration
MK	$\pm 0^m\!.3$	$\pm 0^m\!.6$	[2]
Mt. Wilson	0.5	1.0	[3]
Kuiper	0.3	0.6	[3]
Mt. Wilson (Joy & Abt)		0.55	[1]
Kuiper (Bidelman)	0.3	0.6	Gliese (unpublished)
objective prism classifications:			
MK(Upgren et al.)	~0.5		[1]
MK(Michigan)	not yet done		
Vyssotsky [4]	not possible		
Smethells	0.4	0.9	[5]
Robertson [6]	0.6	0.8	Gliese (unpublished)
Stephenson [7], [8]	0.6		Gliese (unpublished)
Sang-Gak Lee [9]	0.6	1.1	Gliese (unpublished)
photoelectric colours:			
$B - V$	0.4	0.7	[3], [10], [11]
$R - I_{Kron}$	0.4	0.6	[3], [10], [11]

past [3] with a considerably enlarged and improved sample. Based on an earlier YPC-version received in 1985 we have calibrated in recent years several spectral type-luminosity and colour-luminosity relations for late main-sequence stars to derive spectral and photometric parallaxes.

Obviously, the new 1989-system of trigonometric parallaxes deviates slightly from the 1985-version: the average difference is about $< \pi_{89} - \pi_{85} >= -0''\!.001$. This produces somewhat brighter absolute magnitudes $\Delta M = -0^m\!.05$ at the catalogue limit of 25 pc, and the absolute magnitudes of the resulting calibrations will be $0^m\!.02$ to $0^m\!.03$ brighter than before, i.e. an insignificant correction for our present purpose. Therefore, the final calibrations will be postponed until the definite YPC should be available.

Calibrations   In the spectral region of late K and M stars there exist spectral classifications in various systems and we try to apply most of these as luminosity indicators. In our opinion we do not gain very much in reducing one system to another. This means that each system should be calibrated separately for absolute magnitudes independent of related systems.

In the Table 1 all the spectral types used for our nearby star work were compiled together with the estimated standard errors of the absolute magnitudes.

Table 2: Standard errors of spectral type-luminosity calibrations for Kuiper's G0 to M6 dwarf stars after removing the parallax errors

$\sigma(M_V)$	G0 – G9		K0 – M1		M1.5 – M6	
$\sigma(M_V) \leq 0^m.10$	$\pm 0^m.29$	7	$\pm 0^m.31$	53	$\pm 0^m.56$	82
$0^m.10 < \sigma(M_V) \leq 0^m.20$	0.29	24	0.43	69	0.71	61
$0^m.20 < \sigma(M_V) \leq 0^m.30$	0.44	32	0.51	117	0.74	40

For comparison also the standard errors of the resulting absolute magnitudes for the principal broad-band colours $B - V$ and $R - I_{Kron}$ are given in the last two lines.

Since some of these results are still unpublished, let us have a more precise look on 485 Kuiper dwarf stars with spectral types G0 to M6 from Bidelman's [12] compilation having accurately known trigonometric parallaxes from the 1985-version of the YPC. This sample was subdivided into three groups according to the size of the standard error in absolute magnitude. After eliminating the contribution of the trigonometric parallax errors we find for three distinct spectral ranges the *cosmic dispersion* in absolute visual magnitude M_V as listed in Table 2 . In columns 3, 5, and 7 the number of stars are given that were contained in the different subsamples. Table 2 indicates an increase of the cosmic dispersion with increasing parallax errors even after elimination of the dispersion produced by the parallax errors published in the YPC. Therefore, we may ask wether the standard errors given in the YPC are slightly underestimated.

For many objects we know several distance determinations: trigonometric, spectroscopic, and photometric parallaxes. Absolute magnitudes based on various luminosity indicators are more or less correlated with each other. Moreover, a combination of trigonometric parallaxes with other distance estimates as applied in the second edition [13] is not straightforward. Of course, we may give the different parallaxes estimates separately. But, finally, one *resulting* parallax describing the *best* available value should be published for each star to facilitate the application of the catalogue data for further investigations. These requirements make it inevitable that problems of weighted means come into question!

This leads to the conclusion that still some more time is required to end up with a widely acceptable and successful publication of the *Third Catalogue of Nearby Stars.*

References

[1] W. Gliese and H. Jahreiß: 1979, Astron. Astrophys. Suppl. **38**, 423

[2] W. F. Van Altena and J. T. Lee : 1989, in *Star Catalogues: A Centennial Tribute to A. N. Vyssotsky*, eds. A. G. Davis Philip and A. R. Upgren, L. Davis Press, Schenectady, New York, p 83

[3] W. Gliese : 1971, Veröffentl. Astron. Rechen-Institut, Heidelberg, No. 24.

[4] A. N. Vyssotsky : 1963, in *Basic Astronomical Data*, ed. K. Aa. Strand (Univ. of Chicago), Chicago, Chap. 10

[5] W. Gliese : 1981, Astron. Astrophys. Suppl. **44**, 131

[6] T. H. Robertson : 1984, Astron. J. **89**, 1229

[7] C. B. Stephenson: 1986, Astron. J. **91**, 144

[8] C. B. Stephenson: 1986, Astron. J. **92**, 139

[9] S. -G. Lee: 1984, Astron. J. **89**, 702

[10] W. Gliese and H. Jahreiß: 1989, in *Star Catalogues: A Centennial Tribute to A. N. Vyssotsky*, eds. A. G. Davis Philip and A. R. Upgren, L. Davis Press, Schenectady, New York, p 3

[11] H. Jahreiß and W. Gliese : 1989, in *Star Catalogues: A Centennial Tribute to A. N. Vyssotsky*, eds. A. G. Davis Philip and A. R. Upgren, L. Davis Press, Schenectady, New York, p 11

[12] W. P. Bidelman : 1985, Astrophys. J. Suppl. Ser. **59**, 197

[13] W. Gliese : 1969, Veröffentl. Astron. Rechen-Institut, Heidelberg, No. 22.

W. Gliese and H. Jahreiß
Astronomisches Rechen-Institut
Mönchhofstr. 12 –14
D-6900 Heidelberg
Fed. Rep. of Germany

Estimation/Prediction under Dispersive Prior Information in Astronomy

E.W. GRAFAREND

Geodätisches Institut, Universität Stuttgart, FRG

Summary

At first we introduce the nonlinear model $Y = E\{Y\} + e_Y$, $E\{Y\} = F(X)$, $D\{Y\} = Q\sigma^2$ (full rank) of *complete prior information* $\hat{x} = E\{\hat{x}\} + e_{\hat{x}}$, $E\{\hat{x}\} = X - b_0$, $D\{\hat{x}\} = V\sigma^2$ (full rank) of the *unknown non-stochastic parameter vector* X. $E\{\ \}$, $D\{\ \}$ denote the first moment (expectation value) and second moment (variance-covariance matrix or dispersion matrix). $\hat{x}$ is the prior information of the unknown vector which might be *initially biased* by b_0. e_Y contains the error vector of the *stochastic observation vector* Y, $e_{\hat{x}}$ the error vector of prior information. The two-step taylorization is outlined in order to arrive at a linear model $\Delta y = A\Delta x + e_Y = Ab_0 - Ae_{\hat{x}} + e_Y$. The generalized *least-squares method* $e_Y^T P e_Y + e_{\hat{x}}^T W e_{\hat{x}} = \min$ is presented introducing the weight matrix P observation errors as well as the weight matrix W of prior information errors. The corresponding normal equations $(A^T PA + W)\tilde{\Delta x} = A'P\Delta y + Wb_0$ can be interpreted as *'ridge type estimators'*, *'robust estimation'*, *'Tykhonov regularization'* etc. An equivalence lemma is formulated for alternative *best inhomogeneously linear prediction* (inhom BLIP). In case that only *partial prior information* of the unknown vector X is available the two-step taylorization leads to the so called *mixed model* $\Delta y = A_1\Delta x_1 + A_2\Delta x_2 + e_Y = A_1 b_0 + A_2\Delta x_2 - A_1 e_{\hat{x}_1} + e_Y$ in which Δx_1 are *stochastic effects* while Δx_2 are *fixed effects*. The stochastic effects are predicted, the fixed effects estimated. The *extended least squares method* is portrayed in contrast to $\tilde{\Delta x}_2$ BLUUE (*best linear uniformly unbiased estimation*) and $\tilde{\Delta x}_1$ hom BLIP, inhom BLIP and hom BLUP (*best linear unbiased prediction*): Examples in astronomic networks are given. We refer for more details to the review paper by *E. Grafarend and B. Schaffrin*, Zeitschrift für Vermessungswesen <u>113</u> (1988), 79-103.

For the notion of weak and strong definition of bias we refer to A. *Bjer-hammar*: Hyper-Estimators, Publ. Roy. Inst. Tech., Div. of Geodesy, Stockholm 1977, *M.J. Box*: Bias in Nonlinear Estimation; J. Roy. Statist. Soc. B-$\underline{33}$ (1971), 171-201, *D. Majundar/S.K. Mitra*: Statistical Analysis of Nonestimable Functionals; <u>in:</u> W. Klonecki/A. Kozek/J. Rosinski (Eds.), Math. Statistics and Probability Theory, Springer Lect. Notes Statist., Vol. $\underline{2}$, New York 1980, pp. 288-317, *B. Schaffrin*: Zur Verzerrtheit von Ausgleichungsergebnissen, Mitt. Inst. Theoret. Geod. Univ. Bonn, No. 39, Bonn 1975, *B. Schaffrin*: Model choice and adjustment techniques in the presence of prior information, OSU-Report No. $\underline{351}$, Columbus/Oh. 1983, *B. Schaffrin*: New estimation/prediction techniques for the determination of crustal deformations in the presence of geophysical prior information, Tectonophysics $\underline{130}$ (1986), 361-367, *B. Schaffrin*: Das geodätische Datum mit stochastischer Vorinformation, Habilitationsschrift (Stuttgart 1984), Publ. DGK C-$\underline{313}$, München 1985, *B. Schaffrin*: On robust collocation, Proc. of the First Hotine-Marussi-Symp. on Math. Geodesy (Roma 1985), Milano 1986, pp. 343-361, *B. Schaffrin*: Less sensitive tests by introducing stochastic linear hypotheses, <u>in:</u> T. Pukkila/S. Puntanen (Eds.), Proc. of the Second Int. Tampere Conference on Statistics (Tampere/Finland 1987), University of Tampere Press 1987, pp. 647-664, *E. Grafarend and B. Schaffrin*: A unified computational scheme for traditional and robust prediction of random effects with some applications in geodesy, Paper pres. to the First Int. Comf. on Statist. Computing (ICOSCO-I), Cesme-Izmir/Turkey 1987, *B. Schaffrin*: Tests for random effects based on homogeneously linear predictors, <u>in:</u> H. Läuter (Ed.), Proc. of the Int. Workshop on "Theory and Practice in Data Analysis", Akad. der Wiss. der DDR, Report No. R-MATH-01/89, Berlin (Ost) 1989, pp. 209-227, and *B. Middel and B. Schaffrin*: Robust predictors and an alternative iteration scheme for ill-posed problems, Paper pres. at the 7th. Int. Seminar of "Model Optimization in Exploration Geophysics", Free University of Berlin, Febr. 1989.

Fitting Models to Low Count Data Sets

J. NOUSEK

University Park, U.S.A.

SUMMARY

The basic technique of χ^2 fitting becomes mathematically invalid when the number of events per bin becomes very small. We present the results of Monte Carlo simulations which explore the onset of systematic errors when χ^2 fitting is applied to data sets containing bins with small number of events and compare the results of using the C statistic proposed by Cash (1979). We find that the Cash technique is free of systematic bias, even when many bins contain zero or one event.

1   Introduction

The fundamental problem in data analysis for X-ray astronomy is the extraction of the astrophysically important parameters from the data coming from our experiments. Complications arise from the sharply varying instrumental sensitivity, limited and often complex energy resolution, and the effect of photon counting statistic noise. The most common technique to remove these effects has been the application of a non-linear regression to a hypothetical source spectrum convolved with the response of the detector, seeking to minimize the χ^2 value generated by the model and the data set.

Despite the generally satisfactory results of using this technique, we run into problems using it with small numbers of events. An important mathematical advantage of χ^2 fitting is that, for adequately large samples, χ^2 fitting is asymptotically independent of the shape (i.e. distribution) of the events in the sampling bins, under the null hypothesis (Lindgren 1976, for example). (The null hypothesis is the assumption that the only deviation between model and data is random measurement error.) Unfortunately, in many cases in X-ray astronomy, the data samples are not large enough to assure that χ^2 reaches the asymptotic limit. What we have tested is when and how severely χ^2 fitting departs from its asymptotic behavior.

2 χ^2 Fitting

The χ^2 statistic measures the difference between the observed data and the model. χ^2 minimization is equivalent to maximizing the likelihood that the data observed were produced by the model we fit *if* the samples are described by Gaussian statistics. In this case, the probability for observing a given value in a bin is its Gaussian probability, $P_i \propto \exp{\frac{(n_i - m_i)^2}{2\sigma_i^2}}$, where n_i is the observed data, σ_i is the error, and m_i is the model prediction at that point. If, then, the observations are uncorrelated and the model is correct the probability of observing a set of samples is just the product of the probabilities, $\prod P_i$. The log of this probability is χ^2, plus a constant normalization term,

$$-\ln \prod_i P_i = \chi^2 = \sum_{i=0}^{N} \frac{(n_i - m_i)^2}{\sigma_i^2}$$

The minimum value of χ^2 as a function of the model parameters is the best agreement between the model and the data because the minimum χ^2 is the maximum probability for the null hypothesis. The χ^2 value is directly related to the probability that random chance alone will result in a χ^2 value at least that large. Hence large probabilities suggest empirically valid models (because random chance can easily explain the residuals).

If, however, the probabilities of observing the samples are *not* Gaussian, then χ^2 does not properly correspond to the probability of the null hypothesis, and the fitting procedure may not produce the correct result. In these cases the small numbers in each bin (especially zero or one) are very poorly approximated by a Gaussian distribution, and incorrectly weighting such observations may strongly bias the inferred fit parameters.

Cash (1979) points out that the probability distribution appropriate to bins with a small number of counts is Poisson and not Gaussian. He then uses a theorem by Wilkes (1963) to show that a maximum likelihood ratio can be formed for any statistical distribution which itself has a probability distribution like χ^2 (except for a small additive term), and can then be used just like χ^2 for non-linear regression to fit models to data.

His prescription is to minimize the C statistic, where $C = 2\sum_{i=1}^{N}(m_i - n_i \ln(m_i))$ and m_i and n_i are again the number of counts predicted by the model and observed, respectively. Operationally this proceeds exactly as in the χ^2 minimization case, including the inference of confidence intervals on the parameters. (For one parameter the 1 σ confidence interval is found by finding the parameter values where $C = C_{min} + 1$. For more parameters, this becomes $C = C_{min} + f(\nu)$ where $f(\nu)$ can be found in a table in Lampton, Margon and Bowyer 1976). Cash shows that the C statistic goes to χ^2 in the limit of large n, and that it should be more efficient than χ^2.

3   Monte Carlo Simulation

To test the statistical methods, we assumed an ideal hypothetical spectrum that has the form of a power law, $dN = N_o E^{-\gamma} dE$. Over an energy range from E_1 to E_2 we can calculate the expectation value for the number of counts (possibly non-integer): $N = N_o \int_{E_1}^{E_2} E^{-\gamma} dE$. To perform our test we chose $\gamma = 2$ and created a data set by dividing the total range into 15 equal bins. We then calculated N_i, the expectation value in each of the bins.

As we are only concerned with the effect of photon counting statistics the only source of error is the random arrival of photons. This process is described by a Poisson probability distribution, with a mean equal to the expectation value, or N_i. The simulated data, n_i, consist of Poisson random deviates (always integer) drawn from the distributions for each bin.

A power law model was fit to the simulated data. This process of generating randomized data sets and then fitting a model to each data set was repeated 250 times, and the best fit parameters were accumulated. Each fit was provided with identical starting conditions; only the data sets were different. The parameters of the best fits were averaged to minimize the effects of random fluctuations in the simulated data sets. The mean of the minimum χ^2 values and the mean and standard deviation of the best fit parameters were also calculated. The fits that had not converged were excluded from the calculation of mean and standard deviation.

By comparing the mean of our fit parameters to the parameters of our original data we test to see if the fitting technique faithfully recovers the original parameters. On a case by case basis one expects random variations in the data to cause the best fit parameters to deviate from the original parameters used to create the data sets. Indeed the confidence interval on the parameters should be an estimate of the amount of variation. Our experiment is an empirical test of the self consistency of the fitting. Averaged over many data sets the mean of the fits should converge to the original parameters; otherwise we are introducing a systematic error when we perform the fit. Similarly, the standard deviation of the best fit parameters is an empirical measure of the confidence interval on the parameters, which should agree with the confidence intervals generated in the individual fits.

Table I compares the results of using χ^2 and the C statistics. The column, N, corresponds to the total number of counts in the spectrum. The columns, N_{calc}/N_o and γ_{calc}/γ, are the ratios of the mean of the best fit parameters divided by the original value that was used to create the data sets. 'Converged' is the fraction of fits that converged according to the convergence criteria.

Table I. Comparison of χ^2 and C Results

	χ^2 Minimization			C Statistic Minimization		
N	N_{calc}/N_o	γ_{calc}/γ	Converged	N_{calc}/N_o	γ_{calc}/γ	Converged
25	0.709	1.152	0.96	1.269	0.958	0.86
50	0.647	1.134	1.00	1.079	0.998	1.00
75	0.636	1.130	1.00	1.078	0.995	1.00
100	0.673	1.109	1.00	1.053	0.996	1.00
150	0.707	1.094	1.00	1.015	1.005	1.00
250	0.767	1.072	1.00	1.019	1.000	1.00
500	0.863	1.040	1.00	0.997	1.004	1.00
750	0.905	1.025	1.00	0.997	1.002	1.00
1000	0.937	1.017	1.00	1.001	0.999	1.00
2500	0.973	1.007	1.00	1.005	1.000	1.00
5000	0.988	1.003	1.00	0.984	1.003	1.00
10000	0.996	1.001	1.00	1.003	1.000	1.00

The immediate conclusion is that for this problem the Cash statistic introduces virtually no systematic bias to the fit results, while the χ^2 approach is biased. Even as low as 100 counts the C statistic produces relatively accurate results (within 5% on average of the true values).

The principal disadvantage of the C statistic is that there is no value corresponding to the Reduced χ^2 value with which we can measure the goodness of the fit. We can only determine the best parameters by minimizing the function, but we have no criteria to reject the model.[1]

4   Summary

We have confirmed via a Monte Carlo simulation that χ^2 fitting is satisfactory only when the number of counts per bin is large. In our simulation with 15 bins the results of the fits with $N = 75$ suggest that the fit could differ from the ideal value by as much as 50%. For $N = 1000$ a specific fit could differ from the ideal value by as much as 10%. Therefore, depending on the accuracy required of the fit, when the number of counts per bin is low we suggest using the C statistic. As the C statistic is equivalent to χ^2 in the limit of a large number of counts, it can also be used with large data sets, as long as a goodness of fit value is not required.

One final caveat should be observed. This simulation was conducted with a single model and with computer generated data, rather than with large amounts of real data. The amount and nature of the bias introduced by the χ^2 assumption may depend on the form of the model chosen, particularly in cases where only some bins have very small numbers of counts due to low detector efficiency, even though other bins may have large numbers of counts. Thus we recommend against using these results to scale actual fit results. Instead we suggest reanalysis of χ^2 fit results using the C statistic.

5   Acknowledgements

We acknowledge the helpful suggestions of Dr. David Burrows for improving the presentation of our results, and Mr. David Shue for performing many of the calculations. This work was supported in part by NASA Grant 5-100 and Contract NAS-8-36749.

6   References

1. Cash, W. 1979, *Ap.J.*, **228**, 939.

2. Lampton, M., Margon, B. and Bowyer, S. 1976, *Ap.J.*, **208**, 177.

3. Lindgren, B.W. 1976, *Statistical Theory* (3rd ed.: New York: Macmillan).

4. Wilkes, S. 1963, *Mathematical Statistics* (Princeton: Princeton University Press).

---

[1] The reason general tables can not be produced is that the distribution in each bin is different, and depends on the model. This exactly the regime in which χ^2 fails.

One could simulate the effect of random deviations for the given model and observed data set, but the simulation would have to be repeated for each data set and each model.

Criteria for Data Fitting

R.L. BRANHAM, Jr

Centro Regional de Investigaciones, Cientificas y Tecnologicas, Mendoza, Argentina

Abstract. To achieve a unique solution from overdetermined systems one minimizes the p-th norm of the residuals. The norms usually selected are p=2, the method of least squares, and p=1, an L_1 solution. A least squares problem may be solved by normal equations, orthogonal transformations, or the singular value decomposition. Least squares are computationally less demanding than the alternatives, but more subject to ill-conditioning. Extreme ill-conditioning can be ameliorated by scaling the equations of condition, recasting the mathematical model, or performing critical operations with accurate floating-point summation. To achieve robustness one can use iteratively reweighted least squares or calculate an L_1 solution. If errors are present in the equations of condition as well as the vector of observations the total (orthogonal) least squares or L_1 problem must be calculated by modifications of the standard techniques for a least squares or L_1 solution.

1. Introduction.

Fitting or modelling of data is a common task in astronomy. I assume that the term "data fitting" refers to determining the coefficients in a given mathematical model, not to fitting an arbitrary number of parameters, whose total number depends on statistical tests of reduction of variance, to a data set. The latter is useful for finding an empirical function suited for interpolation within the range coverd by the data and for limited extrapolation beyond the range, but inappropriate for the problem considered here: finding the best values for the coefficients in a definite mathematical model.

I also assume that the data being fitted can be represented by a linear system of equations. Many data fitting problems are linear, and a common technique for nonlinear parameter estimation, known to statisticians as the Gauss -Newton method for nonlinear regression and to astronomers as a differential correction, replaces the nonlinear problem by a series of linear approximations that converge, if they converge, to the solution of the nonlinear problem.

Represent the linear system by

$$\underset{\sim}{A} \cdot \underset{\sim}{X} = \underset{\sim}{d} , \qquad\qquad 1)$$

where $\underset{\sim}{A}$ is a matrix, known as the matrix of the equations of condition or data matrix, of size m x n, with m ≥ n, $\underset{\sim}{X}$ an n-vector of the desired solution, and $\underset{\sim}{d}$ an m-vector of the observations or data points. Because of observational error in $\underset{\sim}{d}$, or perhaps in both $\underset{\sim}{d}$ and $\underset{\sim}{A}$, Eq.(1) is inconsistent; no vector $\underset{\sim}{X}$ satisfies all of the equations. Define a vector $\underset{\sim}{r}$ of residuals by

$$\underset{\sim}{r} = \underset{\sim}{A} \cdot \underset{\sim}{X} - \underset{\sim}{d} . \qquad\qquad 2)$$

One may obtain a solution to Eq.(1) by imposing the condition that the p-th norm of the residuals be a minimum,

$$||\underset{\sim}{r}||_p = \{ \sum_{i=1}^{m} |r_i|^p \}^{1/p} = min. \qquad\qquad 3)$$

There is an infinity of choices for p, but in practice only two are used: p=1, the L_1 criterion; or p=2, the least squares or L_2 criterion. (p=∞, the Chebyshev or L_∞ criterion is inappropriate for fitting observational data (Branham, 1982).)

2. Ordinary Least Squares.

When p=2 the solution of Eq.(1) may be found by tak-
ing the gradient of $\underset{\sim}{r}^T . \underset{\sim}{r}$ with respect to $\underset{\sim}{X}$. (Minimizing
$\underset{\sim}{r}^T . \underset{\sim}{r}$ also minimizes $||\underset{\sim}{r}||_2$.) Upon doing this we find

$$\underset{\sim}{A}^T . \underset{\sim}{r} = 0 , \qquad\qquad 4a)$$

or

$$\underset{\sim}{A}^T . \underset{\sim}{A} . \underset{\sim}{X} = \underset{\sim}{A}^T . \underset{\sim}{d} . \qquad\qquad 4b)$$

Eq.(4b) constitutes the normal equations for solving Eq.
(1).

Normal equations possess attractive properties. $\underset{\sim}{A}^T . \underset{\sim}{A}$,
of size n x n, is symmetric: a linear system of size
m x n + m becomes compressed to one of size n(n+1)/2 + n,
a considerable saving of space when m >> n, as occurs fre-
quently in astronomy. $\underset{\sim}{A}^T . \underset{\sim}{A}$ is also positive definite and
may be factored into the Cholesky decomposition,

$$\underset{\sim}{A}^T . \underset{\sim}{A} = \underset{\sim}{C}^T . \underset{\sim}{C} , \qquad\qquad 5)$$

where $\underset{\sim}{C}$ is upper triangular, without row or column inter-
changes. The solution to Eq.(4b) follows from solving two
triangular system and involves half the labor of more general linear equation solvers such as LU decomposition
($n^3/6$ operations rather than $n^3/3$).

Because $\underset{\sim}{C}$ is triangular it is easily inverted and
$(\underset{\sim}{A}^T . \underset{\sim}{A})^{-1}$ found from $\underset{\sim}{C}^{-1} . \underset{\sim}{C}^{-T}$. Given certain assumptions
$(\underset{\sim}{A}^T . \underset{\sim}{A})^{-1}$ may be identified with the unscaled covariance
matrix, allowing us to calculate error estimates for the
components of the solution vector $\underset{\sim}{X}$; the covariance matrix readily converts to the correlation matrix. The assumptions may be summarized as (Branham, 1989a): the ob-
servational error is concentrated in $\underset{\sim}{d}$, $\underset{\sim}{A}$ is error-free;
the observational error follows the normal distribution;
the error components of $\underset{\sim}{d}$ are statistically independent;

the error components of $\underset{\sim}{d}$ do not vary with time. Unless these assumptions are valid, mean errors and correlations from a least squares adjustment may be mere formalities.

Despite their advantages normal equations suffer a defect, albeit not as grave as sometimes claimed. Decompose the matrix $\underset{\sim}{A}$ of Eq.(1) into the singular value decomposition,

$$\underset{\sim}{A} = \underset{\sim}{U} \cdot \underset{\sim}{S} \cdot \underset{\sim}{V}^T \qquad\qquad 6)$$

where $\underset{\sim}{U}$ is m x m orthogonal, $\underset{\sim}{V}$ n x n orthogonal, and $\underset{\sim}{S}$ m x n with lower (m-n) x n part null and upper n x n part diagonal. The elements s_i of $\underset{\sim}{S}$ are $\underset{\sim}{A}$'s singular values. The ratio s_{max}/s_{min} furnishes one definition of the condition number of $\underset{\sim}{A}$, COND($\underset{\sim}{A}$), which varies from unity for perfectly conditioned systems to infinity for singular systems; the higher COND($\underset{\sim}{A}$) the more poorly conditioned the system.

The condition number of the normal equations, COND($\underset{\sim}{A}^T \cdot \underset{\sim}{A}$) = s_{max}^2/s_{min}^2, is the square of COND($\underset{\sim}{A}$); forming normal equations deteriorates the conditioning of the problem. But it is possible to obtain a least squares solution without normal equations. Orthogonal matrices have unit condition number in the p=2 norm, COND($\underset{\sim}{Q}$)=1, and also conserve the p=2 norm of Eq.(3),

$$||\underset{\sim}{Q} \cdot \underset{\sim}{r}||_2 = ||\underset{\sim}{r}||_2 = min. \qquad\qquad 7)$$

One obtains the solution by the least squares criterion upon applying an orthogonal matrix $\underset{\sim}{Q}$ to Eq.(1) such that $\underset{\sim}{Q} \cdot \underset{\sim}{A}$ is upper triangular, the same upper triangular matrix $\underset{\sim}{C}$ that would be calculated by Cholesky decomposition of the normal equations (within the limits of round-off or truncation error). But COND($\underset{\sim}{Q} \cdot \underset{\sim}{A}$)=COND($\underset{\sim}{A}$); we have not squared the condition number.

The matrix $\underset{\sim}{Q}$ is built up from elementary orthogonal

transformations. The Givens transformation and the Householder transformation are used frequently. Golub and Van
Loan (1983) may be consulted for details about these
transformations, but a few brief remarks are in order. A
Givens transformation zeroes one element of a column of $\underset{\sim}{A}$
and must be applied $m(n+1) - (n+1)(n+2)/2$ times to reduce
$\underset{\sim}{A}$ to upper triangular and simultaneously reduce the right
-hand-side $\underset{\sim}{d}$, requiring $2\ mn^2 - 2\ n^3/3$ operations for the
triangularization of $\underset{\sim}{A}$. A Householder transformation zeroes k elements of a column of $\underset{\sim}{A}$ and need be applied only
$(n+1)$ times. Although more complicated than a Givens
transformation, it needs only $mn^2 - n^3/3$ operations to
triangularize $\underset{\sim}{A}$. With neither Givens nor Householder
transformations is it necessary to form an explicit m x m
matrix $\underset{\sim}{Q}$; for m >> n this would be ruinously expensive.
Both transformations, rather, work directly with rows or
columns of $\underset{\sim}{A}$ and achieve the triangularization of $\underset{\sim}{A}$ in no
more space than m x (n+1).

But for m >> n even space requirements of m x (n+1)
may be burdensome. When one uses normal equations one
hardly ever reads in the complete matrix $\underset{\sim}{A}$. Rather, one
uses a one-dimensional array --call it CONDEQ-- to hold a
row of $\underset{\sim}{A}$ and another array --call it $\underset{\sim}{B}$-- to hold the normal equations $\underset{\sim}{A}^T. \underset{\sim}{A}$. If $\underset{\sim}{B}$ is two-dimensional it may be
formed by FORTRAN-77 code such as

```
DØ  I = 1, N
 DØ J = I, N
  B(I,J) = B(I,J)+CONDEQ(I)*CONDEQ(J)
 END DØ
END DØ  .
```

(Because $\underset{\sim}{B}$ is symmetric it may be dimensioned as a vector
with i,j elements located by the mapping $k=j(j-1)/2 + i$
if it is stored in column order; in this way nearly fifty

per cent less space is needed.)

Givens transformations, like the normal equations, accumulate the equations of condition sequentially and need no more space than an n(n+1)/2 array for the triangularized matrix $\underset{\sim}{A}$, an array of size n for the right-hand-side, and some ancillary arrays. Even though they need no more space Givens transformations still require more operations, for m >> n about four times as many operations.

Householder transformations, on the other hand, pay a definite performance penalty for sequential accumulation of the equations of condition. Rows of $\underset{\sim}{A}$ should be read in blocks of k, with k much greater than unity. See Lawson and Hanson (1974) for details. Even for k >> 1 Householder transformations require about twice as much labor as normal equations for a least squares solution. And they need more space, at least double the space even for k=1.

That orthogonal transformations require more operations and, perhaps, more space, obviate some of the putative advantages in their favor. In particular, Lawson and Hanson's (1974) assertion that orthogonal transformations solve a least squares problem in single-precision that needs double-precision when one employs normal equations is untenable for most problems of astronomical data reduction, where m >> n. (Even when m > n but m is not much greater than n the assertion is dubious. See Branham (1987) for details.)

Should one, nevertheless, wish to employ orthogonal transformations, a modification of the Givens transformation, the fast Givens transformation (FGT), permits solving a least squares problem in no more space than that needed by normal equations and with only double the operation count, like Householder transformations with k >> 1. Furthermore, the FGT is even more efficient, in terms of fewer operations, than Householder transformations for sparse $\underset{\sim}{A}$. For problems that merit orthogonal transformations I can recommend the FGT and have given FORTRAN code elsewhere (Branham 1989b).

3. Ill-Conditioning.

Those who recommend orthogonal transformations over normal equations are concerned with the higher condition number of the latter compared with the matrix $\underset{\sim}{A}$. Some, such as Press et al. (1986), go even farther and recommend a least squares solution by the SVD. Given the decomposition of Eq.(6) the least squares solution follows from

$$\underset{\sim}{X} = \underset{\sim}{V} \cdot \underset{\sim}{S}^{-1} \cdot \underset{\sim}{U}^{T} \cdot \underset{\sim}{b} \; . \tag{8}$$

Eq.(8) furnishes a solution when $\underset{\sim}{A}$ is poorly conditioned or even singular. Whether such a solution represents anything worthwhile is another question. And the SVD solution also requires a __minimum__ of four times more labor than a least squares solution.

Often an ill-conditioned $\underset{\sim}{A}$ arises from a poor selection of variables or inappropriate mathematical formulation of the problem. A poor selection of variables may be remedied by postmultiplication of $\underset{\sim}{A}$ by scale factors to reduce the variation among the columns of $\underset{\sim}{A}$. In mathematical terms we form a new matrix $\underset{\sim}{A'}$ by postmultiplying the m x n matrix $\underset{\sim}{A}$ by an n x n matrix $\underset{\sim\sim}{SC}$ of scale factors,

$$\underset{\sim}{A'} = \underset{\sim}{A} \cdot \underset{\sim\sim}{SC} \; , \tag{9}$$

such that $\mathrm{COND}(\underset{\sim}{A'}) < \mathrm{COND}(\underset{\sim}{A})$.

An inappropriate mathematical formulation may be cured by recasting the problem. When differentially correcting minor planet orbits, for example, the osculating elements corrected should have an epoch near the middle of the time span covered by the observation, not an epoch far removed from the earliest or the latest observation.

If the ill-conditioning, however, is intrinsic to the problem then the time has come to gather more data to strengthen the system rather than to worry about solving it. A solution eked out by orthogonal transformations or the SVD, should normal equations fail, is likely a dubious contribution to astronomy.

One may also take measures to assure that the normal equations are less poorly conditioned than the matrix $\underset{\sim}{A}$ than would otherwise be true. One may sequentially accumulate the equations of condition and sum inner products of $\underset{\sim}{A}^T . \underset{\sim}{A}$ in a higher precision than that used for the rest of the calculations. Numerous techniques are available for accurate floating-point summation, the Kahan trick, binary summation, cascaded accumulators (Branham, 1989a), but the simplest uses hardware higher precision. If we are working in single-precision in FORTRAN, for example, we could accumulate the equations of condition by code like

```
    DØ  I = 1, N
     DØ  J = I, N
        B(I,J) = B(I,J)+ DBLE(CONDEQ(I))*DBLE(CONDEQ(J))
      END DØ
    END DØ  .
```

If the precautions of avoiding unnecessary ill-conditioning in $\underset{\sim}{A}$, obtaining sufficient data, and performing critical operations with accurate floating-point summation are taken, normal equations should suffice for most least squares problems. Orthogonal transformations and the SVD should be reserved for refractory problems that require them.

4. Robustness.

Another difficulty with least squares arises not from the method of solution but from the criterion itself: least squares are not robust; a few discordant observations can ruin the solution. Two palliatives are available: robustize least squares, or adopt a more robust criterion, such as the L_1.

Iteratively reweighted least squares (IWLS) achieve robustness by assigning weights to residuals calculated from a preliminary solution, such as a least squares solution with all observations, discordant or not, thrown in. Many weighting functions are possible. Coleman et al. (1980) discuss the method and weighting functions. Let $\underset{\sim}{W}(\underset{\sim}{r})$ be an m x m diagonal matrix of weighting functions. In IWLS we minimize, instead of $||\underset{\sim}{r}||_2$,

$$||\underset{\sim}{W}^{1/2} . \underset{\sim}{r}||_2 = ||\underset{\sim}{W}^{1/2} . \underset{\sim}{A} . \underset{\sim}{x} - \underset{\sim}{W}^{1/2} . \underset{\sim}{d}||_2 = min \qquad 10)$$

We use routines for ordinary least squares, normal equations, orthogonal transformations, or the SVD, with $\underset{\sim}{W}^{1/2} . \underset{\sim}{A}$ and $\underset{\sim}{W}^{1/2} . \underset{\sim}{d}$ in place of $\underset{\sim}{A}$ and $\underset{\sim}{d}$ and keep on iterating until the gradient

$$(\underset{\sim}{W}^{1/2} . \underset{\sim}{A})^T . (\underset{\sim}{W}^{1/2} . \underset{\sim}{r}) = 0 . \qquad 11)$$

The weighting matrix itself depends on what the researcher considers reasonable. A typical example is the biweight,

$$\underset{\sim}{W}(r_i) = \left[1 - (r_i/4.685)^2 \right]^2 \left.\begin{array}{c} \\ \\ 0 \end{array}\right\} \begin{array}{l} |r| \le 4.685 \\ \\ |r| > 4.685 \end{array} \qquad 12)$$

with each individual residual scaled by a factor such as the median residual of $\underset{\sim}{r}$ in absolute value.

How many iterations are necessary? My experience indicates between nine and twenty to satisfy Eq.(11) to five decimals. Given the number of iterations it is important to consider whether one wishes to pay the additional computational price of between two and four times more labor per iteration should one opt for orthogonal transformations or the SVD.

The L_1 criterion, minimize the sum of the absolute values of the residuals, also achieves robustness. An L_1 solution corresponds to a distribution of error of the form

$$f(r_i) = h/2 \; e^{-h|r_i|} \; , \qquad\qquad 13)$$

where h is the modulus of precision. Eq.(13) goes to zero much more slowly than least squares' error distribution $\propto e^{-r_i^2}$. The Table, with h of Eq.(13) equal to $1/\sqrt{2}\,\sigma$ of the least squares error distribution, emphasizes the difference:

Table: Probability of Residual Exceeding Limit

$k\sigma$	L_1	Least Squares
1	0.493	0.317
2	0.243	$0.455 \cdot 10^{-1}$
3	0.120	$0.270 \cdot 10^{-2}$
5	$0.291 \cdot 10^{-1}$	$0.573 \cdot 10^{-6}$
10	$0.849 \cdot 10^{-3}$	$0.152 \cdot 10^{-22}$
20	$0.721 \cdot 10^{-6}$	$0.551 \cdot 10^{-88}$

The L_1 criterion, therefore, is far more tolerant of large residuals than is the least squares criterion and such residuals have little pernicious effect on the solution, unlike the situation with least squares.

Although the L_1 criterion achieves robustness, it does so at a computational price. Practical algorithms, such as Barrodale and Roberts' (1974), work with the full m x n matrix $\underset{\sim}{A}$; it is impossible to compress the matrix to a smaller size as with least squares, particularly when a problem is solved by normal equations or the FGT. For m >> n the computational price can be excessive.

Which of the two methods, IWLS or an L_1 solution, is more convenient depends upon, according to my experience, the size of $\underset{\sim}{A}$. If $\underset{\sim}{A}$ fits comfortably into physical memory, the L_1 solution is calculated quickly and may be used by itself or as a means to filter out discordant observations for subsequent least squares adjustments. For this situation the L_1 method should be faster than IWLS.

But if the size of $\underset{\sim}{A}$ exceeds physical memory, the calculation of an L_1 solution will be laborious, necessitating backing storage on non-virtual memory systems and implying high paging rates, with consequent inefficient CPU usage, on virtual memory systems. For this situation IWLS will calculate the solution more quickly.

5. Total Least Squares and L_1 Solutions.

Whether one uses the least squares or the L_1 criterion, standard solution techniques assume that the observational error of Eq.(1) is concentrated in the vector $\underset{\sim}{d}$; the components of $\underset{\sim}{A}$ itself are assumed error-free. This may or may not be a valid assumption. If invalid the techniques for a least squares or L_1 adjustment presented so far are inappropriate and may lead to biased or at least questionable results. Particularly questionable is the standard practice of calculating error estimates of the $\underset{\sim}{X}$ of Eq.(1) and correlations among the unknowns from the covariance matrix because the derivation of the covariance matrix assumes that $\underset{\sim}{A}$ is error-free (Branham 1989a).

When errors in $\underset{\sim}{A}$ are present the problem is known as a total or orthogonal least squares or L_1 problem. Total because we wish to minimize the total error present in both d and A; orthogonal because in lieu of minimizing $||r||_p$, the vertical distance between a data point and the fit-. ting hypersurface, we minimize the perpendicular distance,

$$||r||_p / (1 + X^T . X) = min .$$ 14)

Eq.(14) furnishes one way to calculate a total solution. Because Eq.(14) is nonlinear any number of techniques for nonlinear parameter estimation could be applied to it. But simpler, linear mechanisms are available. Golub and Van Loan (1980) discuss the total least squares (TLS) problem and show how a solution can be calculated from the SVD. Calculate the smallest singular value s_n of A and the smallest singular value s_{n+1} of A:d, the matrix A with the vector d appended as the (n+1)th column. If $s_n < s_{n+1}$ a TLS does not exist. Otherwise the TLS exists and may be found from

$$X = V . (S^T . S - s_{n+1}^2 I)^{-1} . S . U^T . d ,$$ 15)

where I is the unit matrix. Golub and Van Loan show how s_{n+1} may be calculated from just the SVD of A, thus avoiding an SVD of both A and A:d.

The SVD, nevertheless, is more computationally demanding than normal equations. I have shown how the TLS may be calculated from the normal equations (Branham 1989c). Because the FORTRAN code uses the Cholesky factor C of the decomposed normal equations it could also be applied to this factor obtained from orthogonal transformations such as the FGT.

Späth and Watson (1987) present a technique for calculating a total L_1 solution. Choose a unitary vector v such that $v = e_{n+1}$, where $e_{n+1}^T = (1 \ 0 \ 0 \ ...)$. Find an orthogonal transformation such that

$$Q . v = e_{n+1} .$$ 16)

The Späth and Watson algorithm is iterative, and for the
first iteration take $\underset{\sim}{Q} = \underset{\sim}{I}$. For further iterations define

$$\underset{\sim}{u} = (\underset{\sim}{v} - \underset{\sim}{e}_{n+1})/\ \sqrt{2(1 - v_{n+1})},\ (\underset{\sim}{u}^T \cdot \underset{\sim}{u} = 1)\ . \tag{17}$$

$\underset{\sim}{Q}$ may be constructed as a Householder transformation by
use of Eq.(17) as

$$\underset{\sim}{Q} = \underset{\sim}{I} - 2\ \underset{\sim}{u} \cdot \underset{\sim}{u}^T \tag{18}$$

(notice that $\underset{\sim}{u} \cdot \underset{\sim}{u}^T$ is the backwards product of $\underset{\sim}{u}$ and $\underset{\sim}{u}^T$)
or may be built up from Givens transformations if Eq.(16)
is used.

Having the matrix $\underset{\sim}{Q}$ define

$$\underset{\sim}{Z} \cdot \underset{\sim}{Q}^T = \underset{\sim}{A}{:}\text{-}\underset{\sim}{d}\ , \tag{19}$$

where $\underset{\sim}{A}{:}\text{-}\underset{\sim}{d}$ is $\underset{\sim}{A}$ with $\text{-}\underset{\sim}{d}$ appended as an additional column.
Define a change of variables

$$\underset{\sim}{c} = \underset{\sim}{Q} \cdot \underset{\sim}{d} \tag{20}$$

and solve the standard L_1 problem

$$||\underset{\sim}{A} \cdot \underset{\sim}{x} - \underset{\sim}{d}||_1 = \min.\ , \tag{21}$$

with $\underset{\sim}{c}^T = (\underset{\sim}{x}\ \ 1)$, by an algorithm such as Barrodale and
Roberts'. Iterate until the solution converges within a
specified tolerance. As with TLS a total L_1 solution may
not exist. This occurs when v_{n+1} of Eq.(17) is null within the tolerance of the computer's floating-point arithmetic.

Total least squares and L_1 solutions can differ start-
lingly from their non-total counterparts. Consider the
simple system

$$
\begin{pmatrix} 1 & 1 \\ 1 & 2 \\ 1 & 3 \\ 1 & 4 \\ 1 & 5 \end{pmatrix} \begin{pmatrix} X_1 \\ X_2 \end{pmatrix} = \begin{pmatrix} 1 \\ 1 \\ 2 \\ 3 \\ 2 \end{pmatrix} . \tag{22}
$$

The least squares and L_1 solutions, with their respective errors (the latter found from Branham's (1986) method), are

$$
\begin{pmatrix} X_1 \\ X_2 \end{pmatrix}_{L_2} = \begin{pmatrix} 0.6 \\ 0.4 \end{pmatrix} \pm \begin{pmatrix} 0.663 \\ 0.200 \end{pmatrix} , \quad \begin{pmatrix} X_1 \\ X_2 \end{pmatrix}_{L_1} = \begin{pmatrix} 0.5 \\ 0.5 \end{pmatrix} \pm \begin{pmatrix} 1.333 \\ 0.400 \end{pmatrix} .
$$

The corresponding total solutions, without mean errors because the precepts for calculating them are no longer valid, are

$$
\begin{pmatrix} X_1 \\ X_2 \end{pmatrix}_{L_2} = \begin{pmatrix} 1.649 \\ 0.115 \end{pmatrix} , \quad \begin{pmatrix} X_1 \\ X_2 \end{pmatrix}_{L_1} = \begin{pmatrix} -1 \\ 1 \end{pmatrix} .
$$

Notice that the TLS falls outside the formal error bounds given by the non-total least squares solution as do the X components of the L_1 solution with its non-total counterpart. This example demonstrates how a formally calculated least squares solution with its mean error may bear little relation to the true solution if the approximation by least squares to the true solution should have been calculated by TLS. In such an instance the formal least squares solution will be severely biased.

6. Conclusions.

Least squares by normal equations, especially when

care has been taken to prevent a poorly conditioned matrix
and critical operations are performed with accurate floating-point summation, should be the first choice of reduc-
ers of astronomical data. If ill-conditioning is intrinsic to the problem a solution by orthogonal transformations or the SVD may be indicated. Among orthogonal transformations the FGT may be recommended because it needs no
more space than normal equations and has about the same
operation count as the Householder transformation.

To achieve robustness one may use iteratively reweighted least squares or an L_1 solution. The former must be it-
erated a significant number of times for convergence. The
latter will be computationally more demanding if the matrix of the equations of condition exceeds available physical memory.

If errors are present in both the vector of the right-
hand-side and the equations of condition the proper mathematical tool is a total (orthogonal) least squares or L_1
solution. Although the computational techniques for total
solutions are more demanding than their non-total counterparts, they should be used if indicated because a total
solution can differ significantly from a non-total solution.

References:

Barrodale, I, and Roberts, F.D.K. (1974). Commun. ACM,
    17, p.319.

Branham, R.L.Jr. (1982). Astron. J., 87, p.928.

Branham, R.L.Jr. (1986). Celestial Mech., 39, p.239.

Branham, R.L.Jr. (1987). ACM Signum Newsletter, 22(4),
    p.14

Branham, R.L.Jr. (1989a). Scientific Data Analysis: an
    Introduction to Overdetermined Systems (Springer-
    Verlag, New York [in Press]), Sec. 1.5, Ch.5.

Branham, R.L.Jr. (1989b). Computers in Physics, [submitted] .

Branham, R.L.Jr. (1989c). Computers in Physics, $3(3)$, p.42.

Coleman, D.C., Holland, P., Kaden, N., Klema, V., and Peters, L.C. (1980). ACM Trans. Math. Software, 6, p.327.

Golub, G.H., and Van Loan, C.F. (1980). SIAM J. Numer. Anal. 17, p.883.

Golub, G.H., and Van Loan, C.F. (1983). Matrix Computations (Johns Hopkins U.Press, Baltimore), Sec. 6.2, 6.3.

Lawson, C.L., and Hanson, R.J. (1974). Solving Least Squares Problems (Prentice-Hall, Englewood Cliffs, N.J.), Sec. 19.1, Sec. 27.1.

Press, W.H., Flannery, B.P., Teukolsky, S.A., and Vetterling, W.T. (1986). Numerical Recipes: The Art of Scientific Computing (Cambridge U.Press, Cambridge), pp.515-520.

Späth, H., and Watson, G.A. (1987). Num. Math., 51, p.531.

Centro Regional de Investigaciones
Científicas y Tecnológicas
C.C. 131
5500 - Mendoza, Argentina

Unhomogeneity and Bias in Flare Prediction Problem

M. JAKIMIEC

Astronomical Institute, University of Wroclaw, Wroclaw, Poland

1. INTRODUCTION

In the procedure of short-term solar flare activity prediction the daily characteristics of active regions and the daily flare activity characteristics are used to forecast flare activity for the next day (see Sawyer et al, 1986). The preliminary investigation of such multivariate data sets (the training data sets, TDS) which we use for the building of the predicting algorithms is of great importance. The TDSs should be choosen randomly from a homogeneous population and should be unbiased. We will discuss here some types of the data bias and some sources of the heterogeneity which may be essential for the predictions of solar fare activity.

2. THE HETEROGENEITY PROBLEM

The heterogeneity may be inhered both in the flare population and in sunspot group (or active region) population. Analyzing seven variables characterizing solar flares, Jakimiec (1987) found the flare population heterogeneity which consists in the fact that variable interconnection structures obtained for very faint, stronger and the strongest X-ray flares are different. Bartkowiak and Jakimiec (1986) found that the interrelations between the variables used for predictions are different for sunspot groups of various Zurich classes. Jakimiec (1989) investigated two possible sources of the heterogeneity of sunspot group population: one is connected with the fact that we have to do with sunspot groups in the various phases of the evolution, and the second with the fact that we deal with active regions of the various age. The entire data set

was divided into the pairs of disjoint data subsets
accordingly to the analysed sources of heterogeneity. The
pair of data subsets, for instance the sunspot groups in the
increase and in the decay phase of the evolution, were then
compared by means of several statistical methods. Jakimiec
(1989) analysed the heterogeneity problem in a threefolt
manner: analysis of one-dimensional variables (ODV),
analysis of the variable interrelation structures (VIS) and
regression function (RF) analysis. It was found that the
data subsets for two evolution phases are not strongly
different when we employ the ODV analysis, but the
differences are more pronounced for the results of the VIS
and RF analysis. All three ways of analysis, namely, the
ODV, VIS and RF reveal the heterogeneity connected with the
age of active regions.

3. THE DATA BIAS PROBLEM
Jakimiec (1989) discussed some types of the data bias, which
can be inconvenient for the prediction quality:

A. First is the unrandom choice of the data vectors
describing sunspot groups and their flare activity. Such
unrandom choice may result in the data sample
unrepresentativness.

B. The inconvenient impact of the individual atypical data
vectors on the intrinsic relation structure seems to be also
one important source of the bias. This problem was
comprehensively analysed by Jakimiec and Bartkowiak (1989a,
1989b). They found that the data set used for the
predictions contain some atypical data vectors, which
correspond to sunspot groups with strong flare activity or
with unusual interrelations between considered
characteristics. However, the intrinsic relation structure
of the variables and also the predicting algorithms are not
very strongly disturbed by such atypical data vectors. It
means that the interrelation structure is sufficiently
stable and that the predicting algorithms are defined, in
essence, by the bulk of the data vectors.

C. One of the important types of the data bias arise from

the fact that in the prediction procedure the daily characteristics of solar activity are used and because the flare activity changes strongly in time interval shorter than 24 hours. So, the results of calculation (the predicting function coefficients) are burdened with the effect of the data selection. This problem was discussed by Jakimiec and Wanke-Jakubowska (1988).

D. The unstationarity of solar activity process in the eleven year cycle introduce additional source of the bias in the flare activity prediction procedure. After the estimation of the predicting algorithms for the definite time interval we apply them to forecast the solar flare activity. We should be sure that the algorithms may be used to perform the extrapolation. So, we should know the time distance after which we must estimate anew the algorithms. Jakimiec (1989) found that the interval of half-year is too long for this purpose.

4. CONCLUSIONS

The common factor analysis and the regression analysis methods allowed us to investigate the intrinsic relation structure of the variables used for the predictions. They allow us also to examinate how much the results of calculations are burdened with various kind of the data bias.

The heterogeneity connected with the variable interrelation structures seems to be of great importance for the construction of the predicting algorithms. There is a question whether such heterogeneity entails the necessity of the construction of several conditioned predicting algorithms.

REFERENCES

Bartkowiak,A. and Jakimiec,M., 1986, in Sol.-Terrest. Pred. Proc., Boulder, USA.

Jakimiec,M., 1987, Acta Astr., 37,283.

Jakimiec,M., 1989, Sol.-Terrest.Pred. Workshop, Sydney, 1989.

Jakimiec,M. and Bartkowiak,A., 1989a, Acta Astr.,39,1.

Jakimiec,M. and Bartkowiak,A., 1989b, send to Acta Astr.

Jakimiec,M. and Wanke-Jakubowska,M., 1988, Acta Astr., 38, 431.

Sawyer,C., Warwick,J.W. and Dennett,J.T., 1986, Solar Flare Prediction. Colorado Associated University Press, Boulder, USA.

DISCUSSION

T. Kiang to H. Eichhorn: Is one of your points this: the published error of star positions from photographic astrometry is too small because the published error is calculated from the residuals, based on *estimated* plate constants, and there should be another component due to the departure of the estimated values of the plate constants from their true values.

You have read me correctly. I have pointed out in 1963 (Eichhorn and Williams, 1963, *Astron. Journal*, **68**, 221) that *every* quantity computed with the use of estimated parameters reflects the accidental (and unavoidable) errors of these parameter estimates as a systematic error. These parameter errors seem to be consistently overlooked by many investigators.

J. Pfleiderer to H. Eichhorn: I would be hesitant to consider a box shaped error distribution as an indication of a very good model. I have had examples of the opposite: a distribution of residuals getting narrower on top of a broader distribution or with tails. In these cases, the correct interpretation seemed to be that increasing goodness of the model narrows down the residual distribution except for the original measuring errors which cannot be fitted by any model and stand therefore unchanged, getting increasingly noticeable as "tails".

You are correct, but please note that I am not claiming that *only a good model* can produced a box-shaped distribution function of the residuals. The pattern you describe is not unlike what one would get by a Chebychev-like adjustment as I have described it in Sec. 2; whether the resulting excess of the total distribution is positive or negative will then depend on the standard deviation of the normal Gaussian error function.

F. Murtagh to H. Eichhorn: On a number of occasions you made points about appropriate and recommendable methods. This points to the need for well-documented, easily-available software. This situation here is, in my mind, appalling. Would you agree with this, and do you have any suggestions on what should be done?

I agree not only that there is a lack of well-documented software, but I also think that the task of writing and distributing this is somewhat thankless. Nevertheless, some people have set a good example, e.g., Jefferys with his GAUSSFIT.

H. Eichhorn to W. Gliese (comment): You surely deserve the thanks of the astronomical community for the diligent and careful work you are doing. I also agree with your suggestion that you should list *one* final parallax for each star, because, even within the framework of general relativity, each star can only have one correct parallax at any one time.

A. Upgren to W. Gliese: One datum given in your first viewgraph deserves some comment. It shows the total number of objects in your catalogue rising from 3500 just a few years ago to 4600 now. If many of these are stars newly discovered to be nearby and if this trend continues, the stellar luminosity function will need a revision and the missing mass will no longer be missing. How many of the new objects fall into your sample by fine tuning of the data and how many are new (newly discovered to be nearby)?

You are right, any comparison is not covered. The number of 3500 objects (1988) includes simple stars and double and multiple systems, each taken as one object only, whereas the 4600 objects include not only a fairly small number of stars which are newly discovered to be nearby, but also well separated comparisons now counted separately and whose quantities have to be inspected separately.

C. Jaschek to W. Gliese (comment): I think the role of spectral classification should be simply to assure that the star is normal and dwarf – once one knows that, spectroscopic parallaxes should be abandoned in favor of photometric and trigonometric parallaxes.

H.-M. Adorf (remark, following remark of E. Feigelson to M. Broniatowski): There *are* people in the astronomical community who exploit density estimators. We have used a fairly simple density estimator – that included a scale parameter which we kept control over – in order to search for Young Stellar Objects in the IRAS Point Source Catalog (Prusti, Adorf and Meurs, in preparation).

E. Feigelson to H.L. Marshall (comment): This is a well-motivated effort, analogous to the extensive development of *parametric* survival analysis in the field of industrial reliability. Distribution function estimates that use prior information appropriate to a given experiment (e.g. positive flux. Poisson distribution) should in fact perform better than the nonparametric Kaplan-Meier. A search of the huge literature in the field might uncover relevant material.

J. Pfleiderer to H.L. Marshall: You supposed as prior knowledge that intensities must be positive. There are cases where apparent intensities may be negative, i.e. an absorption hole within extended radio emission which is, as usual, subtracted as part of the baseline. How can such cases be handled in an intrinsically "positive" approach such as maximum entropy?

First, the actual measurement is of a positive definite quantity since one still measures a flux over the background. The "negative" aspect of the problem is due to the relative nature of the scientific question, not the fundamental quantity. Secondly, I only suggest that one includes the physical restrictions in the problem when concerned with estimation, whatever these restrictions may be.

J.D. Scargle to H.L. Marshall: Comment: There are two examples of objects seen in absorption against the cosmic background (formaldehyde in the interstellar medium and clusters of galaxies), so you have to be careful about assuming non-negative observables. Question: Do your results in any way contradict survival statistics?

Concerning the comment, such absorption problems may be assumed to be one-sided, also if no emission along the line of sight is expected. Answering the question, I believe that survival statistics can certainly be used for certain purposes: choose your tool on the basis of the restrictions in your problem. Many problems, however, should not require this method because they have available error distributions: if so, then survival methods would give much weaker results.

R. Branham to E. Grafarend: When you refer to stabilizing the solution do you refer to decreasing the condition number of the matrix?

The structure $A^T P A + W$ of the *extended* normal equation matrix for *hybrid norm* optimization of measurement noise as well as errors in *"prior stochastic information"* may be interpreted as a *regularization solution* of an unstable (ill-conditioned) problem. E.g. stabilizing weight numbers W are added to the diagonal elements of the classical least-squares normal equation matrix $A^T P A$.

H. Eichhorn to E. Grafarend: (1) Can you say something about prior information when only the covariance matrix of a subset of the unknowns, but not estimates of their values, are known? (2) I might comment that what you said underscores very well the statement that sets of estimates published with only their variances but without the covariance matrix lose a lot of their inherent value.

(1) In cases where the model is *linear*, no prior information, no estimate of the unknown quantity itself is needed. Only in *nonlinear* models approximate unknown parameters are necessary when the model equations should be linear – the linearization process – and ready for computer implementations. (2) Certainly I agree.

J. Pfleiderer to E. Grafarend: The linearization of your equations is possible only if the form of the non-linearities is known and only the argument is uncertain. What do you do when you do not know the non-linearity terms?

You are quite right in arguing that the chosen *nonlinear* functional model might be incorrect. But you raise another problem – not treated in my review – actually the problem of *model identification*. In cases where the underlying model is insufficient, there was some hope by applying the technique of *"total least squares"*, but here one only reflected an error in the *linear* model matrix A, namely $y+\delta y = (A+\delta A)(x+\delta x)$.

H.-M. Adorf to J. Nousek: How sensitive are your conclusions to the fact that you have got essentially no *background* in your bins?

If the background is substantial then the number of events will become large enough so that Gaussian statistics will apply. For cases where the source plus background is still small then the background can be handled by including it in the model. (N.B. this assumes that the joint distribution of the source plus background follow a Poisson distribution. If not then a separate evaluation of the procedure would be required.)

H.L. Marshall to J. Nousek: Can you use the Kolmogorov-Smirnov test to examine the fit of your model to the data? Comment: I have an additional problem with the application of χ^2 tests to spectral data. The χ^2 test does not require ordered bins, so that they may be scrambled without changing the value of χ^2. Unfortunately, a model may appear to be well-fitted while in fact the residuals depend on energy (or wavelength). In general, it would seem necessary to apply another test to ensure that the residuals have no correlation with the independent variables.

Yes, and the K-S test has no troubles with zeros, but it is a nonparametric test. One can only approximate a parametric fit by iterative K-S test over a fixed grid.

T. Kiang to J. Nousek: Since the C-statistic includes empty bins, how do you define the set of bins to be included in calculating the value?

Once a specific set of bins are selected they are then fixed throughout the regression procedure. The actual evaluated values of C only occur from bins which have $n_C > 0$. It is up to the data analyst to select the most reasonable set of bins for the problem in hand. (This can include having no more than *one* event in each bin, which is equivalent to unbinned data.)

E. Grafarend to J. Nousek: From my side I have the following comment: replacing the Gaussian distribution within *maximum likelihood* by the Poisson distribution would mean that you are still working in the class of *exponential distributions*. Just I want to emphasize that there exists a general maximum likelihood estimation based on *general* exponential-type distribution models. I certainly agree that the statement "any errors are *asymptotically* Gaussian distributed" is no basis for a real-life probability distribution of events.

J.D. Scargle to R. Branham: This question could as well be addressed to Dr. Eichhorn. Both the incorporation of "prior information" and the rejection of large residuals bias your procedure toward confirming what you think you already know and against finding new and unexpected results. Would it not be preferable to neither blindly include large residuals nor blindly reject them, but rather to inspect them (e.g. look for patterns) and reject them only after understanding them?

One must be careful of outlier rejection. Had Rutherford applied outlier rejection to his observations of scattering of alpha particles, he never would have discovered the atomic nucleus. But when I know my model, I know that large residuals cannot exist. From my experience with minor planets I know that a 60 arc-sec residual from a good orbit is impossible and must be rejected. In any event one should never reject residuals without first examining them: they often contain information of interest, systematic trends and so forth.

H. Eichhorn: Whether an outlier can be (ought to be) rejected depends on how sure one is of the model which is the basis of the adjustment. If the model is correct, an inconsistent outlier is patently suspect. If the model is empirical, an outlier may well suggest that the model is inadequate.

E. Feigelson to R. Branham: (1) When ordinary least squares regression lines are compared with total least squares lines, errors to the total line slope and intercept should also be considered. We present formulas for this error analysis in the work summarized by T. Isobe, and can provide a short Fortran routine upon request. (2) The recommendation that astronomers examine different regression methods (L_2 vs. L_1, ordinary vs. total) for a given problem should be applauded. The choice of which regression should be considered "best" depends in part on the goal of the particular experiment.

I gave no error estimates for my total least squares solutions because the precepts for their calculations from the standard covariance matrix from ordinary least squares are no longer valid. But I agree that error estimates can be given. I also agree with your second comment.

H. Eichhorn to R. Branham/E. Feigelson: With regard to total least squares, it might be worth pointing out that there are two distinct problems: curve fitting and parameter estimation. Total least squares appears to be a valid technique in the context of curve (or straight line) fitting, but if parameters are to be estimated from observations on the basis of a presumed linear relationship between the observables, the covariance matrix – in astronomical practice – is usually known and can be taken into account in such a way that there is no reasonable alternative to using one particular technique of estimating the parameters in this linear relationship.

When one has a definite model one may assume it is error-free in the sense of not incorporating observational error in the sense that the data do. In that case one should not use total least squares. But if the matrix of the equations of condition incorporates coefficients that have themselves been determined from observation, then total least squares is indicated.

F. Murtagh to R. Branham: A point of clarification: in practice, in total least squares, does one use errors on the coordinate values (leading to, e.g., weights defined as $\frac{1}{\sigma}$), where the ratio of errors is not constant?

One can introduce a coupling parameter λ between errors in the vector of data points and those of the matrix of equations of condition. $\lambda = 0$ corresponds to all error in the data points, none in the matrix; $\lambda = 1$ to equally distributed error; and $\lambda = \infty$ to all error in the matrix, none in the data. A more complicated model allows for variations in the distribution of the error in the matrix similar to what we do with the errors in the data.

S. Bhavsar to R. Branham: This is a general philosophical question arising from your asking "What is the true answer?" Though the *true* answer does exist in nature, is it not possible that with the given data we cannot arrive at any *one* number, even with large error bars as the answer, but should just quote the confidence limits, since different schemes give different "true" answers.

I think we should use all of the information possible and quote errors from various reduction methods, least squares, L_1, and others, including confidence limits. This gives us a better idea of the errors present than just error estimates from a standard method like least squares.

E. Grafarend to R. Branham: In cases where you leave the model open, e.g. by physical fit, it is quite well known that e.g. by a proper choice of the *degree* of the polynomial you can obtain a zero error fit. But no-one would use this approach due to the peakedness of the high degree polynomial fit. Thus changing the physical model is a rather sensitive technique, thus definitely *not* being solved by "*linear* total least squares". In addition I would like to propose the definition of a blunder as a sample from a *second* distribution while the "regular noise" originates from a *first* distribution. *Variance-covariance component estimation* would be the proper tool to estimate the parameters of such a *two-component error model*.

I agree.

T. Kiang to R. Branham: What is the actual practice of people working in asteroid positions, do they use the least-square solution or the L^1 solution? (Having had an asteroid named after me, I am an "interested" party.)

My experience has been that they publish an ordinary least squares solution with formal errors given by standard techniques, but repeat that this has been my experience and may not be typical.

J. Pfleiderer to R. Branham: Bias is a concept sparsely treated in text-books. Are you acquainted with any references that treat the subject fully?

I am not aware of any extensive treatment in the sense I wanted to stress here, viz. qualitative aspects.

J.F.L. Simmons to R. Branham: Estimation of parameters cannot really be separated from the question of model testing. If one is trying to interpret data on the basis of a particular model, and so determine the parameters appearing in the model, one should first test the model itself, using a χ^2 test, say, or some test suited to the problem. Evidently, if the model does not fit the data, then any confidence intervals or regions will have no sense.

I agree. Also, one should never reject residuals without first inspecting them. Systematic trends in the residuals can indicate a faulty model.

BIAS

Bias

J. PFLEIDERER

Institut für Astronomie der Universität Innsbruck
Austria

Summary: This paper is intended to show the role of personal judgment and experience in avoiding or reducing bias, in addition to the application of mathematical tools. We illustrate our view by a number of astronomical and non-astronomical examples. Bias is a result of unused but necessary (relevant) information. If that information is not available, the problem consists of recognizing the lack, and using data such that the incompleteness least influences the conclusions. Often, some information is available but not used. We cover the cases of neglecting or low-weighting data of complex quality, of over-weighting observations of standards in the reduction procedures, and of neglecting interdependence of data. We discuss in some detail photographic photometry of stars. An unbiased quantitative measure of structure to be used as smoothing functional in deconvolution is derived.

1 INTRODUCTION

The American Heritage Dictionary describes bias as "a preference or inclination, especially one that inhibits impartial judgment; prejudice" and adds: "Bias has generally been defined as 'uninformed or unintentional inclination'; as such it may operate for or against someone or something."

This definition immediately leads to two groups of bias: (1) the uninformed inclination, where the misjudgment stems from necessary information not being available, hence the bias of missing information, and (2) the unintentional inclination, where such information necessary for impartial judgment is available but apparently not used, hence the bias of neglected information. Both groups play a role in the interpretation of data sets. I shall not and cannot give a review over all possible kinds of bias but shall rather speak mostly on certain aspects of the second kind where mathematical possibilities for at least partially avoiding the bias can be demonstrated. However, some words on the first kind are also appropriate.

This conference has of course a tendency to stress a strict mathematical point of view, with an analytical definition of bias and an analytical bias removal or bias reduction. I would like to stress another side of the bias problem: the role of common sense and of a feeling what might happen, for the understanding of what is occuring in bias. As I shall show in section 4, such thinking can even contribute to a good quantitative and mathematically usable formulation of problems.

We should always remember that our ultimate goal is to obtain not so much a mathematically consistent result but rather a result which we estimate to be as close to the truth as possible, to our best knowledge and judgment. The use of refined statistical methods is an extremely good way, often definitely the best one, but it is certainly not the only way to approach that goal.

Konrad Lorenz, one of the fathers of ethology, coined (1959) the notion of "Gestaltwahrnehmung" which was, as I take from Immelmann (1982), translated into English as "gestalt perception". Animals are able to condense a complex visual impression to an extremely simplified conclusion of one bit or a few bits only: Recognizing a fellow, an enemy, a person, a place, a situation. The phenomenon is actually much more general, including all senses, and thus deserves a more general name: "complex comprehension" or "Komplexwahrnehmung" (M.Pfleiderer 1988). For scientists we can add the ability of condensing complex intuitive impressions, resulting, e.g., in an estimation of the value (not as quantity, but as quality) of a scientific statement. Examples would be: I feel this could be done better, this looks sound, I don't believe this, I reckon there must be some bias, I rather would estimate the error to be larger, I am inclined to trust someone's statement, &c. We would loose a good part of science if we were to neglect our scientific judgment and experience, and were to acknowledge only facts that are sufficiently well understood to be cast into a mathematical form.

2 THE BIAS OF MISSING INFORMATION

Bias often relates to data sets and their interpretation, mostly in the sense of statistical interpretation. Bias comes into the picture not because of missing data but because of an implicit assumption that additional data (as the enlargement of a well-sized random sample) or additional information (as the name of objects) would not appreciably change the statistical properties of the set. The question therefore is how to use a data set such that this assumption has a reasonable probability of being reasonably true. The exact value of the probability or the exact definition of what is meant by "reasonably true" is often less important. A data set is never truly complete but always - somehow non-randomly - selected and therefore can be said to be biased in respect to any question asked or interpretation given which depends on this selection. Or one could say that the selection is or is not a biased one.

There is a general recipe to avoid such biases which is as easy to state as it is difficult to use in a particular case: Use, if possible, other data or other information and, in any case, common sense to answer the question: Would a datum contradicting the conclusion have an increased probability to be not a member of the data set?

I give an example. Lemma: "Most astronomical audiences are male-dominated." This is a statement, and as such it cannot be biased. It can only be true or false. In the case of the audience of this conference, it is certainly true (see list of participants). It is the interpretation which may or may not be biased and, by that, become essentially true or not true. Some interpretations would be:

(a) "There are more male than female astronomers". True (unbiased), because the sample is, at least mainly, selected by job, not by sex. We can safely expect additional information to comply. While a female astronomer does not contradict the conclusion just by her very existence, she nevertheless does at least contribute to the support of the opposite conclusion. Our question is therefore: Would a female astronomer have a higher probability than a male to not attend a conference? Probably not, or nearly not. The conclusion thus makes sense.

(b) "Females are discriminated in astronomy". Mainly wrong (biased) because the missing information on how and why to become astronomer is essential. A datum supporting the conclusion would be a female astronomer who wants to attend a conference but is not allowed to do so. A datum contradicting it is a female astronomer who does not want to attend. Her probability not to be there is of course increased. It follows that the conclusion rests on biased feet only and should not be persued without additional information.

(c) Most attenders of astronomical conferences have spouses, at least in the wider sense of friends. The conclusion that spouses are female-dominated in general would be wrong (biased) because the data set is biased in respect to this question. The conclusion that spouses of astronomers are female-dominated is, however, true.

Another example: Inspection of the Bonner Durchmusterung would suggest that nearly all stars are brighter than the 10th magnitude. True? No, and we all know why not. Similarly, inspection of a UBV catalogue would suggest that nearly all stars have colour indices B-V larger than -0.3. True? Yes, and we also know why.

It is the range and the filling factor (completeness factor) of the relevant parameter space which makes the difference. The space of the magnitude parameter is not limited, i.e., all magnitudes are possible. The accessible range is however limited by the telescope size, and the filling factor of the data set decreases rapidly near that limit (truncation). The space of the colour parameter is limited for theoretical reasons (stars should not be bluer than a black body) while the accessible range for observations reaches beyond that limit such that a filling factor near 1 (or, more strictly, a constant filling factor) can be assumed at and beyond the limit.

Therefore, another form of the bias-avoiding recipe is to ask: Is the accessible parameter range limited? Is the filling factor constant over the observed parameter range? If not, is a more quantitative statement possible?

In our first example, the parameter space consists of two boxes, male and female. The data selection (astronomers, attenders, spouses) may be and indeed is biased as compared to a more general human set; astronomers prefer one of the boxes.

If a data set contains information on more than one parameter (e.g., 3 colours, or magnitude and redshift, or magnitude and type), the uniformly covered parameter ranges may differ from one to another parameter. Then statistics (e.g., averages) should use the well covered parameters rather than the incompletely covered ones.

Consider the magnitude-redshift relation of quasars, m vs. z. Can an average of quasars be used as cosmological standard candles? There is strong selection, for any telescope, in magnitude but little in redshift. The average magnitude $\langle m \rangle$ as function of redshift is strongly influenced by the truncation of the data set at faint magnitudes. The average magnitude is, therefore, little dependent on redshift for all z for which most quasars are very faint. The resulting relation $\langle m \rangle$ vs. z is useless.

On the other hand, selection in redshift is much less severe. When taking an average redshift $\langle z \rangle$ as function of magnitude, one can expect to find a reasonably unbiased relation down to the faintest magnitudes, and another kind of unbiased result, viz. no data at all, beyond the faintest magnitude. While the resulting relation m vs. $\langle z \rangle$ is much too uncertain for a decision between different cosmological models, it fits well into the range of possible models. One would rather expect a different relation if redshifts were intrinsic and due to some unknown effect but independent of distance. The relation has, therefore, been one of the strongest arguments in favour of a cosmological interpretation of quasar redshifts.

For de-entangling parameters into independent ones with independent ranges, a principal-component analysis (PCA) is often useful.

3 THE BIAS OF NEGLECTED INFORMATION
There is no data set for which there is not some information which is neglected in doing statistics with it or handling it otherwise. No harm is in doing so as long as the information is irrelevant to the question asked. A bias can arise if information is neglected or insufficiently considered in a systematic fashion, i.e., systematic in respect to the kind of statistics or handling done. In classifying such bias, it seems to me that three kinds will essentially show the point. First, data may be of different quality. There is not too much difficulty in handling different accuracies (higher or lower quality), but the case of simpler or more complex quality is often answered too simply by inclusion (weight 1) or exclusion (weight 0) in the further treatment. I shall therefore call this the weighting bias. Second, we tend to estimate the quality of a measurement the higher, the more is *a priori* known on the outcome. To give an example of my personal experience: If I do photoelectric observations of stars, I do indeed tend to consider observations of standards to be more accurate than observations of non-standards, even if I know that this is nonsense. The same philosophy is found in many reduction procedures. These procedures are, of course, much simplified if each program star is separately fitted just into the standard-star measurements. This means that standard observations have,

effectively, increased weight. I call this the bias of standards. Third, the outcome of a measurement may depend on the outcome of other measurements creating an interdependent data set. A well-known example is the smearing-out of a star image by atmospheric turbulence (seeing, described in detail by a point spread function). Here the bias is towards smooth features and against sharp gradients. I call this the interdependence bias or, more specifically, the point spread bias.

3.1 The Weight Bias

One problem is to combine data of different quality to find a frequency distribution that is as unbiased as possible. It seems that, *in praxi*, the most relevant case is that of censored data: The exact value of a datum is not known but it must be larger (or smaller) than a certain limit (f.i., the radio flux of a galaxy not detected with a radio telescope). The method proper is called "survival analysis". The case is extensively covered by Dr. Feigelson in this conference.

Here I only want to point out that it is useful to be aware of the possibility of more complex data qualities. For instance, a datum may have a large and non-negligible uncertainty (a photographic magnitude within a set of photoelectric magnitudes). It may be ambigous (more than one value possible, as a variable star with several possible discrete periods), or ranged (fainter than the limit on one plate, visible on another), and so on. Survival analysis is unable to cope with these more general cases. Frequency distributions can nevertheless be found from a set of nonlinear equations (Pfleiderer & Krommidas 1982, Pfleiderer 1983). However, methods of solving these equations systematically have not yet been developed, to my knowledge. Also, as is the case with censored data, the available information may not suffice to solve the problem completely and/or uniquely.

3.2 The Bias of Standards

Most astronomical observations are relative measurements. The absolute position of a star on a plate or the absolute number of photons which have been recorded from a star are not interesting. It is the position or photon number relative to a comparison object called "standard" which is relevant.

Most reduction procedures assume that the relevant data of the standard or the standards are well known, i.e., the uncertainties need not be taken into account. The standards which are measured together with the program objects (non-standards) do however have a measuring error. Also - and this is more dangerous -, the *a priori* knowledge about the standard or its interpretation within the given problem may be wrong (f.i., wrong photoelectric magnitudes). Generally an error of a standard enters the reduction only insofar as discrepancies between the measurements of different standards or repeated measurements of one are present. The system of data with which a non-standard is compared is solely determined from the measurements of standards.

By that procedure, a bias is introduced because any conflicting information which could be derived from non-standards is neglected.

I illustrate this statement by two examples, the first being star positions derived from overlapping plates. The problem of using all available measurements for determining a maximally consistent common scale on all plates can be formulated analytically because the imaging properties of plates can be described analytically by plate constants. H. Eichhorn (1988) was able to show that the resulting system of linear equations for the final differential corrections can be solved explicitly. A similar problem is the establishment of photoelectric sequences over the whole sky. A bias or systematic error is introduced by the fact that, during one night or one year of observations, one has to start with early sequences and go on with later ones, and also that sequences at different declinations have systematically different positions for observations by any one telescope. Many years of observations from many observatories were necessary to empirically solve such problems. Again it turned out that the best results came from a combination of all available observations.

Photographic star photometry: This is a particularly interesting problem for two reasons: First, the non-linearities of the densities as well as of the apparent sizes (diameters) of stars very easily lead to biased (i.e., systematically shifted) results that have made photographic photometry cumbersome and unreliable ever since photography was introduced into astronomy. Second, the exposure of a plate is a measurement that gives quantitatively usable results ("numbers") only after the plate is subjected to a further, "second-generation" measurement (plate measurement). That is, there are two different kinds of errors, the plate error (a star has different magnitudes on different plates), and the plate measuring error (repeated measurements of a star on a given plate yield different magnitudes). In photometry, a third kind of error arises from the uncertain relation between the instrumental and some standard photometric system.

Modern applications of photographic photometry concern the use of unique plate or print material, as the Palomar or ESO surveys, where the plate error is often unaccessible (one plate only, except for overlaps) and therefore generally irrelevant, as well as integrated densities from plate measuring machines where the plate measuring error is generally small enough to be neglected. In more old-fashioned but also more easily accessible applications, as iris photometry, both errors are substantial.

A parametrization of the non-linearities, or a full computer treatment, is difficult. However, manual intervention in computer procedures has in recent years become rather easy, with the implementation of astronomical reduction packages. So one should think again of those old and nearly forgotten means of combining strict mathematical treatment with visual inspection and manual interference.

It is usual to take a set of readings of one plate, determine a scale from the standards, and get a set of magnitudes for the non-standard stars, to be averaged lateron. At this stage arises a bias of standards. Part of the scale error results from plate errors and measuring errors of the standards but is systematically transferred to all non-standard stars of similar brightness. The different errors are completely mixed up in such a reduction procedure.

The existence of three different kinds of errors suggests to do the reduction stepwise. First, get for each plate a set of readings, one per star, with estimated or otherwise determined uncertainties in the instrumental system (here, iris readings). I omit here the complication that repeated iris readings of one star on one plate tend to be non-normally distributed. Second, get from all plates (of course in one colour only) again a set of readings, one per star, with the combined uncertainties of measuring and plate errors.

The average of different sets of readings, $\{R\}$, with mildly non-linear differences, can be found as follows: Start with the plain average, giving a set $\{S\}$. Correct each set linearly (result $\{R^*\}$) such that R^* agrees with S when averaged over all stars. Take the remaining deviations R^*-S to find a non-linear correction for R^* (result R^{**}) such that $R^{**}-S$ is zero not only for the average of all stars but also for groups of stars within a small range of R^{**}. The final result is a set of one effective reading $S^*=\langle R^{**}\rangle$ with internal variance $\langle(R^{**}-S^*)^2\rangle$ for each star.

It is obvious that up to this stage, standards are in no way preferable to non-standards. However, now we have used up all available information within the instrumental system. It is therefore now and not before that the standards come legally into the play. If we take them to get a scale for finding magnitudes, the result will be minimally biased.

Advantages of such procedure are: high-order terms of the non-linearity tend to be different on different plates and even for repeated readings on one plate. They are partly damped by the averaging - the non-linearity is reduced. Therefore, the extrapolation of the scale to faint stars as well as the colour equation are better defined. The same is true for the (internal) total error.

An improvement of the scale is possible if *additional information* is obtained. Particularly awarding seems to be the use of two plates with different exposure times. The case was first treated by K. Schwarzschild (1906) but did not become common practice because at that time there were no photoelectric standards which are necessary to keep the scale from running away. It was re-introduced by Schneider (1976) and applied to iris photometry. He showed that *ceteris paribus* the iris difference on the two plates for any star corresponds, in the average, to a constant magnitude difference (a magnitude "step") between stars having the same two readings on one plate. The step size cannot be derived from the exposure times alone but it follows from (a minimum of two) photoelectric standards. If iris readings are, via average iris differences, translated into numbers of such steps, then the magnitude vs. number-of-steps scale is essentially linear. Scale extrapolation to faint magnitudes thus becomes less uncertain, even if the linearity probably breaks down near the plate limit.

The essential point is that the non-linear magnitude-reading relation which is usually derived from the standards alone is reflected in the relation iris differences vs. readings which can be derived from all stars, thus avoiding a bias of standards.

It is well known that the determination of photographic magnitudes suffers from the possibility that plate sensitivities may depend on the location on the plate. The main reasons are: (1) Change of emulsion properties, (2) different sky backgrounds, (3) different optical images, and (4) tilting of the plate. In first approximation, zero point and scale factor of the magnitude-"step" relation may change but the relation will stay linear. That is, a scale transfer within the plate is possible with a small number of standards.

Another possibility of increasing the available information would be to take iris readings for more than one instrumental setting, thus obtaining two or more different readings per star from each plate. This has not yet been done, nor a theory been developed. The reading differences would reflect the gradual increase of photographic density towards the star centre. Again, a "step number" vs. reading relation can be constructed for all stars and hopefully be used for an improvement of the magnitude scale.

3.3 The Interdependence Bias
The first problem which I would like to discuss in this section is the case of discordant data in a set of otherwise consistent data. The interdependence is given by the fact that the outcome of a measurement is approximately known if it is to be consistent with the other data. The weight that should be given to a discordant result depends on the consistency of the remaining set of data.

Does the discordant datum stem from a random deviation and is thus a legal member of a random sample of the underlying distribution? Even then it may or may not be a good strategy to keep it in the sample. For a reasonably normal or otherwise well-enough known underlying distribution, the probability can be calculated that the average without discordant result is closer to the truth than that with it. However, in many cases, if not in most, a personal judgment is as good as - and faster than - a thorough statistical analysis.

Is the discordant result not a random deviation? Then it may be a good strategy to keep it in the sample in order to demonstrate the possibility of discordances. Or it may be treated separately - f.i., as a hint on variability of an otherwise constant star. Or it may be rejected at all, particularly if there is reason to assume that the measurement plainly went wrong. Whether this is so or not can rarely be decided by statistics but is rather a question of the observer's experience.

Often, a good criterion is whether or not the neglect of the discordant result does improve the internal consistency sufficiently. The fact of discordancy should, however, be remembered. Neglect is not forget.

I give an example which is meant to show how a good knowledge of the interrelation of data can help to find a - hopefully - most unbiased estimate of a resulting quantity. Let us try to estimate the age of a person from a set of data, as height, weight, skin, muscles, teeth, hair colour, &c. People define a distribution in a multidimensional parameter space spanned by these data. The fact that it is often easy to get the age reasonably well from a data subset or a single datum shows that the

effective dimensionality of the distribution is rather small. The same
would follow from a PCA.

Grey hair is normally an indication of old age but it is known that
sometimes young people also have grey hair. Consider a young-looking
person with grey hair. Is it a young grey person, or an older person
looking younger? The method of projecting each datum on the time axis,
according to the experimental distribution, and taking a weighted
average, would give the following result: The best estimate is that the
person is older than it looks otherwise but younger than the hair colour
indicates. However, it may be that all data except the colour would lead
consistently to the same smaller age. Then the colour evidence, being
known to be sometimes misleading, should be dismissed. Indeed, this is
what we would intuitively do. In other words, what we hope to be a most
non-biased estimate takes into account the interdependence of different
data. A datum can increase or decrease the importance, the valuableness,
the worth, the reliability of another datum.

The reason why I have discussed this example so extensively is the
following: As criminologists can tell us, age estimates combine exact
measurements and complex comprehension (intuitive judgment). There are
never enough data known to have a purely mathematical method working
sufficiently reliably.

Interdependency bias can also be found with "half-known" quantities.
H.Eichhorn (1988) showed that the consistency of a fit and its error
estimate can be improved if a reasonably well known variable (f.i., the
focal length of a telescope) is only allowed to vary within a small
range.

Point spread bias: A frequent case of interdependent data is a signal
smeared out by the observation. The signal is convolved with a smearing
function. Examples are the instrument profile in spectroscopy, the
effect of seeing, the beam of a radio telescope. Convolution problems
occur not only in astronomy but as well, f.i., in molecular beam
scattering, in seismic-signal interpretation, in multiple radar imaging,
in tomography, in photography, in spectral analysis, in time series.

Convolution will always decrease contrast. The measurements are biased
against the detection of sharp features. If one tries to recover those
features which are not completely smoothed out (image recovery, image
sharpening), one removes bias. On the other hand, any deconvolution will
either miss some detectable contrast or tend to detect additional
spurious contrast. The problem then is to avoid, as much as possible,
either bias. As with other biases, the question is to use properly and
in an optimum way the information contained in the data or known from
other sources. It should neither be partly neglected nor partly
overemphasized.

In deconvolution, too little structure (too smooth a model) results in
an unsufficient data fit, i.e., the model of the signal will, when
convolved with the point spread function, reproduce the measured data
not well enough. On the other hand, one can, by adding spurious
structure, reproduce the data too well (fitting of noise). Such fit can
be avoided by introduction of a functional that gets large if too much

structure is found. These Lagrange methods of deconvolution thus consist of finding a model which fits the data while leaving the functional at a minimum or, more generally, at an extremum.

4. STRUCTURAL INFORMATION

Reasonable forms of possible functionals have been extensively discussed. The best-known of these is probably the entropy (maximum entropy method = MEM). It has been claimed that the entropy is the most non-committal or most non-biased functional possible. Let me show here that elementary information-theoretical considerations lead to another functional which can be interpreted as the information I inherent in the structure, hence the name "structural information". This is the most unbiased functional I can at present offer (Pfleiderer 1989).

The intuitive meaning of the term "structure" is of qualitative nature. It has a minimum (no structure) and can be ordered (more or less structure). A quantitative measure I of structure should thus be non-negative and be zero only for a completely flat image.

In fig.1, all five images a-e are equal except for having been moved (translated) and turned around (left/right or upside down). One could also say that the same image is looked on from different standpoints and from different directions. The images should, therefore, have equal measures, implying translational and mirror invariance. I can thus depend only on the squares of distances d in location (abscissa x) and f in elevation (ordinate z), or on the unsigned distances. Fig.2 suggests different measures for a and b, a and c. The images of fig.2 could, however, be made coinciding if adequately different scales w in x and v in z, respectively, were used. This implies scale invariance for I.

Note that the concept of "structure *per se*" is invalidated by this last argument, at least as far as quantification of its measure is concerned. Structure is narrow or wide, large or small only in comparison to some naturally provided scale. In other words, one cannot quantify "intrinsic structure" but one can do so with "observed structure". For instance, a polished surface is macroscopically smooth but, as can be seen in a microscope, unsmooth on a smaller scale.

Images are often given in pixels. Then I is made up of two-pixel combinations, and the variables entering I are

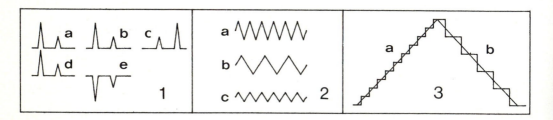

Fig.1: Translational and mirror invariance of I
Fig.2: Scale invariance of I
Fig.3: Pixel-size invariance of I

$X_{1j} = (d_{1j} / w_{1j})^2$, $d_{1j} = x_1 - x_j$,   and $Z_{1j} = (f_{1j} / v_{1j})^2$, $f_{1j} = z_1 - z_j$.

Here, i and j stand for the two pixels. x is a scalar in the one-dimensional case and a two-component vector in the case of two-dimensional maps. w and v may depend on the pixel combination as indicated by the indices.

Structure should not depend on the pixel size s except, of course, for that part which is finer than the pixel grid (same measure for a and b in fig.3). That implies scaling with s (I proportional to s) and defining z as independent of s, e.g., photons per unit area, not photons per pixel.

We add a request for completeness and symmetry. It implies that all two-pixel combinations should be used, that each combination depends only on the location and content of these two combined pixels (or that each combination is strictly two-pixeled), and that there is no cancelling of combinations (i.e., each combination contributes non-negatively to I).

Consider a flat image with noise which can be described by a variance, $\langle f^2 \rangle$. If the measure I of such noise-structured image is to be essentially independent of the exact noise distribution, it must have the same (i.e., a quadratic) dependence on f. One could also say that it is sufficient and even reasonable to use only the first non-vanishing term of a Tayler series in f. Each term (two-pixel combination) is proportional to Z, not to any other function of Z, and contains a function $G_{1j}(X_{1j}) \geq 0$. Then one can also use $p_{1j} = \exp(-G_{1j})$ with $0 < p_{1j} \leq 1$. The function p has the form of a probability.

Putting everything together, we arrive at

$$I = -\Sigma_1 \Sigma_j s_1 s_j Z_{1j} \log p_{1j} (X_{1j})$$

where only the dependence of p (or G) on X is not yet fixed. It cannot be given in general because it reflects the relation (interdependence) between pixels which is problem dependent. The generalization to a continuum distribution (infinitely small pixel sizes, integral representation) is straightforward.

A point source, or an extremely narrow source of a width approaching zero, is a very sharp feature and thus should contribute a large amount to I. Since any practical problem involves finite pixel sizes, one then could - without loss of generality - attribute an infinite value to $G(0)$. On the other hand, an extremely extended source with a width approaching infinity gives a very smooth image and thus should contribute little to I. Correspondingly, a z-difference f for adjacent pixels adds a great deal (large G, or small p, $p=0$ for $d=0$) while the same difference adds little for distant pixels (small G, or p near unity).

The equation is reminiscent of the formula for the information content

$$\text{Inf} = -\Sigma_1 n_1 \log p_1$$

of events i that occur n_1 times with probability p_1. Here, Z (or more strictly $s_1 s_j Z_{1j}$), a measure of variance, represents the frequency (or perhaps better: the strength), and p, a measure of distance, the probability (or perhaps better: the effectiveness) of "structural events".

It is appealing to measure a complex phenomenon by the information it contains. The measure of structure as derived here, is equivalent to the information provided by "structural events" which have low probability (high information content) if occuring in adjacent pixels (much structure) and high probability (low information content) when occuring in distant pixels (little structure). Thus we would argue that not only can structure be measured, but it is measured best by an information expression which we call "structural information".

It should also be noted that the function G need not at all be a monotoneous function of distance. For instance, in an assumedly periodic source distribution, only the structure within a period is relevant. Correspondingly, distances would enter only *modulo* the period, and G would also be periodic.

Extremal functionals used in the deconvolution literature include Σz^2 (essentially corresponding to $G=$const, Cornwell 1983), and the square sum of the second derivative of the intensity distribution (representable by partly negative G). Both cases are thus incomplete versions of I. The entropy (essentially $\Sigma_i z_i \log z_i$) has positional invariance but does otherwise not comply with our requests. It is a very good smoothing function but not a measure of structure.

Direct applications of I, say for the comparison of two given maps, are certainly possible but I do not yet have the experience to report results. It has already been applied to deconvolution (mimimum information method MIM, Pfleiderer 1988).

Acknowledgements: The theory of structural information was devised during a stay at the Department of Astronomy, University of Florida, Gainesville FL. I have to thank many colleagues, particularly E.E.Baart, J.Condon, T.Cornwell, H.Eichhorn, L.Erickson, S.Gottesman, J.Högboom, O.Schneider.

References
Eichhorn H. 1988: Astrophys.J.*334*,465
Immelmann K. 1982: Wörterbuch der Verhaltensforschung, Parey
Lorenz K. 1959: Z.exp.angew.Psych.*4*,118
Pfleiderer J. 1983: pp.3 and 77 in: Proc. Statistical Methods in
   Astronomy Symp. Strasbourg (ESA SP-201)
Pfleiderer J. 1988: Astron.Astroph. *194*,344
Pfleiderer J. 1989: Naturwiss. *76*,297
Pfleiderer J., Krommidas P. 1982: M.N.R.A.S. *198*,281
Pfleiderer M. 1988: p.279 in: Bioastronomy - the Next Steps (G.Marx
   ed.), IAU Coll.99, Astrophys. Space Library *144*, Kluwer
Schneider O. 1976: Mitt.Astron.Ges. *40*,205
Schwarzschild K. 1906: A.N. *172* Nr.4109, 65

Jörg Pfleiderer
Institut für Astronomie
Technikerstr.25, A-6020 Innsbruck, Austria

Bias in Determination of Cosmological Parameters

M.A. HENDRY – J.F.L. SIMMONS

Department of Physics and Astronomy, University of Glasgow
Glasgaw, Scotland, U.K.

ABSTRACT

An illustrative study of the effects of bias on different distance estimators is undertaken. Estimates are inferred from apparent magnitudes for the simple case of a Gaussian luminosity function. The study is applied to a distant sample of ScI galaxies in order to determine Hubble's Constant and the velocity of the Local Group. The results indicate the importance of a careful choice of distance estimator in cosmology.

1. INTRODUCTION

Cosmologists have long been aware of the impact of selection effects on galaxy distance determinations. Furthermore, the consequences of this for the determination of Hubble's constant and other cosmological parameters and testing for isotropy of the Hubble Flow have been an extremely contentious issue over the last decade (e.g. Tammann *et al*, 1979). There is no consensus about how to correct for bias introduced by selection in a magnitude limited sample. In order to illustrate some effects of this bias we consider the most readily understood distance indicator – the apparent magnitude, m. This is related to absolute magnitude, M, and distance, r (in Mpc) by the equation :-

$$m = M + 5\log r + 25 \qquad (1)$$

From equation (1) we may obtain the conditional distribution, $\Phi(m|r)$, of m at a given distance r, in terms of the galaxy luminosity function, $\Psi(M)$. Using this conditional distribution, for a Gaussian luminosity function of mean M_0 and variance σ^2, we have investigated the properties of several different distance estimators. These are described below.

2. DISTANCE ESTIMATORS

Three of our estimators were derived from equation (1) by solving for r, using three different 'effective' absolute magnitudes.

1) 'Naive', $\hat{r}_N$: $M=M_0$, the true mean absolute magnitude of the luminosity function.

2) 'Malmquist', $\hat{r}_{MAL}$: $M=M_0-1.38\sigma^2$, the mean absolute magnitude of visible galaxies in a magnitude limited sample (Malmquist, 1920).

3) 'Feast', $\hat{r}_F$: $M=M_0+0.26\sigma^2$, the value for which the percentage bias of $\hat{r}_F$ tends to zero as r tends to zero. (The derivation by Feast (1987) was, however, somewhat different.)

Notice that these estimators differ simply by a constant scale factor. We considered also the following two estimators :-

4) 'Maximum likelihood', $\hat{r}_{ML}$: defined as the value of r which maximises the likelihood of obtaining the observed apparent magnitude.

5) 'Mean', $\hat{r}_{ME}$: defined as the value of r at which $E(m|r)$ is equal to the observed apparent magnitude. For a Gaussian luminosity function $\hat{r}_{ME}$ and $\hat{r}_{ML}$ are identical, although generally this will not be the case.

Fig (1) shows how the estimators may be obtained from observations. Having defined the risk as $E\{(\hat{r}-r)^2|r\}$, figs (2) and (3) show the percentage bias and risk of the estimators, as a function of distance. All distances are given in units of the 'limiting distance', r_L, at which a galaxy with the mean absolute magnitude, M_0, would appear at m_L, the limiting apparent magnitude. Thus no explicit value of M_0 is required.

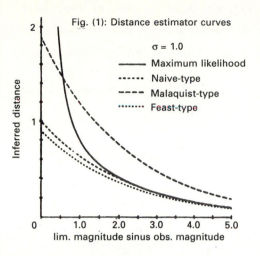

Fig. (1): Distance estimator curves

$\sigma = 1.0$

—— Maximum likelihood
······ Naive-type
- - - Malaquist-type
········ Feast-type

Inferred distance

lim. magnitude sinus obs. magnitude

From fig (1) it can be seen that as $m \to m_L$, $\hat{r}_{ML} \to \infty$. It is found that $E(\hat{r}_{ML}|r) = \infty$, for all r, and so this estimator has infinite bias and risk. One may define a modified estimator with finite bias in several ways: e.g. one may place an upper limit on the value of $\hat{r}_{ML}$ or, as in figs (2) and (3), one may reject galaxies whose magnitudes lie close to m_L (in this case within 0.5 magnitudes). Each of these estimators becomes negatively biased at large distances. The range in which one estimator is preferred over others will depend strongly on σ and m_L: generally $\hat{r}_F$ is better at nearby distances and $\hat{r}_{MAL}$ at large distances, although a suitably modified $\hat{r}_{ML}$ can be better than the latter.

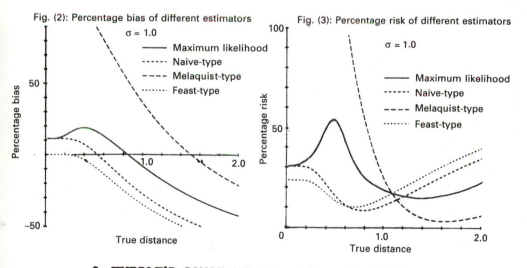

Fig. (2): Percentage bias of different estimators

$\sigma = 1.0$

—— Maximum likelihood
······ Naive-type
- - - Melaquist-type
········ Feast-type

Percentage bias

True distance

Fig. (3): Percentage risk of different estimators

$\sigma = 1.0$

—— Maximum likelihood
······ Naive-type
- - - Melaquist-type
········ Feast-type

Percentage risk

True distance

3. HUBBLE'S CONSTANT AND THE LOCAL GROUP MOTION

The systematic bias of the different estimators will have important consequences for the analysis of real data. As an example we have taken a sample of 184 ScI galaxies, used by Rubin (1976) in studies of the Hubble Flow, and from their redshifts and apparent magnitudes

determined estimates of Hubble's constant and the velocity of the Local
Group with respect to these galaxies. The data fit was then repeated
with Hubble's constant specified. Some results are given in Table (1).

Our analysis was by simple
least squares, taking no
account of the different
errors at different apparent
magnitudes. It would be wrong,
therefore, to attach too much
significance to the actual
values obtained: however, the
strong variation with choice
of estimator clearly indicates
how inadequate treatment of
bias can lead to widely
discrepant notions of the
Hubble Flow.

Table (1) : Inferred Cosmological Parameters

$M_0 = -20$　　　　$m_L = 15$

$\sigma = 0.5$	$\hat{r}_{ML}$	$\hat{r}_N$	$\hat{r}_{MAL}$	$\hat{r}_P$
$\hat{H}_0$ (kms^{-1}Mpc^{-1})	56	86	73	88
$|\hat{V}|$ (kms^{-1})	1011	617	617	617
$|\hat{V}|$ ($H_0 = 50$)	1260	1814	1814	1814
$|\hat{V}|$ ($H_0 = 100$)	2591	1003	1003	1003

$\sigma = 1.0$	$\hat{r}_{ML}$	$\hat{r}_N$	$\hat{r}_{MAL}$	$\hat{r}_P$
$\hat{H}_0$ (kms^{-1}Mpc^{-1})	44	86	73	88
$|\hat{V}|$ (kms^{-1})	745	617	617	617
$|\hat{V}|$ ($H_0 = 50$)	803	1814	1814	1814
$|\hat{V}|$ ($H_0 = 100$)	4915	1003	1003	1003

Since each of the estimators $\hat{r}_N$, $\hat{r}_P$ and $\hat{r}_{MAL}$
differ one from another only by a constant,
it follows (but not trivially) that $\hat{V}$ will be
identical for all three; although, of course,
$\hat{H}_0$ will differ in each case.

4. CONCLUSIONS

The use of different distance estimators inferred from measurements of
apparent magnitude has serious consequences for the determination of
cosmological parameters. The discrepancy between estimators increases
sharply with the dispersion of the luminosity function. It would be
instructive to extend this analysis to include other distance indicators
and other luminosity functions. This we hope to do in future work.

REFERENCES

Feast, M.　　　　　　　　　　Observatory, 1987, **107**, 1080, 185–188

Malmquist, K.　　　　　　　　Medd. Lund. Astr. Obs.　1920, **20**

Rubin, V., Thonnard, N., Ford, W., Roberts, M.　　Astr J.　1976, **81**,
687–718

Tammann, G., Sandage, A., Yahil, A.　Physical Cosmology, 1979, 53–125

Censored Data in Astronomy

Eric D. Feigelson

Department of Astronomy
Pennsylvania State University
University Park PA 16802 U.S.A.

1 MOTIVATION AND BACKGROUND

Consider the following situation: an astronomer goes to a telescope to measure a certain property of a preselected sample of objects. The scientific goals of the experiment might include finding the luminosity function of the objects, comparing this luminosity function to that of another sample, relating the measured property to other previously known properties, quantification of any relation by fitting a straight line, and comparing the measured property to astrophysical theory. In the parlance of statistics, the astronomer needs to estimate the empirical distribution function, perform two- sample tests, correlation and regression, and goodness-of-fit tests. Most astronomers are familiar with simple statistical methods (e.g. Bevington 1969) to perform these tasks. However, these standard methods are not applicable when some of the targetted objects are not detected. In this case, the astronomer does not learn the value of the property, but rather that the value is LESS than a certain level corresponding to the sensitivity of the particular observation. In statistics, these upper limits to the true value are called 'left censored' data points.

'Censoring' due to nondetections in astronomical surveys should be distinguished from the more long-standing bias known as 'truncation'. A truncated astronomical sample is one in which the fainter objects are missing entirely from the sample. A censored sample has all of the objects of interest, some of which are not detected. Recovering from the original truncation, sometimes called the selection or 'Malmquist' bias, can be very difficult because most astronomical samples are truncated in flux, but the variables intrinsic to the objects are often luminosities (flux times distance squared). A simple flux-limited truncation then produces a complicated bias in the luminosity distribution that may depend on the luminosity function, spatial distribution, interstellar reddening, cosmic evolution and so forth. Overcoming truncation bias (see Trumpler and Weaver 1953) is not addressed by the methods discussed here.

Astronomical studies with censoring bias have become common in recent years because of the growth of multiwavelength astronomy. A sample of stars or galaxies originally selected in a flux-limited optical survey often provides the targets for a survey at radio, infrared or X-ray wavelengths. Nondetections arise because some objects are too faint or distant to produce significant signals in the detectors in the allotted exposure time. In principle, one might return to the telescope and continue observation until all objects in the sample are detected. But in practice, the competition for telescope time may be too strong to permit this. Some of the most important telescopes today are borne on satellites costing hundreds of millions of dollars and, once the satellite mission is completed, further observations are nearly impossible to perform. The recent *Infrared Astronomy Satellite* mission, for example, has produced several dozen studies to date with censored data sets.

In the past, astronomers have dealt with censored data in a variety of fashions. First, many recognize the inapplicability of standard statistical methods and refrain from any quantitative analysis of their data. Second, some omit censored points and perform standard tests on the detected points. This is likely to bias luminosity functions and two-sample tests by omitting low luminosity objects, and may also affect correlation and regression results. Third, a few astronomers have artificially changed nondetections to detections and proceeded with standard analysis. This is a falsification of the data and should be discouraged. Fourth, a method arose particularly among radio astronomers during the last decade that performed statistical tests on the ratio of detected-to-nondetected objects. Auriemma *et al.* (1977), for example, obtained heavily censored radio observations of optically selected elliptical galaxies. They computed the 'fractional luminosity functions' based on the detected ratio, fitted analytical functions to them, and compared samples with each other using standard statistical techniques.

The fractional luminosity function was, however, criticized for lack of self-consistency (it is not correctly normalized when all objects are detected) and its failure to take the precise value of each upper limit into account. Astronomers Pfleiderer (1976), Avni *et al.* (1980) and Hummel (1981) independently proposed an alternative: an iterative algorithm that redistributes each nondetection in proportion to the distribution of detected objects at lower luminosities. They established this algorithm is self-consistent and satisfies a maximum-likelihood, minimum-information criterion. This algorithm gained some popularity during the 1980's, particularly among X-ray astronomers using Avni's computer codes. Avni and Tananbaum (1986) provided an additional procedure for computing a linear regression line when the dependent variable is censored. A limitation of these efforts was that the error analysis was inadequately validated or required cumbersome numerical calculations, and no further statistical methods followed.

Independently, Schmitt (1985) and researchers at Penn State (Feigelson and Nelson 1985; Isobe, Feigelson and Nelson 1986) recognized that the Pfleiderer- Avni-Hummel algorithm was identical to Efron's (1967) formulation of Kaplan and Meier's (1958) estimator for the empirical distribution function of a randomly censored variable. The Kaplan-Meier estimator had already been proven to be the unique self-consistent, generalized maximum-likelihood, asymptotically normal estimator of the true distribution function. When presented in the explicit product-limit formulation rather than the iterative redistribution formulation, error analysis and moments of the distribution are readily calculable from simple analytic formulae. The Kaplan-Meier estimator lies at the foundation of broad range of statistical methods developed for biometrical and industrial applications during the past several decades. Two- sample tests, correlation coefficients, regression procedures and goodness-of- fit tests had been extensively studied. This field devoted to the study of censored data is often referred to as 'survival analysis'.

The first explicit use of standard survival analysis in astronomy is that of Tytler (1982) in a study of the redshift distribution of certain Lyman α absorbing clouds found in quasar spectra. Quasars without a detected absorption system were assumed to have one with a redshift beyond the edge of available spectra, and are thereby censored data points. Tytler calculates the Kaplan-Meier estimator of the redshifts, fits it with a parametric model, and used two-sample tests to compare nearby and distant quasars. The statistical tests support a model in which the absorbing clouds are distributed in intergalactic space and are not produced by the quasars. Though this problem has been extensively pursued in recent years, most related studies have not adopted survival analysis statistical techniques.

2 SURVIVAL ANALYSIS IN DIFFERENT DISCIPLINES

The applicability of survival analysis to astronomical problems as it has been developed for other fields bears some scrutiny. Table 1 describes a simple experiment in three fields involving comparison of two univariate samples with censoring. One involves drug testing in biomedical research, the second inquires whether galaxies in rich clusters are stripped of their infrared-emitting interstellar medium, and the last examines the performance of manufactured products in an industrial context. The basic similarity of the statistical issue and structure of the data is clear, and the same two-sample survival analysis tests can be applied to all cases.

Some differences, however, are present. First, astronomy typically has left-censoring rather than right-censoring. This is easily dealt with by changing the sign of the variable. Second, the different experiments exhibit different censoring patterns. The described biomedical experiment has all censored points at a fixed value; this is called Type I censoring and occurs in astronomy if one considers fluxes from flux-limited surveys. The industrial experiment has a fixed fraction of points censored; this is called Type II censoring and rarely occurs in astronomy. The described astronomy

Table 1
Survival Analysis in Different Disciplines

	Biomedical research	Astronomical research	Industrial testing
Problem	Does Drug A cure cholera?	Do spiral galaxies in rich clusters have less interstellar dust than isolated spirals?	Does Engine A have fewer breakdowns than Engine B?
Data	20 rats with cholera, 10 given Drug A and 10 given a placebo.	20 spirals observed, 10 in clusters and 10 isolated.	20 engines operating until 8 break; 10 Engine A and 10 Engine B.
Variable	Survival time	Infrared luminosity	Survival time
Censoring	Type 1, right-censored because some rats are living when experiment ends	Quasi-random, left-censored because some spirals not detected in available exposure	Type 2, right-censored because 12 engines still at end.
Statistical reference	Miller 1981, Kalbfleisch and Prentice 1982	Schmitt 1985, Feigelson and Nelson 1985, Isobe et al. 1986	Lawless 1982

experiment involves satellite observations with different sensitivities of galaxies at different redshifts, so that the censoring pattern is somewhat randomized. Care must be taken to establish that the survival analysis method used is appropriate for the censoring pattern in a given experiment. Third, some survival analysis methods assume a specific parametric or semi-parametric form for the underlying parent population. Some biomedical applications assume a 'proportional hazards' model, many industrial applications assume a Gaussian or Weibull distribution. There are no standard survival analysis procedures for power law distributions often encountered in astronomy.

In addition to this two-sample test, one finds other statistical problems in common between these fields. The epidemiologist may seek a correlation and compute a linear regression between the survival time of cancer patients and their consumption of cigarettes. The industrial statistician may compute a regression between the failure times of steel beams and the stress placed upon them. The astronomer may seek a linear relation between the far-infrared emission and the molecular gas content of spiral galaxies. In all cases, similar techniques dealing with censored data are needed. In certain problems of public health, one even sees combinations of censoring and truncation reminiscent of those encountered in astronomy. Estimation of the mean survival time of AIDS patients involves censored data sets, but estimation of the total population of individuals undiagnosed but infected with the AIDS virus requires evaluation of truncation selection effects.

3 SURVIVAL ANALYSIS FOR ASTRONOMY

The investigation of statistical methods for survival data is well-established. Hundreds of mathematical papers have been written, summarized and synthesized in a dozen modern texts and monographs. Interested astronomers might start with the short overviews by Miller (1981) or David and Moeschberger (1978). More detailed monographs by Kalbfleisch and Prentice (1980), Lawless (1982) and Schneider (1986) are also valuable. A dozen commercial software packages implementing survival analysis methods are reviewed by Wagner and Meeker (1985). Most packages include only a few methods; few if any, for example, implement linear regressions or procedures for censoring in more than one variable.

A selection of survival analysis methods selected for their likely utility in astronomical contexts are described by Schmitt (1985), Feigelson and Nelson (1985), Isobe, Feigelson and Nelson (1986), and Wardle and Knapp (1986). These methods are only briefly reviewed here; the reader is referred to these papers, Isobe (1989) and the monographs listed above for a mathematical presentation. A software package called ASURV (Astronomical SURVival analysis) implementing most of these procedures has been developed by Isobe and Feigelson (1987). A stand-alone version of the package consisting of 10,000 lines of Fortran 77 can be obtained free of charge from the authors. It has also been incorporated into the large IRAF/STSDAS astronomical software system available from the Space Telescope Science Institute in Baltimore.

3.1 Univariate Distribution Function

The Kaplan-Meier estimator is considered to be the best possible estimate of the empirical distribution function of a censored data set. It would be used by astronomers to estimate the shape of luminosity functions and other distributions when upper limits are present. Mathematically, it is the unique maximum-likelihood estimate under the important assumption of random censoring. This means that the censoring mechanism is blind to the value of the data point, not that the censored points appear to be randomly spread throughout the data. The astronomer must have some prior understanding of the cause of the censoring to know whether the assumptions of the Kaplan-Meier are met.

The Kaplan-Meier estimator is calculated as a normalized integral step function. In an uncensored data set, it converges to the obvious solution where the function

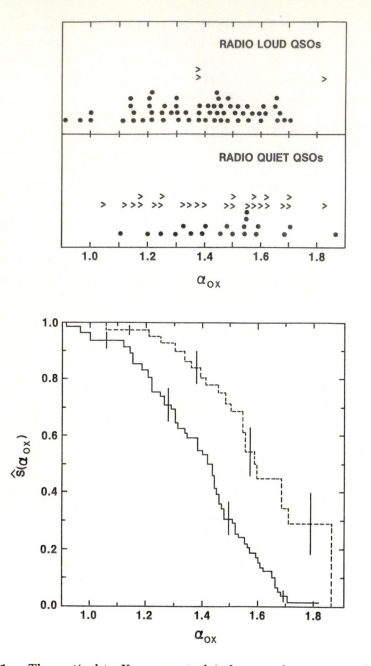

Figure 1. *The optical-to-X-ray spectral index α_{ox} for quasars observed with the Einstein X-ray Observatory by Zamorani et al. (1981). Top panels: Observed data. Bottom panel: Kaplan-Meier estimators for the radio-loud (solid) and radio-quiet (dashed) samples. This variable is right-censored because the nondetections occur in the X-ray flux, which is in the denominator of α_{ox}. Figure from Feigelson and Nelson (1985).*

drops by a step $1/N$ at each of the N data values. In a censored data set, it drops by increasing large steps at each of the detected values as the weight of the nondetections are accounted for. The size of the step and the growing uncertainty in the estimator as more censored points are included, are given by simple explicit formulae developed by Kaplan and Meier (1958). The error analysis is closely related to Greenwood's formula used for many years in actuarial life tables. Figure 1 shows an example of the Kaplan-Meier estimator for two astronomical samples, one lightly and one heavily censored.

The Kaplan-Meier estimator can be also reformulated as a binned, differential distribution function. This approach has intuitive appeal (especially to astronomers who historically draw luminosity functions in this fashion), but has disadvantages: it has an iterative rather than explicit formula for the estimator, and has no formulae at all for the error analysis. It was this differential form of the Kaplan-Meier estimator, originally given by Efron (1967), which was independently rediscovered by the astronomers Pfleiderer, Avni and Hummel. We encourage, however, astronomers to use the original unbinned integral formulation because its mathematical advantages, and because it provides the basis for many other survival analysis statistical tools.

3.2 Comparison of two samples

Another class of statistical tools for univariate data are techniques for comparing two or more samples. Two-sample tests give the significance level measuring whether two datasets are drawn from the same parent distribution. The mathematical challenge has been to formulate statistics with calculable distributions that converge to well-known tests (e.g. F-test for Gaussian data) in the limit of no censored data points. We have presented for astronomical use three nonparametric tests, usually called the Gehan, logrank, and Peto-Prentice test respectively. They are generalizations of the Wilcoxon and Savage nonparametric two- sample tests. The statistics are based on the number of data points in Sample 1 greater or less than each data point in Sample 2. The three tests differ in how they weight the censored points, and consequently have different sensitivities and efficiencies with different parent distributions and censoring patterns. The logrank test is more efficient at discriminating differences in two samples if the parent distribution is exponential and the differences occur at the censored end of the distribution. The Gehan test is more efficient with Gaussian distributions and at the uncensored end of the distribution. The Peto-Prentice test may be best when the censoring patterns differ significantly. All tests may give unreliable significance levels in very small or very heavily censored samples. Since the parent distributions and censoring patterns in astronomical samples are often uncertain, we recommend astronomers apply all three tests and judge any discrepancies in significance levels cautiously. Application to the two samples shown in Figure 1 indicate the probability they are drawn from the same parent distribution is extremely low (P

$\ll 10^{-3}$), even though the distribution of detected objects do not differ greatly.

3.3 Correlation between two variables

Many astronomical problems involve comparing two or more properties of an observed sample of objects, one or more of which may contain nondetections. While truly multivariate survival analysis techniques are not available, a variety of bivariate methods have been developed. We first present two correlation coefficients, which give the significant levels that two measured properties are uncorrelated with each other. One is based on Cox's (1972) proportional hazards model, known as Cox regression. It is extensively used to quantify relations in biometrical survival analysis studies, and is available in many commercial software packages. A simple version of Cox regression can be used to test the null hypothesis that no correlation is present between a censored dependent and an uncensored independent variable. The statistic is based on the measured values of the independent variable and the censoring status of the dependent variable. The significance levels are strictly valid only when the proportional hazards model applies, which may not be true in all astronomical samples.

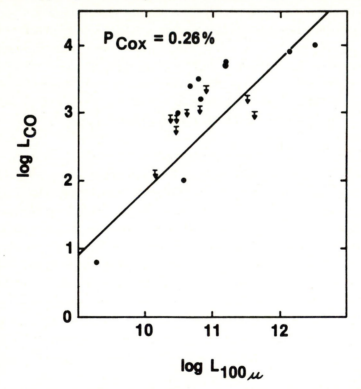

Figure 2. *The empirical relationship between the luminosity of carbon monoxide molecular gas, measured with a radio telescope, and the far-infrared luminosity, measured with the* Infrared Astronomical Satellite, *in a sample of infrared-detected spiral galaxies observed by Young* et al. *(1984). Figure from Isobe, Feigelson and Nelson (1986).*

A second correlation test is a generalization of Kendall's τ rank correlation coefficient described by Brown, Korwar and Hollander (1974) and Oakes (1982). While not widely used in biometrics, it is fully nonparametric and permits censoring in both the dependent and independent variables. Its principal difficulty is computational; calculating the significance level requires $O(N^3)$ operations and becomes prohibitively consuming of computer time when N exceeds a few hundred points. Figure 2 shows an application to an astronomical problem; Cox's regression and the generalized Kendall's τ give P = 0.26% and 0.36% respectively for the significance level of independence.

3.4 Linear Regression

Several similar least squares linear regression methods under the assumption that the data points are normally distributed about the line were independently developed by industrial statisticians (called 'iterative least squares'; Nelson and Hahn 1972), mathematical statisticians (called the 'EM Algorithm'; Dempster, Laird and Rubin 1977; Wolynetz 1979), and astronomers (called 'detections and bounds' regression; Avni and Tananbaum 1986). Though the implementation and error analysis procedures differ, all involve maximizing the likelihood with the nondetections weighted by their distance from a trial regression fit. Buckley and James (1979) suggest a less restrictive regression, in which the distribution of residuals from a trial fit is taken to be the Kaplan-Meier estimator derived from the data points. The Buckley-James fit in Figure 2 gives the line $L_{CO} = -7.73 + (0.96 \pm 0.21)L_{IR}$; the fully parametric EM algorithm gives almost the same result with slope (0.98 ± 0.21).

Astrophysicist Schmitt (1985) contributed a new linear regression procedure that, like the generalized Kendall's τ described above, allows censoring in both the independent and dependent variables. The X-Y plane is divided into a grid and a two-dimensional Kaplan-Meier estimator is calculated by redistributing censored points in perpendicular directions simultaneously. The regression line is then fit by least squares to the grid of Kaplan-Meier densities. The method has the advantage, important in many astronomical applications, of dealing with doubly-censored data. But it suffers two limitations. First, error analysis based on analytical formulae or derivitives of a likelihood has not been derived for this situation; Schmitt suggests instead performing bootstrap simulations of the data set. Second, the user must arbitrarily specify the grid spacing. If the grid is too coarse, the regression line is inaccurate; if it is too fine, the two-dimensional Kaplan-Meier estimator becomes ill-defined.

4 COMMENTS ON SURVIVAL ANALYSIS IN ASTRONOMY

Given the fundamental similarity between problems frequently arising in multi-wavelength astronomy, and those encountered in biostatistical and industrial reliability contexts, the use of survival analysis methods for astronomy seems desirable. Survival analysis is not by itself capable of recovering information lost through truncation; *i.e.*, sample selection effects. But it does provide mathematically rigorous and optimized procedures for recovering from upper limits in surveys of previously defined samples.

There is little doubt that using survival analysis generally gives better results than previous treatments based on detected fractions or detected objects only. The fear that survival analysis somehow gives 'something from nothing' and overinterprets the data is not well founded. A good statistical method is one that optimally uses what information is available, and the failure to detect an object is indeed meaningful information. Survival analysis methods generally have the excellent characteristics of converging on standard results when the censoring fraction approaches zero, and giving increasingly weak results (*e.g.*, poor significance levels in two-sample and correlation tests; large uncertainties in regression coefficients) as the censoring fraction increases.

However, astronomers MUST carefully consider whether a given experimental situation or dataset satisfies the mathematical assumptions of a given survival analysis method. Magri *et al.* (1989) provide a good discussion of the applicability of random censoring to a specific astronomical dataset. Though it is difficult to generalize, the performances of survival analysis statistics tend to deteriorate under the following circumstances: very high censoring fractions (exceeding, say, 60-70%); non-random censoring patterns, particularly when censoring is correlated with the variables of interest; significant fraction of identical values (ties); very small samples ($N \leq 20$). The results under these circumstances are not necessarily wrong, but do bear close scrutiny.

Isobe, Feigelson and Nelson (1986) have tested bivariate survival analysis procedures for a typical astronomical situation: comparison of the luminosities at two wavebands of a sample of objects distributed uniformly in space. Simulated parent populations of objects with a power law luminosity function were numerically constructed and randomly located in space. A sample was then obtained by simulating a flux-limited survey in one wave band; this is the truncation step and results in samples with a clear 'Malmquist bias'. Luminosities at a second waveband where then synthesized and 'observed" by a telescope with different sensitivities. When the luminosities were constructed to be uncorrelated, survival analysis correlation and regression procedures correctly reported no correlation, though standard methods using detected points along found a spurious correlation (Figure 3). When a linear relation between the luminosities was simulated, survival analysis recovered the intrinsic relation. As expected, the derived regression coefficients were increasingly inaccurate as the censoring fraction increased, but the calculated regression uncertainties commensurately. Cox regression, normal (EM algorithm) and semi-nonparametric (Buckley-James)

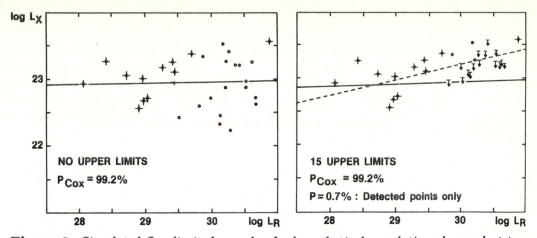

Figure 3. *Simulated flux-limited sample of a hypothetical population observed at two wavebands. The luminosities are assumed to be uncorrelated. The plusses are nearby objects and filled circles are distant objects. Left panel: All objects are detected. Right panel: Half of the sample is censored at the second waveband. Dashed line shows spurious correlation if nondetections are omitted. From Isobe, Feigelson and Nelson (1986).*

linear regressions thus performed with considerable success in this astronomical simulation, even though the data and censoring patterns were not fully random.

5 DIRECTIONS FOR FUTURE RESEARCH
Since survival analysis methods were developed primarily for applications in other fields, a number of limitations and deficiencies are notable when astronomical problems are considered. Following are five areas where further research is needed. Work on some of these problems is being pursued at Penn State University, but the involvement of other statisticians and astronomers is strongly encouraged.

• Perhaps the most distinctive difference of astronomical censored samples is that the censoring levels are usually not known precisely, as assumed in standard survival analysis. They usually arise due to the failure to detect a signal in the astronomical detector. Typically the detection limit is defined to be some multiple (e.g. 3 or 5) of the root mean square level of a Gaussian noise distribution in the detector. In recent years, photon counting detectors have been developed and the underlying distribution of signal and noise is assumed to be Poisson. The problem is that the scientist must subjectively decide the confidence limit at which the censored value is reported. In biometrical or industrial applications, there is little or no subjectivity in determining when the patient dies or product fails. Thus, astronomical survival analysis methods that fully account for the parametric form of the noise is needed. Since the noise is also present in the detections, the methods should also include measurement errors in the uncensored values. Several astronomers have recently addressed this problem. Turner (described in Wardle and Knapp 1986) attempted an analogue of the Kaplan-Meier estimator with these properties, but it is not self-consistent. Walsh *et al.* (1989) develop a procedure that uses measurement uncertainties to "smooth" the two-dimensional Kaplan-Meier distribution in bivariate problems. Marshall (1990) develops a promising formalism based on the Poisson model. Existing parametric models from the industrial reliability field may also be relevant.

• Practitioners of survival analysis must often deal with the possibility that the underlying distribution and/or censoring pattern of a dataset are not consistent with the assumptions of a given statistical method. Astronomers, however, in some situations have prior knowledge of these distributions. The distribution function of normal star luminosities, for example, can be approximated as a truncated power law relation of the form $N(L_1 \leq L \leq L_2) \propto L^{-3.5}$, and more complicated tabulated distributions are available. The luminosity function of normal galaxies appears to follow a Schechter (1976) function of the form $N(L_1 \leq L \leq L_2) \propto (L/L_*)^\alpha \, exp(-L/L_*)$. Censoring patterns are also sometimes predictable. The flux sensitivity of a given observation at the telescope is always known, and the relation between flux and luminosity is usually known (errors in this relation are due to uncertain interstellar absorption or cosmological model). Thus, analytical predictions and numerical simulations similar to those in Figure 3 can be performed to test or develop statistical tools most appropriate for a given astronomical problem.

• Related to this point, simulations of realistic astronomical surveys and survival analysis tests can lead to the design of maximally efficient astronomical surveys. Major surveys with censored results can occupy significant fractions of the available time on large ground- based or satellite-borne telescopes, with an effective cost of millions of dollars to national research agencies. Planning tools could be developed that indicate, for example, what ratio of observing time should be spent on distant compared to nearby objects in order to most accurately constraint the Kaplan-Meier estimator of the luminosity function. Unlike scientists in the social or biological sciences, astronomers have no tradition in quantitative experimental design.

• Most survival analysis techniques treat situations in which only one variable is censored in only one direction. Astronomers are increasingly confronting problems in which a sample of objects is observed at several wavelengths or properties, any of which could be censored. The *Infrared Astronomical Satellite* active in the early 1980's, for example, measured the luminosities of thousands of galaxies with high detection rates in the 60μ band, but with increasing censored fractions in the 100μ, 25μ and 12μ bands. Scientific problems require knowledge of the infrared colors of various galaxy samples, measured by the ratio of fluxes in the four bands. These ratio variables will thus have right-censored, left-censored and doubly-censored data points in a single sample. Bivariate and multivariate plots of the data will have censoring in all variables. More effective correlation and regression techniques are needed to treat such situations. The ultimate desiderata is a fully multivariate survival analysis. A principal component analysis permitting censoring in any variable in any direction, for example, would be needed to attain a complete statistical model of the multiwavelength datasets that are being compiled in the 1980-2000 era.

• Finally, any advanced methods appropriate for astronomical data analysis must be meaningfully communicated to the practicing astronomer. The great majority of astronomers have no formal education in statistics, do not read statistical journals, and are unfamiliar with the major software packages such as SAS and BMDP. Yet they have considerable need for sophisticated statistical techniques. This poor situation can be rectified by statisticians providing clear guidelines and software to the astronomical community, and astronomers making stronger efforts to learn from the statistical community.

Acknowlegements: The author would like to thank colleagues Paul I. Nelson (Dept. of Statistics, Kansas State University) and especially Takashi Isobe (Center for Space Research, Massachusetts Institute of Technology), without whom this work would not have been accomplished. Michael Akritas and G. Jogesh Babu (Dept. of Statistics, Penn State University) provided helpful comments on the manuscript. This work was supported under NASA's IRAS Data Analysis Program funded through the Jet Propulsion Laboratory, and by a National Science Foundation Presidential Young Investigator Award with matching funds from TRW, Inc., and Sun Microsystems, Inc.

REFERENCES

Auriemma, C., Perola, G.C., Ekers, R., Fanti, R., Lari, C., Jaffe, W.J., and Ulrich, M.H., 1977, *Astron. Astrophys.*, **57**, 41.

Avni, Y., Soltan, A., Tananbaum, H., and Zamorani, G., 1980, *Astrophys. J.*, **238**, 800.

Avni, Y. and Tananbaum, H., 1986, *Astrophys. J.*, **305**, 57.

Bevington, P.R., 1969, *Data Reduction and Error Analysis for the Physical Sciences*, N.Y.:McGraw-Hill.

Brown, B.W.M., Hollander, M., and Korwar, R.M., 1974, in *Reliability and Biometry*, ed. F. Proschan and R.J. Serfling (Philadelphia:SIAM), p. 327.

Buckley, J. and James, I., 1979, *Biometrika*, **66**, 429.

Cox, D.R., 1972, *J. Roy. Stat. Soc. B*, **34**, 187.

David, H.A. and Moeschberger, M.L., 1978, *The Theory of Competing Risks*, N.Y.:Macmillan.

Dempster, A.P., Laird, N.M. and Rubin, D.B., 1977, *J. Roy. Stat. Soc. B*, **39**, 1.

Efron, B., 1967, *Proc. 5th Berkeley Symp. Mathematical Statistics*, **4**, 831.

Feigelson, E.D. and Nelson, P.I., 1985, *Astrophys. J.*, **293**, 192.

Hummel, E., 1981, *Astron. Astrophys.*, **93**, 93.

Isobe, T., 1989, Ph.D. dissertation, Penn State University.

Isobe, T., Feigelson, E.D. and Nelson, P.I., 1986, *Astrophys. J.*, **306**, 490.

Isobe, T. and Feigelson, E.D., 1987, *ASURV:A Software Package for Statistical Analysis of Astronomical Data with Upper Limits*, Rev. 0.0. Available from authors.

Kalbfleisch, J.D. and Prentice, R.L., 1980, *The Statistical Analysis of Failure Time Data*, N.Y.:Wiley.

Kaplan, E.L. and Meier, P., 1958, *J. Am. Stat. Assoc.*, **53**, 457.

Lawless, J.F., 1982, *Statistical Models and Methods for Lifetime Data*, N.Y.:Wiley.

Magri, C., Haynes, M.P., Forman, W., Jones, C. and Giovanelli, R., 1988, *Astrophys. J.*, **333**, 136.

Marshall, H., 1990, This volume.

Miller, R.G. Jr., 1981, *Survival Analysis*, N.Y.:Wiley.

Nelson, W. and Hahn, G.J., 1972, *Technometrics*, **14**, 277.

Oakes, D., 1982, *Biometrics*, **38**, 451.

Pfleiderer, R.A., 1976, *Mitteil. Astron. Ges. Hamburg*, **40**, 201.

Schechter, P., 1976, *Astrophys. J.*, **203**, 297.

Schmitt, J.H.M.M., 1985, *Astrophys. J.*, **293**, 178.

Trumpler, R.J. and Weaver, H.F., 1953, *Statistical Astronomy*, Berkeley:U. Cal. Press.

Tytler, D., 1982, *Nature*, **298**, 427.

Wagner, A.E. and Meeker, W.Q. Jr., 1985, in *Proc. Stat. Computing Section, Amer. Stat. Assoc*, 441.

Walsh, D.E.P., Knapp, G.R., Wrobel, J.M. and Kim, D.-W., 1989, *Astrophys. J.*, **337**, 209.

Wardle, M. and Knapp, G.R., 1986, *Astron. J.*, **91**, 23.

Wolynetz, M.S., 1979, *Applied Stat.*, **28**, 185.

Young, J.S., Kenney, J., Load, S.D. and Schloerb, F.P., 1984, *Astrophys. J. Lett.*, **287**, L65..

Zamorani, G., et al., 1981, *Astrophys. J.*, **245**, 357.

DISCUSSION

J.D. Scargle to J. Pfleiderer: Two technical points: (1) Entropy does have the proper dependence on scale, contrary to what you stated. If the variable is discrete, then the entropy of its distribution can be easily seen to be scale invariant. For a continuous process, a scale change by a factor of S produces an entropy change of $-\log S$, but the entropy of a continuous process in infinite – so here too entropy is scale invariant. (2) Translation invariance does not imply that the quantity depend only on coordinate differences. For example, as you pointed out, entropy – because it is an integral over the whole space – is translation invariant but does not depend on coordinate differences.

(1) The vertical scale in entropy as used in MEM is fixed by the condition that the normalized pixel sum $\sum p$ equals unity. MEM never uses a continuous process. I was not discussing entropy in general. (2) I meant that dependence on coordinates themselves (not only on differences) contradicts translational invariance. In this respect, no dependence on any coordinates does not contradict and is thus a special case of what is needed.

J.D. Scargle to J. Pfleiderer: Your discussion of a new measure of structure is very interesting, but I think it is unfair to compare it to entropy and information in the way you did. Entropy is *not* a measure of structure – for example, it is not changed if the pixels are scrambled. Your quantity is like a 2-dimensional entropy. David Donoho and I are currently working on a chaos analysis technique which makes use of an analogous quantity.

MEM deconvolution claims to give the most unbiased recovery of convolved structure but, it seems to me, for that it should have more of a true measure of structure.

E. Grafarend to J. Pfleiderer: (1) In the statistical literature (C.R. Rao, S.K. Mitra, A. Bjerhammar, ...) there are clear *analytical* definitions of *bias* and *unbiasedness* I have missed in your presentation. (2) Your have based your discussion verbally on *Information Theory*, e.g. N. Wiener's notion of *Entropy*. I just wanted to mention that *maximum entropy* estimations coincide with "best" or "*minimum variance*" estimators (the classical estimators of Point Estimation Theory) in mathematical statistics in case that your events enjoy Gaussian probability distributions or "exponential-type probability distributions". Thus there is a strong relation between N. Wiener's Information Theory and Standard Estimation Theory.

(1) It is my strong opinion that bias has an essential qualitative component that is not well covered by an analytic definition. (2) The notion of entropy I used is that of one group of maximum entropy people. Wiener was not restricted in this way. Entropy is a powerful tool but not the most non-biased one in the recovery of structure.

M. Kurtz to J. Pfleiderer: I did not understand how you would use your method to learn something from the CfA Slice.

The structural information in these data as a function of pixel size and spatial correlation width and form might (or, perhaps, might not – I do not know yet) contribute to our knowledge on what the data might mean.

S. Bhavsar to J. Pfleiderer: An additional comment to the one just made on the CfA slice – J.R. Gott and his group are looking at the topology of the distribution of this data, by smoothing it on even larger scales. The point is that raw data can have lots of information, in this case smoothing it on larger scales gives us information on the topology of the initial fluctuations of matter in the universe.

It is useful to look at as many aspects of these data as possible, with as many methods as possible. Every result will give some hint how to understand the data.

T. Kiang to M. Hendry: The luminosity function of galaxies is certainly non-Gaussian, it is exponentially increasing, with an undefined mean. Hence work based on Malmquist bias is not likely to be useful. On the other hand, even for an exponential luminosity function, the luminosity function in an apparent magnitude-limited sample will be one-peaked and will have well-defined means and standard deviations. Perhaps one should concentrate on such apparent magnitude samples and making as little assumption as possible regarding the (space) luminosity function.

Your comment is a criticism which I would accept if one attempts to apply this work to the general luminosity function of galaxies of all types. I must say, however, that if one considers only galaxies of a particular morphological type – as in Rubin's ScI sample – then in many situations a Gaussian has been found to be a *good* approximation to the observations. In any case a Gaussian has proved useful as a means of illustration, even for the general luminosity function, although we would hope to explore other analytic forms (and non-parametric methods) in the near future.

E. Grafarend to M. Hendry: Your have shown beautifully the *dual role* of (minimum) bias and (minimum) mean square errors for various estimators. Even for a linear fixed effects model there are minimum variance, biased estimators which are *better* than BLUE ("better" being measured in terms of variance of the estimators). Would you have some preference for some of *your* estimators in this *multicriteria optimality problem*?

I would certainly agree that the choice of "best" estimator must take into account this "dual role" of minimum bias and variance. I think that which property is regarded as more important must ultimately depend on the context. I might add, however, that the question of whether one should use the mean or median bias seems to us to be an open one. In this work one has a different conditional distribution at each distance, r, so one cannot consider one's sample of galaxies – however large – as drawn from the same distribution. How meaningful therefore is the bias of the "*expected value*" of our estimator, when only one evaluation is made?

M. Kurtz (comment): If one restricts oneself to using only gE or spiral galaxies, according to Bingalli et al. (Virgo cluster data) the luminosity function is nearly Gaussian.

SPECIAL TOPICS

Directional Statistics in Astronomy

P.E. JUPP

University of St. Andrews

SUMMARY

Positional astronomy is concerned with points on the celestial sphere. For statistical analysis of such points, standard linear methods can be inadequate and the special techniques of *directional statistics* must be employed. This paper aims to give an introduction to the topic of directional statistics and to show how it can be useful in astronomy. Both formal parametric inference and informal graphical methods are considered. Astronomical illustrations include applications to cometary orbits, binary systems and double radio sources.

1 INTRODUCTION

In various astronomical contexts distances are unknown or irrelevant and the data of interest are *directions,* i.e., points on the celestial sphere. If these points are concentrated in a small area then there is negligible error in using the customary linear statistical methods. However, if the points cover an appreciable area of the celestial sphere then naive application of standard linear methods is inadequate and the techniques of *directional statistics* are required.

The topic of directional statistics is concerned with observations which are not the familiar counts, real numbers or (unrestricted) vectors but instead are typically either directions in 2- or 3-dimensional space or rotations of such a space. A direction can be regarded as a unit vector, i.e., a vector $\mathbf{x}$ satisfying $\mathbf{x}.\mathbf{x} = 1$, or equivalently as a point on the sphere of unit radius. Given a system of spherical polar coordinates (θ, ϕ), where θ denotes colatitude and ϕ denotes longitude, we have (regarding $\mathbf{x}$ as a column vector)

$x = (\sin\theta \cos\phi, \sin\theta \sin\phi, \cos\theta)^T$.

The aim of this paper is to give an introduction to the topic of directional statistics and to show how it can be useful in astronomy. An excellent guide to current methods for the practical analysis of spherical data (data on the sphere) is the recent book by Fisher, Lewis & Embleton (1987). See also Chapter 10 of Upton and Fingleton (1989). The underlying theory can be found in Chapters 8 and 9 of Mardia (1972). Some material on practical data analysis on the sphere is given in Chapter 1 of Watson's (1983) book but the later chapters are very mathematical. A comprehensive review of the theory of directional statistics is given in Jupp & Mardia (1989).

2 INSPECTING AND SUMMARISING DATA

2.1 Plots

Perhaps the best way of getting a feel for any data set is to plot it. Spherical data can be presented in the form of an equal-area plot obtained by equal-area projection of the sphere onto the plane. One way of obtaining such plots is to map the point with spherical

coordinates (θ, ϕ) to the point in the plane with polar coordinates (r, ϕ), where

$r = 2\sin(\theta/2)$ for $0 \le \theta \le \pi$ (with angles measured in radians). Alternatively, this projection can be applied separately to the upper and lower hemispheres.

To bring out the important features of data, contour plots can be helpful. These show contours of an estimate of the probability density thought to have generated the data. A convenient computer program for producing contour plots has been written by Diggle & Fisher (1985).

Example 1 Binary systems

Figure 1 is an equal-area plot of the positions of 206 binary systems (those of quality "a" or "b" in the catalogue of Batten, Fletcher & Mann, 1978). It is clear that these positions are denser in the northern hemisphere (corresponding to the inner disc) and so are not spread uniformly over the celestial sphere. □

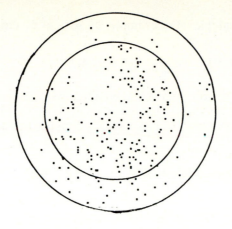

Figure 1 Equal area plot of binary systems data of Example 1.   The centre is at the north celestial pole and the boundary circle is to be identified with the south pole.

2.2 Summary Statistics

If observations $x_1, \ldots, x_n$ on the unit sphere are given, they can be summarised by their vector mean (or "centre of mass"), $\bar{x} = n^{-1} \sum_{i=1}^{n} x_i$. It is convenient to express

x in polar form as $\bar{x} = \bar{R}\,\bar{x}_0$ where $\bar{R} \geq 0$ and $\bar{x}_0$ is on the unit sphere.   Then

$$\bar{R} = \|\bar{x}\| \qquad\qquad\qquad \bar{x}_0 = \|\bar{x}\|^{-1}\bar{x}, \qquad\qquad (1)$$

where $\|\bar{x}\|$ denotes the length of the vector $\bar{x}$. $\bar{R}$ is called the *mean resultant length* and satisfies $0 \leq \bar{R} \leq 1$; $\bar{x}_0$ is called the *mean direction*.   In directional statistics $\bar{x}_0$ plays the role of sample mean and measures the location of the sample, whereas $1-\bar{R}$ gives a measure of its concentration: $\bar{R} = 1$ occurs only when $x_1 = \ldots = x_n$, whereas $\bar{R} = 0$ arises when $x_1, \ldots, x_n$ are dispersed "uniformly" on the sphere.

Another useful summary of spherical data is the "moment of inertia" or second moment matrix

$$M_n = n^{-1} \sum_{i=1}^{n} x_i x_i^T.$$

The eigenvalues τ_1, τ_2, τ_3 of M_n and the corresponding eigenvectors u_1, u_2, u_3 give useful information about the sample. For example, $\tau_1 \approx \tau_2 \approx \tau_3$ suggests that the sample

is close to uniform; $\tau_3 \approx \tau_2 > \tau_1$ indicates that the sample is concentrated near a great

circle normal to $\mathbf{u}_1$ with approximate symmetry about $\mathbf{u}_1$, whereas if $\tau_3 > \tau_2 \approx \tau_1$ and

$\bar{R} \approx 0$ then the sample is clustered near $\pm\mathbf{u}_3$ and is fairly symmetrical about the $\mathbf{u}_3$-axis.

3 PROBABILITY DISTRIBUTIONS ON THE SPHERE

A standard method of drawing conclusions from a sample of data is to assume that the data form a random sample from a *population* of possible values, having a probability distribution in some specified parametric family. For spherical data, distributions of random unit vectors $\mathbf{x}$ are required. Replacing $\bar{\mathbf{x}}$ in (1) by the expected value $E[\mathbf{x}]$ of the random unit vector $\mathbf{x}$ leads to the population analogues of $\bar{R}$ and $\bar{\mathbf{x}}_0$. These are the *mean*

resultant length ρ and the *mean direction* μ defined by

$$\rho = \|E[\mathbf{x}]\| \quad \text{and} \quad \mu = \|E[\mathbf{x}]\|^{-1}E[\mathbf{x}].$$

3.1 The Uniform Distribution

The simplest probability distribution on the sphere is the *uniform distribution*. This is the unique distribution which is isotropic, i.e., invariant under rotations, in the sense that the random vector $\mathbf{x}$ has the same distribution as $\mathbf{U}\mathbf{x}$, for any fixed rotation matrix $\mathbf{U}$. With

respect to spherical polar coordinates (θ, ϕ) the uniform distribution has probability density function

$$f(\theta, \phi) = (4\pi)^{-1}\sin \theta.$$

Thus, for any subset A of the sphere,

$$P(\mathbf{x} \in A) = (4\pi)^{-1}\iint_A \sin \theta \, d\theta \, d\phi = (4\pi)^{-1}\iint_A dS,$$

where $dS = \sin \theta \, d\theta \, d\phi$ is the element of area. Then the probability that the random

vector $\mathbf{x}$ lies in A is proportional to the area of A. For the uniform distribution $\rho = 0$ and

μ is indeterminate.

3.2 The Fisher Distribution

Because of its mathematical convenience, the most widely-used distribution on the sphere is the Fisher distribution. This has probability density functions

$$f(x; \mu, \kappa) = \kappa(\sinh \kappa)^{-1}\exp\{\kappa\mu.x\} \tag{2}$$

with respect to the uniform distribution, where $\kappa \geq 0$ and $\|\mu\| = 1$. Then

$$P(x \in A) = (4\pi)^{-1}\iint_A f(x; \mu, \kappa) \, dS(x) = (4\pi)^{-1}\iint_A f(x; \mu, \kappa) \sin \theta \, d\theta \, d\varphi.$$

The parameter κ is called the *concentration* and μ is the mean direction. The distribution has a mode at μ and is symmetrical about the axis through μ. Taking $\kappa = 0$ in (2) gives the uniform distribution. As $\kappa \to \infty$, the Fisher distribution (2) becomes more and more concentrated around μ.

If spherical polar coordinates (θ, ϕ) are chosen such that μ is given by $\theta = 0$, then

ϕ and θ are independent,

ϕ is distributed uniformly on $(0, 2\pi)$,

$2\kappa(1 - \cos \theta) \sim \chi^2_2$ approximately for large κ,

where χ^2_2 denotes the chi-squared distribution with 2 degrees of freedom.   For large κ the random vector x is close to μ and we can regard x as having approximately a bivariate normal distribution

$$\kappa^{1/2} (\eta - \mu) \sim N(0, I - \mu\mu^T)$$

in the tangent plane at μ to the sphere and standard multivariate statistical methods can be used. These approximations are acceptable for $\kappa > 5$.

4 HYPOTHESIS TESTING AND ESTIMATION

4.1 Testing uniformity

A simple and intuitively appealing test for uniformity of a spherical distribution is the Rayleigh test. This rejects uniformity for large values of the mean resultant length $\bar{R}$. Critical values of $\bar{R}$ for small sample sizes can be found in Mardia (1972) and Fisher, Lewis & Embleton (1987). If the sample size, n, is large, the limiting distribution

$$3n\bar{R}^2 \sim \chi^2_3 \qquad\qquad n\to\infty$$

can be used.

Example 2 Planetary orbits

In 1734 D. Bernoulli asked in his essay on the orbits of the planets whether the similarity of the planets' orbits could have arisen by chance. For the 9 planets now known the mean resultant length of the directed unit normals is $\bar{R} = 0.79$. As the 1% critical value of $\bar{R}$ for n=9 is 0.624, the Rayleigh test provides strong evidence that the orbital planes do not form a random sample from the uniform distribution.   □

Example 3

For the data on 206 binary systems in Example 1 we find that $3n\bar{R}^2 = 86.2$. As $P(\chi^2_3 > 86.2) < 10^{-3}$, uniformity of the positions is strongly rejected.   □

Example 4 Cometary orbits.

For the perihelion directions of the 240 comets with parabolic and nearly parabolic orbits listed in Marsden's (1982) catalogue the value of $3n\bar{R}^2$ is 213.0. As

$P(\chi^2_3 > 213.0) < 10^{-3}$, uniformity of the perihelion directions is strongly rejected. This data set is considered in more detail in Example 8. Equal-area plots and probability plots (see Section 5) of various other data sets concerning cometary orbits are given by Watson (1983, pp. 28-35).   □

Many other tests of uniformity are available, appropriate for a variety of alternative hypotheses. References to these can be found in Section 5.6 of Fisher, Lewis & Embleton (1987).

4.2 Estimation and Testing

For a sample $x_1, \ldots, x_n$ from the Fisher family (2), the values of μ and κ can be estimated by their maximum likelihood estimates $\hat{\mu}$ and $\hat{\kappa}$ which are given by

$$\hat{\mu} = \|\bar{x}\|^{-1}\bar{x} \qquad \qquad \coth\hat{\kappa} - \hat{\kappa}^{-1} = \bar{R},$$

where $\bar{x}$ and $\bar{R}$ are respectively the sample mean and the mean resultant length of the observations. For $\bar{R} \approx 1$ the approximation $\hat{\kappa} \approx (1-\bar{R})^{-1}$ can be used.

Example 5

For the data on binary systems in Example 1 we have $\hat{\kappa} = 1.23$ and $\hat{\mu} = (0.28, -0.35, 0.89)$, corresponding to $\hat{\alpha} = 22\text{h } 36\text{m}$, $\hat{\delta} = 64°$. □

It is useful to have not only an estimate of the mean direction μ but also some idea of the accuracy of this estimate. This is provided by *confidence cones* of the form $\{\xi \mid \hat{\mu}.\xi > c_\alpha\}$. The constant c_α is chosen so that, for confidence cones constructed from many independent samples, $100(1-\alpha)\%$ of them would contain μ. Further, if μ_0 is a given hypothetical population mean direction we can test $H_0: \mu = \mu_0$ by rejecting H_0 if μ_0 lies outside the confidence cone. For details see Mardia (1972, pp. 260-261) and Fisher & Lewis (1983, pp. 131-134).

4.3 Large-sample Inference

For large samples it is possible to make inference without assuming that the data come from a Fisher distribution. For example, if it is assumed only that the underlying distribution is symmetrical about some axis, then an approximate $100(1-\alpha)\%$ confidence cone for μ is

$$\left\{ \xi \mid \hat{\mu}.\xi > 1 - \frac{[1-n^{-1}\sum(\hat{\mu}.x_i)^2]\chi^2_{2;\alpha}}{(4n\bar{R}^2)} \right\},$$

where $\hat{\mu} = \|\bar{x}\|^{-1}\bar{x}$ and $\chi^2_{2;\alpha}$ denotes the upper $\alpha\%$ point of the χ^2_2 distribution. See

Watson (1983, pp 138-140).

Example 6

For the data on 206 binary systems in Example 1 the 95% confidence cone for μ is

$\{\xi \mid \hat{\mu}.\xi > 0.968\}$. If, for example, we were to take μ_0 to be the direction of the galactic

pole, we would have $\hat{\mu}.\mu_0 = 0.496$ and so would reject the hypothesis that the mean

direction of the binary systems is μ_0.   □

5 INFORMAL METHODS

It is good statistical practice to check that the fitted distribution does indeed provide a good
fit to the data and in particular to look for *outliers*, unusual observations which look as if
they may not belong to the fitted model. Although such checks can be carried out by
formal parametric methods, it is convenient and often more informative to use instead
informal graphical methods, such as probability plots.

Probability plots for spherical data can be produced as follows. Given a sample of n

points on the unit sphere, choose new spherical polar coordinates (θ', ϕ') with the sample

mean direction as pole $(\theta' = 0)$, and let $\theta'_{(1)} < \ldots < \theta'_{(n)}$ be the observed values of θ'
placed in increasing order. It is necessary to choose also another set of spherical polar

coordinates (θ'', ϕ'') such that the sample mean direction is at $(\pi/2, 0)$ and such that ϕ''

has range $[-\pi, \pi]$. Then the probability plots are

(i)   "colatitude plots" of $1 - \cos \theta'_{(i)}$ against $-\log\{1-(i-1/2)n^{-1}\}$,

(ii)   "longitude plots" of $\phi'_{(i)}$ against $(i-1/2)n^{-1}$,

(iii)   "two-variable plots" of the ordered values of $\phi''_i(\sin \theta''_i)^{1/2}$ against quantiles of
        the standard normal distribution .

For a Fisher distribution these plots should be approximately linear with slopes
approximately κ^{-1}, 1, and $\kappa^{-1/2}$ respectively and should pass near the origin. (The

approximation is acceptable for $\kappa > 5$.) Thus they provide graphical tests of goodness of fit to a Fisher distribution, quick estimates of the concentration parameter, and a method of detecting outliers. These plots are useful for pointing out possible departures from the assumed statistical model; formal tests of goodness of fit can then be used to test the significance of such discrepancies between model and data. See Fisher, Lewis & Embleton (1987, pp. 117-125, 204, 208).

Example 7   Binary systems

Probability plots for the data from Example 1 on positions of binary systems are given in

Figure 2. Although the estimated value of 1.23 for κ is too small for the plots to be considered a reliable assessment of Fisherness, all three plots suggest that a Fisher

distribution with $\kappa \approx 1.4$ might give a rough fit to the data. However, the colatitude plot appears to detect too many points far from the sample mean direction and the longitude plot finds some asymmetry about this direction, whereas the two-variable plot indicates no

association between θ' and ϕ'. Formal tests confirm these impressions.   ☐

Figure 2 Probability plots for checking goodness-of-fit of the Fisher distribution to the binary systems data of Example 1.

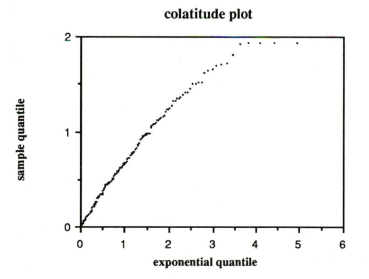

colatitude plot

sample quantile

exponential quantile

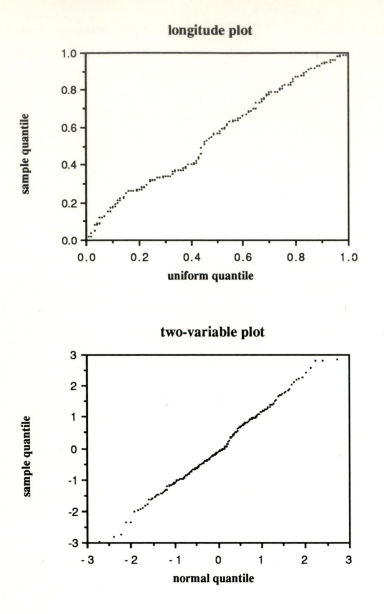

6 ROTATIONS

Sometimes each data point consists not of a single direction but of a pair (x_1, x_2) of orthogonal directions.   Then there is a unique direction x_3 such that the matrix $X = (x_1, x_2, x_3)$ satisfies $X^T X = I_3$, the 3x3 identity matrix, and $\det(X) = 1$, i.e., X represents the rotation taking the vectors $(1, 0, 0)^T$, $(0, 1, 0)^T$, $(0, 0, 1)^T$ into x_1, x_2, x_3, respectively.   The group of rotations of 3-space is denoted by SO(3).

The simplest probability distribution on SO(3) is *the uniform distribution,* the unique probability distribution on SO(3) which is invariant under rotations, i.e., the random rotation **X** has the same distribution as **UX**, for any fixed rotation **U**.   If **X** = (x_1, x_2, x_3) and **X** has the uniform distribution on SO(3) then each of x_1, x_2 and x_3 has the uniform distribution on the sphere.

Uniformity of rotations can be tested by a matrix version of the Rayleigh test. This rejects uniformity for large values of $\bar{R}$, where $\bar{R}^2 = \text{tr}(\bar{X}^T\bar{X})$ and $\bar{X} = n^{-1}\sum_{i=1}^{n} X_i$. If the sample size, n, is large, the approximation

$$3n\bar{R}^2 \sim \chi^2_9 \qquad\qquad n\to\infty$$

can be used.

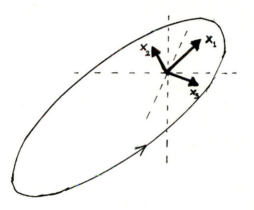

Figure 3 Rotation matrix (x_1, x_2, x_3) determined by a cometary orbit.

Example 8 Cometary orbits
Consider again the orbits of long-period comets discussed in Example 4.   Two natural questions are:
(i)   are the perihelion directions uniformly distributed?
(ii)   are the orbital planes of comets with a given perihelion direction distributed
       uniformly about that axis?

Taking x_1 as the perihelion direction, x_2 as the unit normal to the plane of the orbit (with direction specified by the sense of rotation) and x_3 to complete a right-handed orthonormal frame, the orbit of a comet can be represented by a rotation matrix $X = (x_1, x_2, x_3)$. See Figure 3. $\square$

The Rayleigh test statistic for uniformity of X decomposes as

$$3n \; tr(\overline{X}^T \overline{X}) = 3n \; tr(\overline{X}_1^T \overline{X}_1) + 3n \; \overline{x}_1.\overline{x}_1. \tag{3}$$

The two terms on the right hand side are Rayleigh test statistics for uniformity of $X_1 = (x_2, x_3)$ and x_1, respectively.   Under the null hypothesis, the large-sample distributions of the terms in (3) have limiting chi-squared distributions, namely

$$3n \; tr(\overline{X}^T \overline{X}) \sim \chi^2_9, \; 3n \; tr(\overline{X}_1^T \overline{X}_1) \sim \chi^2_6, \; 3n \; \overline{x}_1.\overline{x}_1 \sim \chi^2_3 \qquad n \to \infty.$$

For this data set of 240 long-period comets, we obtain $3n\overline{x}_1.\overline{x}_1 = 213.0$ and

$3n \; tr(\overline{X}_1^T\overline{X}_1) = 11.8$.   As $P(\chi^2_3 > 213.0) \approx 10^{-3}$, uniformity of perihelion directions is strongly rejected, as in Mardia (1975a, p. 363-365) and Jupp & Mardia (1979).   As

$P(\chi^2_6 > 11.8) > 0.05$, symmetry of orbital planes is accepted at the 5% level.   See Jupp & Spurr (1983). $\square$

7 DISTIBUTIONS ON THE CIRCLE

Sometimes each data point consists of a single angle, e.g., R.A., in the range $(0, 2\pi)$. Such an angle can be regarded as a point on a circle of unit radius.   The simplest distribution of such a point is the isotropic or *uniform distribution* for which all directions are equally likely.   The probability density function is

$$f(\theta) = (2\pi)^{-1} \qquad\qquad 0 \le \theta \le 2\pi.$$

The analogue on the circle of the Fisher distribution on the sphere is the von Mises distribution.   This has density function

$$f(\theta;\ \kappa,\ \mu_0) = \{2\pi I_0(\kappa)\}^{-1} \exp[\kappa \cos(\theta - \mu_0)\} \tag{4}$$

$$= \{2\pi I_0(\kappa)\}^{-1} \exp[\alpha \cos\theta + \beta \sin\theta)\}$$

where $I_0(\kappa)$ denotes the modified Bessel function of the first kind and order 0. The distribution is symmetrical about its mode at μ_0. The parameter κ measures concentration, as in the Fisher distribution.

8 REGRESSION

Sometimes measurements are made of pairs (x_1, x_2) of (not necessarily orthogonal) directions and interest lies in the way in which the distribution of x_2 depends on x_1. In statistical language, the problem is one of *regression* of x_2 on x_1 and a common question is whether or not x_1 and x_2 are independent.

Example 9 Asymmetry of double radio sources

A possible asymmetry of high luminosity double radio sources was reported by Birch (1982). Let x be the position of such a source and Δ the discrepancy between the position angles of the elongation and of the magnetic field vector. Birch noticed that Δ tends to be positive in one half of the sky and negative in the other half and remarked that this might be evidence for rotation of the universe. Kendall and Young (1984) gave a careful statistical analysis of this problem. See also Kendall (1984). Here is an outline of the paper of Kendall and Young.

First note a technicality: as the elongation is undirected, Δ is indistinguishable from $\Delta + \pi$. This ambiguity can be removed by considering 2Δ (since $2(\Delta + \pi) = 2\Delta + 2\pi$, which is equivalent to 2Δ).

If the distribution of Δ were symmetrical about 0 and did not depend on the position x, then it would be reasonable to model 2Δ by a von Mises distribution (4) with $\mu_0 = 0$, so that Δ would have density function

$$f(\Delta; \alpha) = \{2\pi I_0(\alpha)\}^{-1} \exp(\alpha \cos 2\Delta). \tag{5}$$

In order to model possible dependence of Δ on $\mathbf{x}$, model (5) can be generalised to

$$f(\mathbf{x}, \Delta; \alpha, \lambda) = g(\mathbf{x})\{I_0([\alpha^2 + (\lambda.\mathbf{x})^2]^{1/2})\}^{-1} \exp(\alpha \cos 2\Delta + \lambda.\mathbf{x} \sin 2\Delta). \tag{6}$$

The parameter λ describes the asymmetry and (5) is the special case of (6) with $\lambda = \mathbf{0}$.

Given observations $(\mathbf{x}_1, \Delta_1), \ldots, (\mathbf{x}_n, \Delta_n)$, a test of the hypothesis $H_0: \lambda = \mathbf{0}$ can be obtained from the general approach of likelihood ratio tests. For testing H_0 against the alternative $\lambda \neq \mathbf{0}$, the likelihood ratio statistic is

$$w = \frac{\max\left\{\prod_{i=1}^{n} f(\mathbf{x}_i, \Delta_i; \alpha, \lambda)\right\}}{\max\left\{\prod_{i=1}^{n} f(\mathbf{x}_i, \Delta_i; \alpha, \mathbf{0})\right\}},$$

where the maximum in the numerator is over all possible values of α and λ, while that in the denominator is over all values of α. The likelihood ratio test rejects H_0 for large values of w. If H_0 is true then the large-sample distribution of w is

$$2 \log w \sim \chi^2_3 \qquad\qquad n \to \infty. \tag{7}$$

For the data set of 134 sources considered by Kendall and Young, the observed value of $2 \log w$ is 16.04. As $P(\chi^2_3 > 16.04) \approx 10^{-3}$, H_0 can definitely be rejected, assuming that $n = 134$ is large enough for (7) to hold. That this approximation is valid here is one of the conclusions of the more detailed analysis given by Kendall and Young. Thus there is very strong evidence of asymmetry.

9 CONCLUDING REMARKS

Only the simplest aspects of directional statistics have been outlined here. Important distributions on the sphere other than the Fisher distribution include the Watson and Bingham familes which are discussed in Fisher, Lewis & Embleton (1987). Distributions on the sphere which are suitable for modelling observed directions for which the detection probability varies with declination were introduced by Mardia & Edwards (1982) and illustrated with data on cosmic rays. Rotations (such as those in Example 8) can be modelled using the matrix version of the Fisher distribution, introduced by Khatri & Mardia (1977). Directional versions have been developed of nearly all the usual statistical concepts and techniques, in particular, 2-sample methods, analysis of variance, correlation, curve fitting, analysis of residuals, etc.

Bootstrap methods play an important method in the analysis of spherical data (as indeed in many areas of statistical analysis). The basic idea of the bootstrap is to sample many times *without replacement* from the observed data points and to compute each time the value of the appropriate estimate or test statistic. This provides *from the data alone* an assesssment of the variability of the estimate or test statistic and means that (possibly unrealistic or intractable) parametric assumptions can be avoided. Details of bootstrap methods for spherical data can be found in Fisher, Lewis & Embleton (1987) and Fisher & Hall (1989).

It should be stressed that almost all statistical techniques rely on some underlying assumptions, in particular that the observations form a random sample from some population. The user should be aware of the problem of selection effects. See, e.g., Pensado and Lahulla (1976) who point out that the greater the inclination of a visual binary the less likely it is to be detected and to have well-determined orbits.

REFERENCES

Batten, A.H., Fletcher, J.M. & Mann, P.J. (1978). Seventh catalogue of the orbital elements of spectroscopic binary systems. *Publ. Dom. Astrophys.Obs.* **15**, no. 5, 121-295.

Birch, P. (1982). Is the universe rotating? *Nature* **298**, 451-454.

Diggle, P.J. & Fisher, N.I. (1985). SPHERE A contouring program for spherical data. *Computers and Geosciences* **11**, 725-766.

Fisher, N.I. & Hall, P. (1989). Bootstrap confidence regions for directional data. *J. Amer. Statist. Assoc.* (to appear).

Fisher, N.I., Lewis, T. & Embleton, B.J.J. (1987). *Statistical Analysis of Spherical Data.* Cambridge: Cambridge Univ. Press.

Jupp, P.E. & Mardia, K.V. (1979).   Maximum likelihood estimation for the matrix
    von Mises-Fisher and Bingham distributions. *Ann. Statist.* **7**, 599-606.

Jupp, P.E. & Mardia, K.V. (1989).   A unified view of the theory of directional statistics,
    1975-1988.   *Int. Statist. Rev.* **57** (to appear).

Jupp, P.E. & Spurr, B.D. (1983). Sobolev tests for symmetry of directional data.   *Ann.
    Statist.* **11**, 1225-1231.

Kendall, D. G. (1984).   Statistics, Geometry and the Cosmos. (The Milne Lecture,
    1983). *Q. J. R. astr. Soc.* **25**, 147-157.

Kendall, D. G. & Young, G.A.   (1984).   Indirectional statistics and the significance of an
    asymmetry discovered by Birch.   *Mon. Not. R. astr. Soc.* **207**, 637-647.

Khatri, C.G. & Mardia, K.V. (1977).   The von Mises-Fisher distribution in orientation
    statistics.   *J. R. Statist. Soc.* B **39**, 95-106.

Mardia, K.V. (1972). *Statistics of Directional Data.*, London: Academic Press.

Mardia, K.V. (1975a). Statistics of Directional Data (with discussion). *J. R. Statist. Soc.*B
    37, 349-393.

Mardia, K.V. & Edwards, R. (1982).   Weighted distributions and rotating caps.
    Biometrika **69**, 323-330.

Marsden, B.G.   (1982).   *Catalogue of Cometary Orbits,* 4th edition.   International
    Astronomical Union.

Pensado, J. and Lahulla, J.F.   (1976).   Orientación de los planos orbitales de las
    inclinaciones de las órbitas de las estrellas dobles visuales. *Bol. astr. observ.
    Madrid* **9**, 21-35.

Upton, G. J. G. and Fingleton, B.   (1989)   *Spatial Data Analysis by Example,
    Vol 2: Categorical and Directional Data.*   Wiley.

Watson, G.S. (1983).   *Statistics on Spheres.   University of Arkansas Lecture Notes
    in the Mathematical Sciences* 6,   New York: Wiley.

Department of Mathematical Sciences
University of St. Andrews
St. Andrews
Fife   KY16 9SS
U.K.

STATISTICAL SOFTWARE

F. MURTAGH[1]
Space Telescope – European Coordinating Facility
European Southern Observatory
Karl-Schwarzschild-Straße 2
D-8046 Garching bei München

and

Ph. NOBELIS
Département de Mathématique
Université de Strasbourg
11, rue de l'Université
F-67000 Strasbourg

1 INTRODUCTION

1.1 Commercial Packages

The statistical software market is a large and mature one. In Table 1, a number of more widely-used packages are listed, with prices and number of sites. This information must be treated with caution: firstly, the market-place is fast-moving, and this information dates from 1987; secondly, it is in the nature of such surveys that information is not homogeneous, – multiple licences at one site may or may not be repeatedly counted, costs generally vary with the power of the machine on which the package is installed, annual maintenance charges are not explicitly mentioned in most cases here, and so on. Table 1 is derived from Cable and Rowe (1987): this survey of statistical software is updated at approximately 4-yearly intervals, and is available at very little cost. Addresses of all vendors may also be found in this reference. Some other such surveys are referenced in Dallal (1988), and Beaumont et al. (1985) can be referred to for forecasting software.

In Table 1, we note the inclusion of IMSL and NAG which are essentially mathematical subroutine libraries. The general-purpose packages, SAS, SPSS, and BMDP, are widely available. SAS, which contains significant graphics capabilities, is well-established in the business world. SPSS's base was initially the social sciences, and BMDP's was biomedical computing, but these three packages have

---

[1]Affiliated to Space Science Dept., Astrophysics Div., European Space Agency.

Package	Number of Sites	Price (Please Confirm with Vendor!)
BMDP	N.A.	$1200–$3850 (mainframe)
		$495–$2100 (PC)
Genstat	360	From £990 (PCs)
		From £450 with annual licence (mainframes)
IMSL	3000	N.A.
Minitab	2500+	$1200–$2200 (mainframe)
		$900 (PC); Discounts available
NAG	1500+	£864–£1620; Discounts available
S	Approx. 300	$8000 (commercial/source licence)
		$400 (academic/one-time/source)
SAS	9000+	From £1875; Academic discounts available
SPSS/PC+	16000+	£760 (base) + £200–£340
		for each additional module (PC)
SPSS-X		£6900 + £3500/year (mainframe)
		Academic/non-profit discounts available
Statgraphics	20000	$795 (PC only)
SYSTAT	7500 + 40 site licences	$595

Table 1: Some statistical packages. Data valid circa February 1987.

come to be used across all disciplines. Genstat is more widely known in the agricultural world. Minitab is very popular for teaching purposes. The graphics and statistics language S will be further discussed below.

While it is a definite advantage for comparison purposes to have a comprehensive range of packages available for the general user,. and while additionally the contents of packages may well be complementary, it is nonetheless clear that many installations will not have all packages mentioned in this article. The situation in Strasbourg is sketched out in Box 1. A number of important French packages are surveyed (with details of where available). Among other packages discussed is RS/1 (Macon, 1988).

1.2 Astronomical Requirements

Are there particular astronomical requirements that are not catered for by current statistical software packages? Such a question is not easily answered. However, it is our contention that more influence could be exerted by astronomy on developments in the statistical software area, purely by increased use of – and preference for – such packages. Statistical software packages evolve both from the point of view of algorithms implemented in them; and from the point of view of the user interface and the environment presented to the user. The latter issue will be looked at further, below. In this section, we will be concerned with the former issue, – the contents of statistical packages.

Box 1: Statistical Software in Strasbourg

A number of different statistical software packages are available in different laboratories of the Université de Strasbourg. Their comprehensiveness and complementarity allow users and statisticians to apply a wide range of statistical methods to almost any kind of data. Among such packages, the most used are those available on the mainframe of the Centre de Calcul de Strasbourg-Cronenbourg (funded by the CNRS). Apart from the well-known SAS and BMDP, the other statistical packages, ADDAD[1], MODULAD[2] and SPAD[3], are mainly devoted to multivariate data analysis (ADDAD and MODULAD are program libraries rather than packages). The latter are written in Fortran and can be used in the SAS environment, taking advantage of the graphical capabilities of the latter. In addition to these packages, one should add the statistical programs contained in the different mathematical libraries of the Centre (IMSL, NAG, and others).

With the advent of PCs, the use of statistical packages has become easier. Every laboratory, however small, has a PC and, as soon as data has to be processed, acquires statistical software. Considering the multiplicity of PC packages, an informal group of users was created in the Université de Strasbourg, called GSMI, with the aim of helping researchers in their choice of statistical procedure. After a phase of extreme dispersion, there is currently stabilization around a few products, which are now presented briefly.

Beyond the micro-versions of SAS, BMDP and SPAD, the following commercial packages are currently accessible in Strasbourg: SYSTAT, SPSS, Statgraphics and RS/1[4]. The latter has a mainframe version under MVS and has all the capabilities of a spreadsheet editor, together with usual statistical functions (testing, anova, anacova, fitting, and so on). RS/1 contains a large, easily-used graphics module. It incorporates a programming language, which makes for its wealth and originality; this language is derived from PL/1 and C and allows one to extend the package's statistical procedures and adapt them to every problem and environment. However, full use of these features requires an appreciable amount of learning.

There are also packages maintained by educational or research institutes, which are cheaper and sometimes more adapted to specific applications. Among the most used are CHADOC VS[5], LEAS[6] and STATITCF[7]. The latter, created by an agronomical institute, is a very complete package both in classical statistics and in multivariate data analysis. It contains comprehensive help for each procedure, with bibliographical references, and it works interactively by multiple selection menus. It is very easy to use but in an inflexible way. In particular the data are restricted to 1500 units and 60 variables, on average, and the outputs are very succinct but limited. There are no graphics. This last disadvantage will be bypassed in the new release of CHADOC VS. This package, devoted to multivariate data analysis, is well structured and allows the processing of large data files. It is written in Pascal, and has the advantage of allowing procedures in batch mode and obtaining the results on screen, on file, or on a printer. Different procedures can be executed in sequence without any intervention. Finally LEAS is an appropriate package for teaching statistics. It is comprehensive, but very limited in size of data files allowed.

[1]ADDAD: Association pour le Développement et la Diffusion de l'Analyse des Données, 22 rue Charcot, F-75013 Paris, telephone + 33 1 47070889.
[2]MODULAD: Club MODULAD, INRIA, Domaine de Voluceau, Rocquencourt, BP 105, F-78153 Le Chesnay cedex, telephone + 33 1 39635511.
[3]SPAD: Système Portable pour l'Analyse des Données, CESIA, 82, rue de Sèvres, F-75007 Paris.
[4]RS/1: BBN Software Products Corp., 10 Fawcett Street, Cambridge, MA 02238 USA, telephone + 1 617 873 5000.
[5]CHADOC VS: Institut Universitaire de Technologie, Département d'Informatique, 41, boulevard Napoléon III, F-06041 Nice cedex, telephone + 33 93837116.
[6]LEAS: Unité de Biométrie de Montpellier, INRA-USTL-ENSAM, 9, place Pierre Viala, F-34060 Montpellier, telephone + 33 67612428.
[7]STATITCF: Institut Technique des Cereales et des Fourrages, Service des Etudes Statistiques et Informatiques, Boigneville, F-91720 Maisse.

A number of aspects of astronomical data analysis are not well represented in the major statistical packages. Among these are the following (– a necessarily subjective view).

- Small sample sizes feature strongly in astronomical work (cf. Gehrels, 1986; Watson, 1989), yet there is little specific support of this in major software packages.

- Bootstrap significance testing is often a convenient way to rapidly (in terms of human time) assess significance. Albeit not in astronomy, a package which is available at nominal cost is described in Lunneborg (1987), and is to be further expanded.

- Regression with errors in both (or all) variables is commonly appreciated as a problem in astronomy. Murtagh (1989) discusses software and bibliographic references, here. A symbolic mathematics package, with special relevance for least squares estimation, has been described in Jefferys et al. (1988a). A User's Manual (Jefferys et al., 1988b) is also available.

- Survival analysis, or handling data with limits, is dealt with by the ASURV package (Isobe and Feigelson, 1987).

- Statistics has not played a fundamental role in the education of the present-day astronomer. There is a need, therefore, for greater on-line help and constructive decision aids to be closely coupled with statistical software. The use of expert system front-ends to statistical packages is an ongoing research field, although it has not yet resulted in generally-accepted, marketable products.

A number of similar topics are dealt with in Feigelson (1989). These include spatial analysis, for the galaxy clustering problem; analysis of periodic time series in noisy data; and image restoration techniques (an area in which radio astronomy has been significantly in advance of observational astronomy).

1.3 Current Information
The statistical software field, it has been mentioned above, is a rapidly evolving one. In this section, a few possibilities for keeping abreast of developments are reviewed.

Software reviews, just as book reviews, can be valuable. *The American Statistician*, a quarterly publication of the American Statistical Association, contains two regular sections entitled "Statistical Computing Software Reviews" and "New Developments in Statistical Computing". Comprehensive product announcements are to be found, together with thorough reviews, often for PC products. Both commercial and research products are covered. Advertisements in *AMSTAT News*, published 10 times per year by the American Statistical Association, can also usefully indicate market developments. Package reports (generally coming from the vendor) are to be found in *Computational Statistics and Data Analysis*, published 4 times per year by North-Holland. A range of relevant information is also contained in *Chance*, publised 4 times per year by Springer-Verlag, and *Statistical Software Newsletter*, published 3 times per year by the Gesellschaft für Strahlen- und Umweltforschung (GSF), Munich (although this is expected to change, shortly), for the International Association for Statistical Computing (IASC). The latter organises the biennial COMPSTAT (Computational Statistics) conferences in Europe. Both this and the annual Computer Science and Statistics Symposia on the Interface, held in the US, are convenient venues to see demonstrations of the major packages. *Applied Statistics*, published 3 times per year by the Royal Statistical Society, contains a secion on "Statistical Algorithms", often at a high level of sophistication which would not appeal to applied researchers. Many books also contain software listings, or have associated diskettes (e.g. Murtagh and Heck, 1987; Brockwell and Davis, 1987; Rousseeuw and Leroy, 1987).

Among surveys to which reference can be made are: Carpenter et al. (1984) for PC-based statistical software; and Lehman (1987) for Macintosh products.

2 STATISTICS AND ASTRONOMICAL IMAGE PROCESSING

Ever larger amounts of astronomical data are being collected and the rate of collection and storage is certain to speed up in coming decades. The Space Telescope, for instance, will send down over an estimated lifetime of 15 years the equivalent of 6×10^{12} bytes of information. Hence the now accepted centrality of image processing methods in astronomy for aiding bulk processing of data; hence too the growing number of astronomical archives and databases, the objective of which is to allow maximum use of expensively collected data; and hence also the accelerating interest in statistical procedures for their ability to aid in the interpretation and understanding of data.

Substantial effort has been dedicated by the astronomical community to the development of image processing systems to solve the critical problem of handling and displaying large amounts of data. Among the most well established systems are the following:

1. AIPS (Astronomical Image Processing System) from the National Radio Astronomy Observatory (NRAO), Charlottesville, Virginia, which was developed initially for radio astronomy data from the VLA (Very Large Array) telescope, New Mexico.

2. IRAF (Image Reduction and Analysis Facility), from Kitt Peak National Observatory (KPNO), Tucson, Arizona, and the National Optical Astronomy Observatories (NOAO), Tucson, Arizona. IRAF uses as a subsystem the Science Data Analysis System (SDAS), developed at the Space Telescope Science Institute (STScI), John Hopkins University, Baltimore, Maryland.

3. STARLINK is not a homogeneous system but rather a wide-ranging collection of software (including graphics, database management, and applications) maintained at the Rutherford Appleton Laboratories, Oxfordshire, England. It is accessible over the astronomical research network of the same name.

4. MIDAS (Munich Image Data Analysis System) is developed and distributed by the European Southern Observatory, with its focus on optical astronomy.

For a general survey of image processing systems, see Preston (1981) and Adams et al. (1984), where the systems considered belong to areas other than astronomy. A short appraisal of what is available in Strasbourg is to be seen in Box 2. It is not easy to specify a "representative" institute; however, it is important to note that general image processing systems may offer considerable capabilities for the statistical treatment of data.

For image processing and data analysis in astronomy, see Di Gesù et al. (1984, 1986, 1989). All the systems mentioned above have special characteristics making them suitable for particular areas of application. Apart from MIDAS, very little statistical software is available in these packages. MIDAS contains routines for multivariate statistical analysis, and period detection in time varying data, and further developments are ongoing.

One of MIDAS's data structures is the *table*, a collection of n-tuples which represent the properties of objects in a structured form. They are used to store results of the analysis of images, and to keep catalogues and small databases inside the image processing environment. MIDAS tables ressemble certain aspects of what have been called "scientific and statistical database management systems" (see further, below); they also contain aspects of spreadsheet programs. The MIDAS table subsystem grew out of 2-dimensional photometry: scanning digitised images for objects (e.g. stars, galaxies, or other objects) led to lists of coordinates, and associated flux or intensity values. Both during an image "reduction" session, and as an end-product, tables are needed for storing data ranging from stellar magnitudes, through galaxy isophotal diameters, to spectral line intensities. Most image processing systems use tables in some form, whether visible or otherwise to the user. An efficient table system is therefore highly desirable. It must also interface

Box 2: Image Processing and Statistical Software in Strasbourg

A very powerful interactive image processing environment is at the researcher's disposal on the mainframe (IBM 3090) of the Centre de Calcul de Strasbourg-Cronenbourg: the IBM Image Access eXecutive (IAX). This environment works under VM. In the Centre de Calcul's configuration the use of graphic terminals is necessary, but IAX is independent of such devices. On such a terminal, the images, elementary variables of IAX, and any transforms of them, can be displayed as pictures. In addition to its native functions and help screens, IAX allows the procedures of others to be used, and, in particular, the macro-commands of the Restructured EXtended eXecutor (REXX). Beyond displays and the manipulation of images, IAX's functions are calculus of histograms, means, ordinary numerical transforms, thresholding, and so on. Images are handled as matrices of integers, reals, or complex numbers. So, by using them outside of IAX, one can process images with other statistical software packages, such as SAS.

Another convenient package implemented in the Centre de Calcul is the Subroutine Package for Image Data Enhancement and Recognition (SPIDER[1]). This package is extensive; it contains the usual numerical and structural transformations of images, smoothing, restoring and analysis of contours, lines and textures, and description and geometrical modifications by numerical calculations, such as relaxation on two-dimensional Markov processes.

From a statistical point of view, one should consider the package IMAGIC[2] used by microscopists of the Laboratoire de Génétique Moléculaire. These programs allow, for biological macromolecules, three dimensional reconstruction and pattern recognition from plane images. This latter is brought about by multivariate data analysis. On a set of aligned images, where each pixel is taken as component of a vector, with numerical value the observed intensity, a correspondence analysis allows the relevant information to be summarized by the first eigenvectors, or so-called eigen-images, and an agglomerative hierarchical clustering permits the most representative forms of molecules to be obtained.

Finally, there are various possibilities for image processing in the contexts of physics, electronics and remote sensing at the Centre de Télédétection, Ecole Nationale de Physique, Département d'Informatique and at the Laboratoire de RMN en Biophysique Médicale of Université de Strasbourg.

[1]SPIDER: Electrotechnical Laboratory, 1-1-4, Umezono, Sakura-mura, Niihari-gun, Ibaraki-pref., Japan.
[2]IMAGIC: Dr. M. van Heel, Fritz Haber Institute, Max Planck Society, Faradayweg 4-6, D-1000 Berlin-Dahlen, FRG.

effectively with graphics. Finally, the provision of methods to analyse tabular data is clearly an important aspect.

The MIDAS table subsystem is not a DBMS (it does not have full relational functionality). It does not have the objectives of a spreadsheet, either. It is rather an intermediate data structure, designed for the best possible linking of the image processing part of the system with statistical and mathematical applications software.

3 STATISTICS AND ASTRONOMICAL DATABASES

The centrality of astronomical image processing is closely followed by astronomical database systems. Both areas require statistical techniques. The objective in the first part of this section is to point out the increasing convergence of statistical software, in general, and database management systems; and to proceed from this to briefly explore central research directions in this common area. In the second part of this section, we will overview developments in the area of interactive graphics.

3.1 Databases: Scientific/Statistical Versus Commercial

Current statistical analysis packages are widely used from the business to the scientific arenas, and encompass ever more extensive database capabilities. For example, SAS ("Statistical Analysis System") has been embraced under the 4GL, "Fourth Generation Language" heading, in the same context as the high level interface to ADABAS, NATURAL. Hext (1982) discusses the data storage and manipulation features of many statistical and database systems.

Ochsenbein (1986) and Murtagh (1988, 1989) overview issues in this field in astronomy. Ponz and Murtagh (1986) describe a proto-relational system used in astronomical image processing. This latter study is pursued in Ponz and Murtagh (1987), where this "table system" is related to SSDBMS research issues. The system described is part of an image processing system, designed to complement the image handling capabilities.

The mainstay of most traditional DBMSs lies in business applications. Scientific and statistical data often present different characteristics. Härder and Reuter (1985) include the following areas among those where "non-standard" DBMSs are required: image and speech processing systems (where image frames and spectra are to be stored); text and information retrieval; geographic information systems; office automation, involving forms systems (hierarchically structured user interface units of data); expert system knowledge bases; together with scientific and engineering (CAD/CAE, – computer aided design and engineering) databases.

Some of the characteristics of scientific and statistical databases (Shoshani and Wong, 1985; see also Cubitt and Westlake, 1987; Eurostat, 1986; Rafanelli et al., 1988; Härder and Reuter, 1985; and Hartzband and Maryanski, 1985) are the following.

1. In traditional commercial applications, capabilities for frequent updating and deleting must be supported, whereas in the scientific domain, the experimental (or observational or statistical) results are rarely updated but rather are "reduced" as they are given. Additionally, concurrency of write-access is often an important issue in commercial systems whereas it is not generally important in the scientific area. It results that the concept of transaction is not at all the same in these different fields.

2. Ancillary data (calibration data, instrument modes, experiment and environment details) are often stored for later use; while inextricably part of the data, metadata (or "data about data") nonetheless gives rise to added complexity.

   For example, an image may have ancillary information associated with individual pixels and with the entire image; metadata may be fixed format such as for observing conditions and it may be open-ended such as for processing history.

3. Scientific and statistical data are often available is very large quantities. Efficient compression techniques may be required for voluminous, sparse data. Transposed data matrices may be required for efficiency in addition to the usual "horizontal" record structure.

4. Special data types (temporal data, spectral data, survey data, irregularly sampled measurements) may need to be supported. (For temporal databases, see *inter alia* the bibliography of McKenzie, 1986; or Snodgrass and Ahn, 1986).

   While "entities" are often the basic building blocks of traditional systems, "cases" with many values may be more relevant for scientific and statistical data. Examples include vectors of time-varying values, geographical or astronomical coordinates, spectra, or closely associated textual information.

5. Special operations may need support (e.g. for statistical sampling and aggregation; interpolation and estimation; for handling various types of missing data values; or for data reliability estimates).

   Multidimensional data structures (e.g. grid files, multidimensional binary search trees, etc.) and search methods (e.g. best match or nearest neighbour searching) may be needed.

More "intelligence" may be needed for checking on data integrity producing summary tables, and so on.

With extra programming, such requirements can be handled on top of vendor-supplied DBMSs. But this is inconvenient for the individual scientist or statistician, and effectiveness is compromised by less than suitable storage structures, query optimization methods and buffer management strategies. A consequence of the requirements looked at is that DBMSs and query languages for scientific and statistical databases may in the future need to go beyond what is desirable in more traditional (business and commercial) fields.

3.2 Query Languages and Metadata

As a result of the special needs in the scientific and statistical area, suitable query languages are a subject of much interest. The query language is, of course, often the only user-perspective of the database management system. A survey of issues relating to query languages for statistical databases is available in Ozsoyoglu and Ozsoyoglu (1985). Before sketching out their categorization of work in this area, studies relating to summary tables will be mentioned.

Summary tables are important for statistical data: macro data summaries, in the form of frequencies or counts, supplement the given micro data. As discussed by Ghosh (1986), deletion and updating are not often practised in the statistical database context: outlier handling or the use of robust procedures are made use of instead. The stage of summarizing an often voluminous questionnaire (formally, very similar to an astronomical catalogue), for instance, leads to a table of aggregates which is of most interest to the statistical analyst. Extensions of widely-used relational database query languages, which deal with summary tables, are described by Ghosh (1986).

Approaches to scientific and statistical database query languages can involve one of the following strategies.

1. Extra features may be added to standard DBMSs to cater for special requirements. Alternatively, interface systems can connect together DBMSs with statistical packages and graphics software.

2. The relational data model can be used, with added capabilities. Access and retrieval may be expedited, for example, by better storage features.

3. Formal extensions to the relational model may be defined. Non-first normal form relations use both atoms and also arbitrarily nested sets of these as tuple

object	observations					
	data					
	date	exposure	spectra			
			start	step	counts	
SN 1990a	Jan 20	300	3400.	1.5	412. 460. 510. ...	
	Jan 28	300	3400.	1.5	402. 420. 510. ...	
	Feb 2	200	3100.	1.9	152. 154. 462. ...	
SN 1992b	Apr 1	300	3400.	1.5	342. 390. 415. ...	
	Apr 8	300	3400.	2.5	472. 520. 610. ...	
	Apr 16	275	3570.	1.1	452. 420. 410. ...	

Figure 1: Example of non-first normal form where attribute values are themselves relations.

components. Extended non-first normal form models allow the handling of various data types (e.g. ordered sets or matrices; see, e.g. Pistor and Traunmueller, 1986).

3.3 Interactive Graphics for Analysing Data

In this section we will be as much concerned with *how* one carries out statistical analyses, as with *what* one does.

Graphical techniques in statistics could be viewed as growing out of (i) the work of John Tukey at the Institute of Advanced Studies, Princeton, whose "Exploratory Data Analysis" described a comprehensive range of graphical and representational techniques for the description and interpretation of data; and (ii) multivariate data analysis, which encompasses a large number of techniques for the parsimonious description of data. It is nonetheless the case, however, that these fields have evolved rather separately.

Although dynamic, interactive graphics have been in existence for some time, it could be seen as one particular type of "visualization" of scientific data (see special issue of *Computer Graphics*, **21**, Nov. 1987, on "Visualization in Scientific Computing"; or the articles in *Computer*, **22**, Aug. 1989). For statistical analysis of data, very recommendable articles on new environments and strategies are to be found in Oldford and Peters (1988), and McDonald and Pedersen (1988). New ideas in interactive graphics are dealt with in Becker et al. (1987) and in a number of articles in the June 1987 (Vol. 82, No. 398) issue of *Journal of the American Statistical Association*. The focus of these articles on integrated, and interactive, programming environments is a clear pointer towards the way that statistical analyses are increasingly carried out.

A widely available tool for the graphic display of multivariate data is the Macintosh-based MacSpin (Donoho et al., 1988). Making use of a 3-dimensional view of the data points, rotation and labelling can be used effectively to analyse the data in an exploratory way. Features include subset selection, highlighting of selected points, identification and labelling, animation through rotation, and transformation of variables. Another feature available in the S language (see below) is to "brush" a certain part of a two-way plot, and to see the "brushed" subset highlighted in other two-way plots. The MacSpin manual includes a discussion of analysing a subset of the Center for Astrophysics Survey galaxy distribution data.

The MacSpin manual includes a short but comprehensive survey of work in the field of dynamic graphics. Initial research (in addition to the MacSpin manual, see Friedman et al., 1987) was on hardware which was outside the reach of most research groups. With powerful Macintosh-type personal computers, and scientific workstations, this is no longer true. In addition to MacSpin, two other systems are well-known in this area: ISP (Interactive Scientific Processor), initially VAX and Apollo-based, and associated with P. Huber (see, e.g., Ostermann and Degens, 1986); and the Unix-based S language, developed at ATT Bell Laboratories (Becker et al., 1988).

4 STRATEGIES FOR THE FUTURE

The very mature statistical software marketplace has been looked at in this article. So too have the exciting and fast-moving interface fields between statistical software and image processing systems, on the one hand, and statistical software and database management systems, on the other. Astronomers may choose not to attempt to influence developments in these different areas, but they cannot remain unaffected by such developments.

It is our wish that astronomers would, rather, intervene positively in their own long term interest. We have noted in section 1 that particular problems relating to astronomical data analysis are not adequately catered for by current statistical software packages. It is quite clear that astronomers cannot compete with very mature disciplines. Manpower and expertise which can be set to work in any of today's major astronomical institutes just cannot attempt to replicate what is already available – reliably, and if one considers all relevant financial aspects, at very moderate cost.

We will not address, here, the place of statistics in the modern astronomer's education (which, empirically, we feel to be rather limited). The entire thrust of this article has been to bring the attention of astronomers to the field of statistical software, in its widest sense. What more should, or could, be done in the foreseeable future?

The need for closer collaborative work between professional astronomers and statisticians is evident. Again on empirical grounds, it may be asserted that current organisational structures do not help this, notwithstanding available funding schemes. Necessary convergence and symbiosis surely must come about, however, given the great amount to offer by, and the great needs of, both sides.

REFERENCES

1. J.R. Adams, E.C. Driscoll, Jr., and C. Reader, "Image processing systems", in M.P. Ekstrom (ed.), *Digital Image Processing Techniques*, Academic Press, New York, 1984, pp. 289–360.

2. Ch. Beaumont, E. Mahmoud and V.E. McGee, "Microcomputer forecasting software: a survey", *Journal of Forecasting*, **4**, 1985, 305–311.

3. R.A. Becker, W.S. Cleveland and A.R. Wilks, "Dynamic graphics for data analysis", *Statistical Science*, **2**, 1987, 355–395.

4. R.A. Becker, J.M. Chambers and A.R. Wilks, *The New S Language*, Wadsworth and Brooks, Pacific Grove, CA, 1988.

5. P.J. Brockwell and R.A. Davis, *Time Series: Theory and Methods*, Springer-Verlag, New York, 1987.

6. D. Cable and B. Rowe, *Software for Statistical and Survey Analysis 1987*, Prepared for the Study Group on Computers in Survey Analysis, Published by British Informatics Society, Ltd., 13 Mansfield St., London W1M OBP, on behalf of The British Computer Society and The Study Group on Computers in Survey Analysis, 1987, 33 pp.

7. J. Carpenter, D. Deloria and D. Morganstein, "Statistical software for microcomputers: a comparative analysis of 24 packages", *Byte*, **9**, April 1984, 234–264.

8. R. Cubitt and A. Westlake (eds.), "Report on the Third International Workshop on Statistical and Scientific Database Management, Luxembourg, 22–24 July 1986", *Statistical Software Newsletter*, **13**, 3–27, 1987.

9. G.E. Dallal, "Statistical microcomputing – like it is", *The American Statistician*, **42**, 1988, 212–216.

10. V. Di Gesù, L. Scarsi, P. Crane, J.H. Friedman and S. Levialdi, *Data Analysis in Astronomy*, Plenum Press, New York, 1984.

11. V. Di Gesù, L. Scarsi, P. Crane, J.H. Friedman and S. Levialdi, *Data Analysis in Astronomy. II*, Plenum Press, New York, 1986.

12. V. Di Gesù, L. Scarsi, P. Crane, J.H. Friedman, S. Levialdi and M.C. Maccarone, *Data Analysis in Astronomy. III*, Plenum Press, New York, 1989.

13. A.W. Donoho, D.L. Donoho and M. Gasko, "MacSpin: dynamic graphics on a desktop computer", *IEEE Computer Graphics and Applications*, **8**, 51–58, 1988.

14. A.D. Elliman and K.M. Wittkowski, "The impact of expert systems on statistical database management", *Statistical Software Newsletter*, **13**, 14–18, 1987.

15. Eurostat, *Proceedings of the Third International Workshop on Statistical and Scientific Database Management*, Commission of the European Communities Statistics Office, Luxembourg, 1986.

16. E.D. Feigelson, "Astronomy, statistics in", in S. Kotz and N.L. Johnson, *Encyclopedia of Statistical Sciences, Suppl. Vol.*, Wiley, New York, 1989, pp. 7–14. (Also in *Newsletter of Working Group for Modern Astronomical Methodology*, Issue 7, Sept. 1988, pp. 8–15.)

17. J.H. Friedman, J.A. McDonald and W. Stuetzle, "An introduction to real-time graphical techniques for analyzing multivariate data", *Computer Physics Communications*, **45**, 161–167, 1987.

18. N. Gehrels, "Confidence limits for small numbers of events in astrophysical data", *The Astrophysical Journal*, **303**, 336–346, 1986.

19. S.P. Ghosh, "Statistical relational tables for statistical database management", *IEEE Transactions on Software Engineering*, **SE-12**, 1106–1116, 1986.

20. T. Härder and A. Reuter, "Architektur von Datenbanksysteme für Non-Standard-Anwendungen", in A. Blaser and P. Pistor (eds.) *Datenbank-Systeme für Büro, Technik und Wissenschaft*, Informatik-Fachberichte Nr. 94, Springer-Verlag, Heidelberg and New York, 253–286, 1985.

21. D.J. Hartzband and F.J. Maryanski, "Enhancing knowledge representation in engineering databases", *Computer*, **18**, 39–48, 1985.

22. G.R. Hext, "A comparison of types of database system used in statistical work", in H. Caussinus et al. (eds.) *COMPSTAT 1982*, Physica-Verlag, Vienna, 272–277, 1982.

23. T. Isobe and E. Feigelson, "ASURV: A Software Package for Statistical Analysis of Astronomical Data with Upper Limits", User Manual, Department

24. W.H. Jefferys, M.J. Fitzpatrick and B.E. McArthur, "GaussFit – a system for least squares and robust estimation", *Celestial Mechanics*, **41**, 1988a, 39–49.

25. W.H. Jefferys, M.J. Fitzpatrick, B.E. McArthur and J.E. McCartney, "Gauss-Fit: A System for Least Squares and Robust Estimation. User's Manual", Department of Astronomy and McDonald Observatory, The University of Texas at Austin, 1988b, 75pp.

26. R.S. Lehman, "Statistics on the Macintosh", *Byte*, **12**, July 1987, 207–214.

27. C.E. Lunneborg, *Bootstrap Applications for the Behavioral Sciences. Vol. 1*, Manual, 185 pp., Department of Psychology, University of Washington, Seattle.

28. H.P. Macon, "RS/1 research system", *Byte*, Feb. 1988, 172–176.

29. J.A. McDonald and J. Pedersen, "Computing environments for data analysis III: programming environments", *SIAM Journal of Scientific and Statistical Computing*, **9**, 380–400, 1988.

30. E. McKenzie, "Bibliography: temporal databases", *SIGMOD Record*, **15**, 40–52, Dec. 1986.

31. F. Murtagh, "Scientific databases: a review of current issues", *Bull. Inform. CDS*, No. 34, May 1988, 3–33.

32. F. Murtagh, "Large databases in astronomy", in A. Kent and J.G. Williams, *Encyclopedia of Computer Science and Technology*, Vol. 21, Marcel Dekker, New York and Basle (in press).

33. F. Murtagh, "Linear regression with errors in both variables: a short review", in this volume.

34. F. Murtagh and A. Heck, *Multivariate Data Analysis*, Kluwer Academic, Dordrecht, 1987.

35. F. Murtagh and A. Heck (eds.), *Astronomy from Large Databases: Scientific Objectives and Methodological Approaches*, European Southern Observatory, Garching bei München, 1988.

36. F. Ochsenbein, "Data storage and retrieval in astronomy", in V. Di Gesù, L. Scarsi, P. Crane, J.H. Friedman and S. Levialdi (eds.), *Data Analysis in Astronomy II*, Plenum Press, New York, 1986, pp. 305–313.

37. R.W. Oldford and S.C. Peters, "DINDE: towards more sophisticated software environments for statistics", *SIAM Journal of Scientific and Statistical Computing*, **9**, 1988, 191–211.

38. R. Ostermann and P.O. Degens, "ISP as tool for EDA-experts", in W. Gaul and M. Schader (eds.), *Classification as a Tool of Research*, Elsevier, Amsterdam, 1986, pp. 361–368.

39. G. Ozsoyoglu and Z.M. Ozsoyoglu, "Statistical database query languages", *IEEE Transactions on Software Engineering*, **SE-11**, 1071–1081, 1985.

40. P. Pistor and R. Traunmueller, "A database language for sets, lists and tables", *Information Systems*, **11**, 323–336, 1986.

41. D. Ponz and F. Murtagh, "Image processing, databases and statistical software: the common interface in MIDAS", *Statistical Software Newsletter*, **12**, 129–132, 1986.

42. D. Ponz and F. Murtagh, "MIDAS Tables: present status and future evolution", in F. Murtagh and A. Heck (eds.), *Astronomy from Large Databases: Scientific Objectives and Methodological Approaches*, European Southern Observatory, Garching bei München, 1988, pp. 441–446.

43. K. Preston Jr., "Image processing software – a survey", in *Progress in Pattern Recognition*, L.N. Kanal and A. Rosenfeld (eds.), Amsterdam: North-Holland, 1981, pp. 123–148.

44. M. Rafanelli, J.C. Klensin and P. Svensson, *Statistical and Scientific Database Management*, Springer-Verlag, Heidelberg, 1988.

45. P.J. Rousseeuw and A.M. Leroy, *Robust Regression and Outlier Detection*, Wiley, New York, 1987.

46. A. Shoshani and H.K.T. Wong, "Statistical and scientific database issues", *IEEE Transactions on Software Engineering*, **SE-11**, 1040–1047, 1985.

47. R. Snodgrass and I. Ahn, "Temporal databases", *Computer*, **19**, 35–4, 1986.

48. A.A. Watson, "Some statistical problems encountered in the analysis of ultra high energy cosmic ray and ultra high energy gamma ray data", in Di Gesù et al. (eds.), 1989, pp. 335–349.

DISCUSSION

J.F.L. Simmons to P.E. Jupp: I liked this talk very much and can see there could be many applications of the analysis of spherical data in astronomy. For example, many properties of close binary stars depend on the orientation of the orbital plane. Often we have to deal with velocity fields or magnetic fields that depend on direction. In relation to this last point, has much work been done by statisticians on the distribution of vectors (and tensors) over the sphere?

Very little: I have plans for some work on this – from the theoretical viewpoint!

S.P. Bhavsar to P.E. Jupp: The perihelia of comets are non-uniform. Does this mean that they are clumped in some direction or could they still be uniformly distributed in a plane? In general, they have some symmetry, though not spherical.

Non-uniformity does not imply clumping and could allow distribution near a plane, great circle or small circle. I have not investigated the cometary perihelia in greater detail.

E. Grafarend to P.E. Jupp: Comment: Besides the knowledge of the distribution of spherical data/data on the sphere, there are also "normal" distributions available for data on other manifolds like an ellipse, ellipsoids, etc. This comment especially relates to your example of an elliptical orbit. Question: Assume a regression model on *spherical data*. In order to fit the model to the observed spherical data what would be the objective or risk function corresponding to "*least squares*"?

Given observations $x_1, x_2, \ldots, x_n$ on the unit sphere and corresponding fitted values $\hat{x}_1, \ldots, \hat{x}_n$. the usual objective function is $\sum_{i=1}^{n} \hat{x}_i . x_i$ and is equivalent to the usual $\sum_{i=1}^{n} \| \hat{x}_i - x_i \|^2$ of least squares. Alternatively, the spherical distance function $\sum_{i=1}^{n} \cos^{-1}(\hat{x}_i . x_i)$ can be used. Distributions on even more exotic manifolds (hyperboloids. Stiefel manifolds, Grassmann manifolds) have been considered!

R. Branham to P.E. Jupp: According to the Oort cloud theory, the distribution of cometary perihelia should be random. Your data shows that this is not the case. Suppose I propose the hypothesis that the distribution is random. but we are seeing a selection effect in the Marsden catalog. Is there a way to test this hypothesis?

I am not aware of such a method. I do not see how one could distinguish between (i) non-uniformity. observed without selection, and (ii) uniformity, observed selectively.

C. Jaschek to P.E. Jupp (comment): It is surprising to see that spherical methods. introduced fifty years ago are not used in astronomy, whereas they are widely in geophysics.

J. Cuypers to P.E. Jupp (comment): I think that there exists a recent paper in the astronomical literature on the orientation of the orbital plane of binary stars in the solar neighbourhood using explicitly statistical methods on spherical data.

E. Feigelson to F. Murtagh and Ph. Nobelis: The trend in which individual or small groups of astronomers develop software that is applicable beyond a local environment is likely to continue during the 1990s. NASA in particular has initiated two significant programs, the Astrophysics Data System, and Astrophysics Software and Research Aids. that promote the broad distribution of databases and software. The problem I perceive is inadequate coordination of such efforts. There is no forum to announce the availability of a product; no clear norms for software documentation or distribution format, portability, etc. I suggest that some organisation – an IAU Commission, the Working Group for Modern Astronomical Methodology, the AAS Working Group on Astronomical Software, or whatever – take a stronger leadership role in such matters.

H. Eichhorn to F. Murtagh and Ph. Nobelis: The problem of more cooperation between package producers and astronomers is extremely complex. I believe the key is the example of successful astronomers' example to their colleagues that more attention to proper and clean statistics is very worthwhile. But this cannot be imposed "from above", as if a university administration would decree that two departments now have to form a successful interdisciplinary program.

GENERAL DISCUSSION

General discussion

M.C.E. HUBER

ESA/ESTEC, Space Science Department
Noordwijk, The Netherlands

In introducing the discussion the Chairman presented a "pedestrian" summary of the Colloquium, consisting of two points :

1. During the Colloquium the statement was made repeatedly that a numerical result has no meaning unless it is accompanied by an uncertainty. Over the last two decades it has become customary for astronomical results to be quoted with an uncertainty, but most data bases do not yet allow storage of uncertainties.

2. The three topics listed in the title of the conference can be put into the following context :

- assumptions	result in bias
- experimental realities	result in uncertainties
- stochastic variations	result in errors

While the papers at the Colloquium dealt most extensively with the formal aspects of errors, the importance of a "mixed model" - comprising systematic uncertainties as well as stochastically generated errors - was also stressed.

It should be realised that the input for assessing the uncertainty consists of estimates, which in fact are based entirely on judgement. The same is true for the bias. This is the reason why the outcome of scientific experiments (or computations) is often not a clear black-and-white situation. In many cases, the question of whether two results differ (and therefore may support different theories) can only be answered by careful assessment of the assumptions that result in the uncertainty estimates given. Thus, in the end, a sound scientific judgement is often just as important as technically and mathematically sound measurements and data analysis !

In the discussion following this introduction, it was agreed that the Chairman's original presentation should be amended by the following additions :

- the choice of the method for measuring, data analysis or computing also introduces bias, thus, method and assumptions result in bias, and

- the co-variance matrix should also be given, as this data set defines the Kalman filter used in analysing the data.

There was general agreement on the importance of a proper understanding of all stages from data taking through the analysis of errors, bias and uncertainties, and on the dominating role of judgement in the scientific process.

    The Chairman indicated that the following discussion should concern itself primarily with practical questions and with how the message of this Colloquium (namely the need for sound treatment of bias, errors and uncertainties in serious astronomical work) can be carried to the "outside world". For this reason some time was reserved to discuss whether a summer school and/or another conference of this kind should be organised.

(**Note** : The book list is not mentioned here, because it will be presented in the Proceedings. The "Hipparcos update" is also not covered, as it did not really concern the Colloquium.

GENERAL DISCUSSION

The following is a record of issues raised during the general discussion. It is based only on the written versions provided by speakers. The order should be considered as approximate.

S.P. Bhavsar: The environment in which science is done in the USA does, to an extent, lead to sloppiness and careful statistical analysis is a prime example. At the post-doc stage one is looking for a job every two years. During the tenure-track level one is trying to show productiveness in terms of papers published and grants obtained. These pressures, and the way one's "productiveness" is judged lead to neglect of certain approaches to problems and a sloppiness in a thorough analysis of a problem. These are practical, sociological issues which enter into how the science is done.

H.L. Marshall: I think that bias in estimation may be introduced by the *method* of estimation, which is not related to the assumption of the model or the taking of data. Here bias is defined by $E(\hat{\theta} - \theta)$ when $\hat{\theta}$ is the estimate of the true value, θ. The model can be correct but bias would occur due to the particular method used to obtain $\hat{\theta}$.

H. Eichhorn: A bias (difference: estimator minus true) is *always* introduced by the fact that a *complete* model is never available.

J. Pfleiderer: Scientific judgement is an essential estimator in statistics and should always be used together with other estimators. Our final goal should not be to use statistics but to get a result as close to the truth as possible for which statistics is a tool we judge to be very efficient.

H. Eichhorn: Not only should uncertainties be stated, but one throws away a lot of information by not stating the covariance matrix of the results. One always runs (during filtering) into the problem of inverting $A + BC^{-1}D$. The inverse equals $A^{-1} - A^{-1}B(C + DA^{-1}B)^{-1}DA^{-1}$ and can thus be computed if A^{-1} is known. A plays the role of the covariance matrix, so that this "inversion lemma" can be applied only if A is provided.

C. Jaschek: I think that the remedy consists basically in introducing statistics in astronomy courses – on the same importance level as differential equations, kinetic theory of gases. Visiting scientists at big institutions can be a partial remedy, but they cannot substitute the basic training. Additional help can come from summer schools or meetings at a local level or a national level.

C. Worley: It is not always true that we don't know the correct model. In the case of binary star orbits, we know the model – an ellipse and Keplerian motion (excluding such effects as possible third bodies or relativistic effects which are, of course, non-detectable). The problem is the observations, and particularly how to judge and weight the older data. There have been many cases where absolute nonsense has been produced because the investigator(s) failed to evaluate the data correctly.

H.L. Marshall: As Dr. Eichhorn points out, one needs the covariance matrix for the proper analysis of a given dataset. This view is a straightforward generalization of the requirement for uncertainties – these are merely the diagonal elements of the covariance matrix. Requiring the covariance matrix is therefore sufficient as well as necessary.

C.S. Cole: In order to reach our goal we must understand the physics of the situation and we must understand the statistical theories and assumptions we use, and we must also understand our data. Error estimates are not unique. Knowledge and judgement must be used with the theory, the statistics and the data in order to do good science.

R. Branham: I would like to emphasize the importance of mathematical judgement or at least numerical analysis judgement. We can tackle a least squares problem by orthogonal transformations or normal equations. The latter are twice as efficient but more subject to problems of ill-conditioning. Steps can be taken, however, to improve their conditioning, and then we have a more efficient algorithm. The steps to be taken in this and other algorithms depends on mathematical judgement. To improve the problem of lack of communication between astronomers and statisticians or other groups, I suggest that we consider publishing in other than astronomical journals. Recently I published an article in *Computers in Physics*, admittedly not a statistics journal, but at least it gives exposure of our ideas to other than the strictly astronomical community.

H.L. Marshall: Let me take the view, temporarily, of a scientist who is under pressure to publish results quickly. As long as the methods used are "roughly correct", there is little motivation to spend the time to learn techniques that might be considered "sophisticated" or "advanced". Improving the result could be left to later scientists as a "clean-up crew". How does one confront this point of view?

H. Eichhorn: In many fields (e.g. astrometry), applying the "best" method and squeezing out another decimal is not up to the "cleanup crew". There is no cleanup crew, the astrometrist who does not produce the best data at the first attempt has simply not done the most competent job. (There are legitimate exceptions to this statement, though.)

E. Grafarend: Comment one: as a "mathematical statement" there is the lemma that all *empirical science* need statistics since they deal with measurements, observations. And astronomy is a subset of the set of empirical sciences. Therefore in astronomy regular courses in statistics are obligatory. A more delicate question would be the choice of topics for these courses. Comment two: there has been raised the point of view that in case an astronomer needs advice for data analysis she/he should contact the statistician. Certainly this is the right way for first aid, but one should not underestimate the situation that *not* all astronomical data analysis problems are solved. And we saw some examples of this type in our conference. Definitely some inherent statistical problems in astronomy are original and have to be solved independently. A fine example is the invention of the probability of *circular data* by R. von Mises: he was studying the periodicities in the system of chemical elements. Here the proper probability distribution was not available beforehand.

E. Feigelson: In addition to improved education of young astronomers, a change in attitude of our community's intellectual and administrative leaders is also needed. Our best scientists do not skimp on careful observing and physical interpretation, our leading instrument developers do not skimp on quality engineering, yet sophisticated data analysis methodology is not adequately emphasized or budgeted.

H.-M. Adorf: Just a brief comment on Prof. Jaschek's previous statement, with which I completely agree: Not only should statistics be taught, but it should be *multivariate*; and *decision theory* should be included.

C.S. Cole: In the US, statistics departments do provide service courses at the graduate level for other departments, e.g. biomedicine and agriculture. Astronomy graduate departments should petition their statistics departments to provide a graduate level course(s) specifically for astronomers.

H. Eichhorn: There seems to be agreement that astronomers are not exposed to enough statistics during their training. Jaschek has suggested a summer school. It would be useful, I think, to discuss this and make a recommendation.

J. Cuypers: There seems to be lack of communication between statisticians and astronomers, especially on meetings that could be of interest to both communities. Astronomers should try to contact international statistical societies to be aware of conferences in related fields. For a next meeting I would encourage astronomers to "feed" statisticians with their problems in order that they can report on their findings.

C. Jaschek: Summer schools should only be accepted if they leave something behind. Perhaps the lecture notes are an approximation to the textbook which should be written.

M.J. Kurtz: It was a short course by Albert Bijaoui which taught me multivariate analysis for classification.

F. Murtagh: As someone who has spoken at summer schools at Goutelas, Luminy and Capri – there will be another at Naples early next year – I would point out that such courses have been and are being held.

M. Broniatowski: There will be a statistical conference at the Department of Mathematics, Université Louis Pasteur, end of May 1991. We will organize a session on Statistics and Astronomy if there are enough contributed papers. Mail to M. Broniatowski, Math. Dept., 7 rue René Descartes, 67000 Strasbourg, France.

C. Jaschek: We will be glad to organize another meeting on statistical matters in three years here at Strasbourg.

F. Murtagh: A word about the lineage of this conference: it follows from the 1983 event, hosted here, and the proceedings of which were published by ESA. In 1987, the conference on "Astronomy from Large Databases" was seen as a follow-up, hosted in Garching. Hence Garching is also a possibility for a follow-up to this meeting. The "Working Group for Modern Astronomical Methodology" is open to contributions at all time. This is an informal working group with an approximate 100 membership. André Heck and I have been sending out a newsletter approximately 3 times per year for the last few years.

M.J. Kurtz: There is no graduate level book on astronomical statistics at the level of a graduate course in statistical mechanics. While a summer school is a possibility to create this, another way is for someone (in this room?) to teach such a course and publish the lecture notes.

F. Murtagh: The list of recommended texts in statistics, a preliminary version of which has been circularized, is open for suggestions.

M.J. Kurtz: Volumes 1 and 3 of the Handbook of Statistics are also very nice, – the whole set is quite good.

POSTERS

Uncertainties in spectral line identification due to rotational broadening using WCS method

S.G. ANSARI

Observatoire Astronomique, Centre de Données Stellaires, Strasbourg, France

Introduction

A large number of HgMn stars possess rotational velocities greater than 50 km/s (Wolff and Wolff, 1975). These stars have thus been ignored in spectral analysis studies. Most of the chemical abundances derived for HgMn stars originate primarily from studies made on stars having rotational velocities less than 20 km/s. How well could we determine the presence of a chemical element in stars having higher rotational velocities ? To answer this question I have tried to quantify the uncertainties brought about by spectral line broadening. In order to demonstrate this; the sharp lined co-added spectrum of κ Cnc with a rotational velocity of 6 km/s between 3700 and 4700 Å was used (Adelman, 1987) at a reciprocal dispersion of 2.4 Å/mm. By convolving the original spectral profile with a calculated rotational profile we obtain a rotationaly broadened spectrum (Cowley, 1970, Ansari, 1988). Adding random noise gives the spectrum a realistic touch. By doing so all instrumental and stellar parameters remain constant; the only free parameter being the rotational velocity. I have calculated rotationally broadened spectra between 10 and 50 km/s with an increment of 5 km/s as well as two spectra at 80 and 100 km/s. Of course, the number of identifiable lines diminish with increasing velocity. To demonstrate this Figure 1 shows $v \sin i$ vs. the number of identifiable lines. At 100 km/s only a handfull of spectral lines were identified. And normally, due to blending and noise, only a very small fraction of those may be used in a chemical analysis.

Wavelength Coincidence Statistics (see Hartoog et al., 1973, Ansari, 1987) (WCS) was developed based on a Monte Carlo method to identify the strong or weak presence of a chemical element in a star, depending on a cross examination of laboratory and stellar wavelengths. The significance parameter helps establish a scale of measuring this presence. However, if only very few lines in the lab list exist, it may give a high significance of something that may not be present. Another thing that one must be aware of is the fact that the significance level may give a high value of a chemical element that may not be analysable, since it may not have unblended lines, a problem usually encountered as one goes on to analyse more exotic species at higher velocities.

In the following section a summary of WCS results of the broadened spectra and a short discussion on elements showing an overall high significance is made.

WCS results:

Wavelengths were measured by fitting parabola to the spectral lines and thus determining the central depth.

For the broadened spectra between 10 and 30 km/s a tolerance level for identification of 60 mÅ was chosen. For the higher velocities the tolerance level had to be increased by double the amount (i.e. 120 mÅ) since blending becomes common and central depth wavelengths become more susceptible to misidentification.

The following is a summary of the results found for each element with an overall high significance level. Figure 2 shows the results of the WCS investigation.

Mg Mg II consists mainly of weak lines. This prediction can be made, on account of the significance level, which drops rapidly with increasing velocity. However, Mg II lines may be primarily blended, because at a higher search tolerance the significance begins to increase.

Si This element shows a strong presence throughout, consisting mainly of strong lines, not blended with others. Even at a larger search tolerance the significance level does not remarkably change. Si III behaves much like Si II, but at higher velocities significance levels drop, confirming that Si III lines in κ Cnc are weak.

P A typical case of a decrease in significance due to rotational broadening. P II shows a systematic decrease, which may be caused by the disappearance of weak lines at high rotational velocities. At a larger search tolerance, no increase in the significance parameter is detected. P III shows a significance level below 4 in the original spectrum, however fades away as velocities increase until 45 km/s, where a sudden high significance is noted. This is presumably due to a misidentification of a blend as a P III line.

Ca Since the H and K lines are identifiable throughout, and only 4 lines are used in the WCS search, it is clear why this element shows a high significance throughout. Of course at high velocities it becomes very difficult to identify the K line blended with the Hϵ Balmer line.

Ti Curiously titanium begins to disappear rapidly with an increase in rotational velocity, indicating again a sharp decrease in the number of identifiable lines.

Cr The significance of chromium is well below the critical significance level of 3. Adelman (1987) determined an abundance of 5.84 (H=12) for Cr II for κ Cnc indicating that this element is weak in this star. WCS results reflect this fact.

Mn The most abundant element in most HgMn stars shows a clear depletion for velocities greater than 20 km/s. However, Mn I is far too weak compared to Mn II. Again this is due to weak lines, which begin to rapidly disappear at higher velocities. Mn II is highly significant. This is reflected accross all velocity values upto 100 km/s.

Fe Fe I behaves very much like Mn I and is undetectable at higher velocities pointing out that it is mainly represented by weak lines in this star. Fe II is strong all along. However, at 100 km/sec it becomes too weak to detect.

Ga One of the most interesting elements encountered in HgMn stars gallium is definitely present. Again, due to bad statistics and a limited number of laboratory lines significance levels fluctuate around 30 km/s. Ga II is strongly present at all velocity levels.

Sr Rather weak in the original spectrum, Sr II significance level begins to increase with velocity. Again, this shows that as the number of blends increase uncertainties in the central wavelengths increase the chance of misidentifying a line.

Xe Xenon will probably only be identified in sharp-lined stars, since it is mainly represented by Xe II spectral lines with equivalent widths around 10 mÅ. Results were very unsatisfactory in this star at velocities greater than 30 km/s, which appears to be a critical velocity for the detection of exotic species.

Hg Only 3 Hg II lines are used in the WCS investigation to identify the presence of mercury, whereby one of them is the prominent line in almost all HgMn stars, λ3984. This line is identifiable at all velocities.

The WCS method seems to work well for elements having a large number of lab lines. I must also point out that the quality of stellar spectra and the dispersion at which they are observed play an important role in determining central wavelengths of spectral lines. In the case of spectral line broadening, blending and noise are the main factors for the difficulties encountered in spectral line identification.

References

Adelman, S.J.: 1987, *Mon.Not.R.Astron.Soc.* **228**, 573.

Ansari, S.G.: 1987, *Astron.Astrophys.* **181**, 328.

Ansari, S.G.: 1988, *A chemical Analysis of the HgMn stars HR 205 and HR 2844*, Ph.D. thesis, Vienna Univ.

Cowley, C.R.: 1970, *Theory of Stellar Spectra*, Gordon and Breach, N.Y.

Hartoog,M.R., Cowley,C.R., Cowley,A.P.: 1973, *Astrophys. J.* **182**, 847.

Hill, G., Fisher, W.A.: 1976, *Publ. Dominion Astrophys. Obs.* **XVI**, No. 13.

Wolff, S.C., Wolff, R.J.: 1975, *Physics of Ap-Stars*, IAU Coll. 32, eds. W.Weiss, H. Jenkner, H.J. Wood, p. 503.

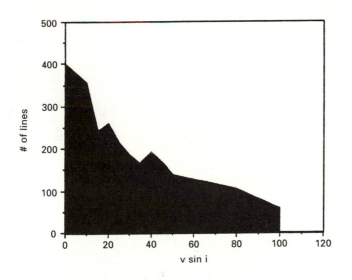

Figure 1

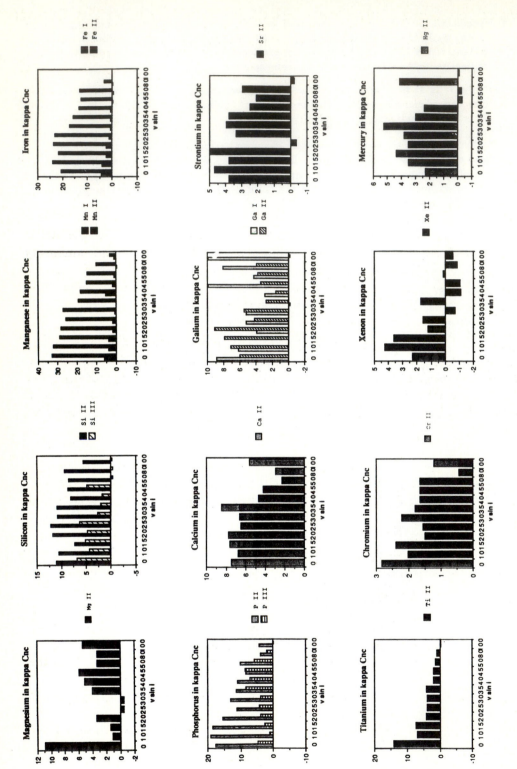

Figure 2

Effects of Uncertainties in Stellar Parallaxes on the Spectral Type Distribution in the Solar Neighborhood

BENEST D. – **GIUDICELLI M.**

O.C.A., Observatoire de Nice, France

1 INTRODUCTION

A significant part of galactic studies is based upon our knowledge of the stellar density, and particularly of the repartition of spectral types and the luminosity function. It is well-known that only in the vicinity of the Solar System can the stellar set be considered as a complete sample. But to which limiting distance? And how accurate are our models of the Solar Neighborhood?

Let us consider here only the Main Sequence stars. Traditionally, the Limiting Distance for the Completeness (hereafter noted the LDC) of our stellar neighborhood (down to $M9$, excluding thus the so-called "Brown Dwarfs") is taken of the order of 5 pc (Lippincott, 1960; Van de Kamp, 1971). Several approaches of the problem have been undertaken. For example, one may use the visual magnitude as a test (see e.g. Gliese, 1981; Bok and Bok, 1981; Upgren, 1985a,b).

For our study, we consider the percentage of a given spectral type (for stars on the Main Sequence) versus distance from the Sun (Benest, 1983). For the M-type stars, this gives us an estimation of the LDC which is in fairly good agreement with the traditional value as well as with the value obtained by other methods (Upgren, private communication). Then, we take into account the uncertainties in the stellar parallaxes, in order to evaluate the uncertainty in our value of the LDC. We detail the method in Section 2, and we present the preliminary results in Section 3.

2 METHOD

Let us assume that the distribution of each spectral type (M, K, G, ...) is isotropic and homogeneous in the large scale in the solar neighborhood.
Let us set in order the nearby stars according to their increasing distance from the Sun.
For the N first stars, let us compute the percentage $P(N)$ of stars of a given spectral type (for stars in the Main Sequence) over all observed stars.
The curve $P(N)$ tends ideally to the exact value of P when N tends to very great values. But, due to the finite power of our instruments, the curve $P(N)$ has a

maximum for $N = N_{max}$, and decreases for greater values of N. This value N_{max} corresponds to a limiting value for the radius of a Sun-centered sphere, inside which the set of observed stars of this spectral type may therefore be considered as a complete sample; beyond this limit, the decreasing apparent brightness of stars of this spectral type allows less and less detection with the present day instruments, introducing more and more observational bias.

Now, let us introduce the uncertainties on the stellar parallaxes.
When the error bars (parallaxe plus or minus delta) of two stars intersect, the places of these two stars (in the arrangement in distance from the Sun) may be exchanged. Let us take an example: in the arrangement cited above, stars 6 (Sirius) and 7 (L 726-8, UV Ceti) have parallaxes respectively 377 (delta=4) and 367 (delta=9); taking into account the uncertainties, it is therefore not impossible that the actual parallaxes might be 377-4=373 and 367+9=376, i.e. the parallaxe of Sirius may be less than the parallaxe of L 726-8, so that the actual places of the two stars in the arrangement may be 7 for Sirius and 6 for L 726-8; their places have been exchanged!
The number of such possible permutations increases very quickly with the number of considered stars, because the uncertainties in parallaxes -hence the error bars- increase with the distance; for example, we have detected several thousands possible permutations for the 300 nearest stars.
We can compute therefore the corresponding changed arrangements; in each of these changed arrangements, the value of N_{max} may be slightly different than the value of N_{max} for the unchanged arrangement. We can finally, from the set of values of N_{max} determined as said above, construct a histogram which will give the error bar for N_{max}.

3 PRELIMINARY RESULTS
We present here only a preliminary report.
In a first stage, we are testing the method using Gliese's Catalogue (1969), for practical reasons.

The figure 1 shows that the abscissa N_{max} of the maximum of the curve $P(N)$ (unchanged arrangement) for the 300 nearest M-type Main Sequence stars is indeed about 50, corresponding to a LDC of about 16 light years $\simeq$ 5 pc, of the order of the traditional value.

The figure 2 shows a sample of curves $P(N)$ corresponding to the first 1000 computed changed arrangements. N_{max} varies mainly between 45 and 55, with an isolated value near 60. This seems to confirm that the actual mean value might be effectively very near 50; but any definitive conclusion could be stated only when we will have computed all the changed arrangements, and furthermore when we will have constructed the histogram.

Of course, we are aware that since the publication of the 2nd Gliese's Catalogue (1969), many nearby stars have been discovered. Therefore, when we will have succeeded in the tests, we will apply our method to the most recent data bases, for example the forthcoming 3rd Gliese's Catalogue (see Jahreiss and Gliese, these proceedings), or the future catalogue issued from the HIPPARCOS experiment. We intend also to apply our method to other spectral types, particularly K and G.

The catalogue file has been furnished by the C.D.S..
The computation is performed on the VAX 11/785 of the Observatory of Nice.

4 REFERENCES

Benest, D., 1983, 'Spectral Type Distribution in the Close Solar Neighborhood', in *The Nearby Stars and the Stellar Luminosity Function*, eds. A.G.Davis Philip and A.R.Upgren, L.Davis Press pp.25-28

Bok, B.J., Bok, P.F., 1981, *The Milky Way*, Harvard Univ. Press, Chap.3 'The Sun's Nearest Neighbors; Stellar Populations'

Gliese, W., 1969, '2nd Catalogue of Nearby Stars', Veröff. Astron. Rechen Inst. Heidelberg **22**

Gliese, W., 1981, 'New Data on Nearby Stars and Problems in the Solar Neighborhood', *Bull. Inform. C.D.S.* **20**, 4-12

Lippincott, S., 1960, 'A Model of our Stellar Neighborhood', *Astron. Soc. Pacific Leaflet* **377**

Upgren, A.R., 1985a, 'The Effect of Uncompleteness among Nearby Stars', *Bull. A.A.S.* **17**, 553

Upgren, A.R., 1985b, 'Defining a Bias-Free Sample of Nearby Stars', *Bull. A.A.S.* **17**, 624

Van de Kamp, P., 1971, 'The Nearby Stars', *Ann. Rev. Astron. Astrophys.* **9**, 103-126

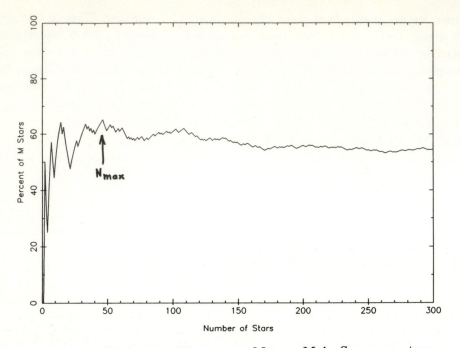

Figure 1. $P(N)$ for the 300 nearest M-type Main Sequence stars (unchanged arrangement).

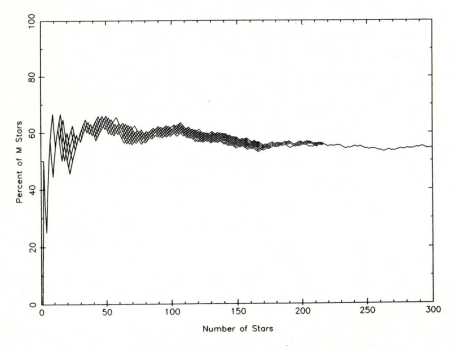

Figure 2. $P(N)$ for 1000 changed arrangements.

Multivariate mixture distributions in stellar kinematics: Statistical and numerical stability of the SEM algorithm

Bougeard M.L.[1][2], Arenou F.[3]

(1) Univ. Paris X, IUT, 1 chemin Desvallières, F-92410 Ville d'Avray
(2) URA 1125 CNRS, Observatoire de Paris, F-75014 Paris
(3) URA D0335 CNRS, Observatoire de Paris-Meudon, F-92195 Meudon

Summary :*In this paper, we are concerned with the problem of estimating the parameters of a gaussian mixture density. Here we tackle the problem of analysing the convergence stability of the SEM process by performing several independent runs. Then, the results of the most stable SEM solution are compared to classical clustering and classification techniques. The method is applied to samples of A type population I stars.*

Keywords: - *Gaussian mixture - Maximum likelihood - Stellar kinematics*

1. Introduction, notations and statistical background

Of interest in this paper is the parametric family of mixture of k normal multivariate densities, i.e the family of density functions of the form

$$f(x,\theta) = p_1 f_1(x|m_1,\Sigma_1) + ... + p_k f_k(x|m_k,\Sigma_k), \quad x \in \mathbb{R}^n$$

where the proportions p_i, i=1-k are constrained to be nonnegative and to amount to one. Each component f_i, i=1-k of the mixture is a n-multivariate gaussian density with n-vector mean $m_i \in \mathbb{R}^n$ and (n x n) covariance matrix Σ_i.. Much has been written on methodology for estimating the unknown parameters $\theta(k) = (p_i, m_i, \Sigma_i, i=1-k)$. For a review, we refer to (Titterington & al 1985; Redner &Walker 1984). A class of iterative procedures for numerically approximating maximum likelihood estimates is known as EM algorithm.which is an algorithm used for incomplete data problems (Dempster & al 1977).

It acts as follows for a N-sample of observations $(x_j, j=1-N)$, N>>k, $x_j \in \mathbb{R}^n$. Let θ^c = $(p_i^c, m_i^c, \Sigma_i^c, i=1-k)$ be a current approximate maximizer of the log-likelihood function of the sample $L(\theta)$, and θ^{c+1} the next one.The **Expectation** step computes for i=1-k , j=1-N,

$$p^c(i,x_j) = p_i^c f_i(x_j|m_i^c, \Sigma_i^c) / f(x_j,\theta^c)$$

that is an estimate of the posterior probability that x_j belongs to the ith component given the approximate estimate θ^c. Then, the **Maximization** step of the EM algorithm yields θ^{c+1} maximizing L(.), given by

$$p_i^{c+1} = (1/N)\sum_{j=1}^{N} p^c(i,x_j)$$

$$m_i^{c+1} = \left\{ \sum_{j=1}^{N} p^c(i,x_j).x_j \right\} / \left\{ N p_i^{c+1} \right\}$$

$$\Sigma_i^{c+1} = \left\{ \sum_{j=1}^{N} p^c(i,x_j).(x_j - m_i^{c+1})(x_j - m_i^{c+1})^t \right\} / \left\{ Np_i^{c+1} \right\}$$

The **EM** algorithm (Redner & Walker 1984) possesses several attractive properties (low computational cost, convergence, **constraints** on θ satisfied) compared to that of several alternative methods (Newton, scoring, quasi-Newton) for numerically approximating maximum likelihood estimates. The SEM algorithm, described in (Celeux &Diebolt, 1986), is a recent improvement of this algorithm : it incorporates a Stochastic step to accelerate the (a priori low) convergence. Nevertheless, even in the context of gaussian univariate mixture, the resulting likelihood surface is littered with **singularities** (Titterington & al 1985, ex. 4.3.2, p83). However, with a *good initialization* - obtained, for instance through graphical methods (Bougeard & al, 1989b)- and reasonable *sample size*, one can expect a (S.)E.M. iteration sequence to converge to a *local maximizer* of the log-likelihood function.

Of interest here is the convergence stability of the SEM algorithm in application to the study of the 3 dimensional x=(U,V,W) velocity distribution of 2 samples of A type population I stars, defined in Grenier & al (1985): an A2V sample (N=97) and an Ap sample (N=36).

2. Pertinence of a gaussian mixture model for the A2V sample
On a bidimensional graph UxV, one can foresee the presence of a potential mixture of two populations. The pertinence of the use of a parametric *gaussian* mixture model was shown in (Bougeard & al, 1989c) using the U component of the velocity. Nevertheless, it is known to be unsufficient to check univariate k' gaussian mixture for each variable U,V,W in order to be able to reject the possibility of a k mixture, k>k' (for example, see Titterington & al, 1985, fig 4.10, p68). So, a **multivariate** analysis has to be performed.

3. Numerical stability of the SEM algorithm
Assuming that the distribution of the (U,V,W) velocity sample is a mixture of multivariate gaussian components, we study the convergence stability of the SEM algorithm by performing several runs : each run represents a 200 iteration sequence.

3.1. Firstly, 31 runs of the SEM algorithm have been performed for the A2V sample using an initialization with **K=3** as upper bound of the number of components. Fig 1 shows the respective estimates in U found at each run. A 3 mixture solution appears as very unstable : 19 runs lead to a two mixture solution, 6 runs find no mixture.
3.2. At this stage, the same process has been performed by initializing SEM with **K=2**. The results are summarized on Fig 2 for the proportions ($p_1 > p_2$) found at each run and on Fig 3 for the distributions in U,V,W.Table 1 gives the most stable solution (21 runs over 31) .Due to the fact that it is a multidimensional analysis, the result is slightly different in the U estimations from those obtained in an univariate context by Soubiran & al (1989), Bougeard & al (1989c). For interpretation in terms of star formation bursts, see Gómez & al (1989).

4. Statistical stability of the SEM estimates
The SEM algorithm provides also the probability for each star to belong to one of the estimated components (see Section 1). We compare the resulting classification with the results of classical multivariate data analysis methods.
Firstly, a Principal Component analysis (PCA) was performed on the correlation matrix (variables:U,V,W), by which it became apparent that the first axis (53% of the variance) was highly correlated with U,V lying in the galactic plane. Axis 2 (33.5% of the variance) is correlated with W perpendicular to this plane.The centers of the two gaussian components, projected as supplementary points, are highly correlated with the first axis and 7 stars are not in the same class if we perform a SEM univariate classification only on the U component.
A hierarchical classification was also performed with the reciprocal neighbour algorithm (Lebeaux, 1986; Lebart & al, 1984). The two clusters obtained by the top-level of the hierarchy are in good agreement with the SEM clusters (Bougeard & al, 1989a,b).

Table 1 : *Sample A2V (U,V,W) - SEM stable solution*

Component #1			Component #2		
Proportion : 0.64			Proportion : 0.36		
m_1(u)	m_1(v)	m_1(w)	m_2(u)	m_2(v)	m_2(w)
-20.8	-14.3	-6.8	11.4	1.7	-7.4
variance-covariance matrix			variance-covariance matrix		
176.9	10.8	17.5	50.2	3.8	-13.2
10.8	103.4	-21.6	3.8	33.6	-3.8
17.5	-21.6	75.3	-13.2	-3.8	51.2

Fig. 1: *Sample A2V(U,V,W)- SEM results per run, initialization with K=3*
Graph of m_i(u) per run, the error bar is the square root of the respective variance v_i(u) in Σ_i

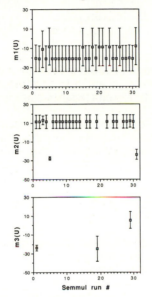

Fig. 3 : *Same as Fig 1, initialization with K=2*

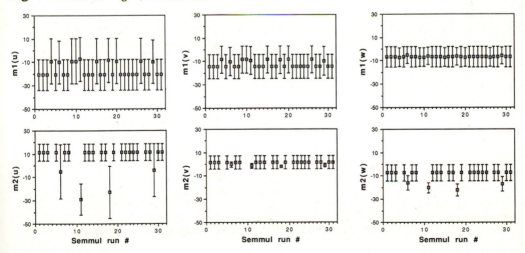

Fig.2 :**Sample A2V**(U,V,W), SEM
(p$_i$) results per run, initialization with k=2

Fig.4:**Sample Ap**(U,V,W),
SEM (p$_i$) results per run,
initialization with K=3

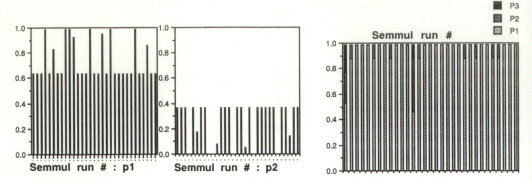

Finally, a linear discriminant analysis based on Fisher linear discriminant function and Mahalanobis distance was also performed to assess the discrimination between the two groups found by SEM . Only 15 stars which were on group 1 according to SEM are affected to group 2 ; this yields to an agreement of 84.5% of well classified stars.

5.   Sensitivity of the SEM algorithm to the sample size
Finally, 31 SEM runs have been performed on the Ap sample (N=36), using an initialization with K=3 as upper bound of the number of the components. Two components were expected (Gómez & al,1989), but Fig 4 shows a high instability (no mixture is found in 20 runs over 31). The main reason is that the sample size is far too small and components are overlapping too much. We note, in the studied application, that a 3 (resp. 2) mixture model yields a SEM estimation of 2+3x3+3x6=29 (resp. 19) unconstrained parameters.

6.   Conclusion
If the sample size is large enough and if the components are well separated, the SEM algorithm has been seen to provide a reasonable good convergence in the estimation of the parameters of gaussian mixtures in stellar kinematics. But it cannot be used rashly in other cases. In the particular case of the A2V sample studied here, SEM results have appeared as nearly stable and in good agreement with other clustering and classification techniques.

7.   Acknowledgements
Principal component analysis, hierarchical classification and discriminant analysis were performed with SAS-ADDAD software on an IBM computer at CIRCE (F-Orsay). We thank Dr Celeux and Dr Diebolt (INRIA, F-Rocquencourt) for allowing the use of SEM software.

8.   References
Bougeard M.L., Arenou F., Gómez A.; 1989a Bull.47th Int. Stat. Inst.: contr. papers, vol1,p161-162 , Paris
Bougeard M.L., Arenou F., Gómez A.;1989b, "Mélanges gaussiens en cinématique stellaire. Une approche
      comparative de méthodes paramétriques et non paramétriques"(in preparation)
Bougeard M.L., Arenou F., Soubiran C., Gomez A., Grenier S.;1989c, (this issue)
Celeux G., Diebolt J.;1986, Revue Stat. Appl., 34, n°2,
Dempster A., Laird N., Rubin D.;1977,J. Royal Stat. Soc. B, 39, p1-38
Gómez, Delhaye, Grenier, Jaschek, Jaschek 1989: Astr. &Astroph. (soumis)
Grenier S. ,Gómez A., Jaschek C., Jaschek M., Heck A.: 1985 Astron. & Astrophys. 145, 331
Lebart L., Morineau A., Warwick K.;1984," Multivariate Descriptive Statistical Analysis", Wiley
Lebeaux M-O.;1986 Manuel de référence ADDADSAS, CIRCE, Orsay, france
Redner R., Walker H.;1984 SIAM,26,n°2, p195-239
Soubiran C., Bougeard M.L., Gomez A., Arenou F.; 1989 (this issue)
Titterington D., Smith A., Makov H.;1985, "Statistical Analysis of finite mixtures", Wiley

Separation de Mélanges Gaussiens Unidimensionnels en Statistique Stellaire
Les Méthodes Graphiques

Bougeard M.L.[1][2], Arenou F.[3], Soubiran C.[3], Grenier S.[3]

(1) Univ. Paris X, IUT, 1 chemin Desvallières, F-92410 Ville d'Avray
(2) URA 1125 CNRS, Observatoire de Paris, F-75014 Paris
(3) URA D0335 CNRS, Observatoire de Paris-Meudon, F-92195 Meudon

Summary . In this Paper, we present several graphic methods available in the context of Gaussian univariate mixture, with application to stellar kinematics. We prove they are pertinent to give indication of the type of the mixture and to exhibit at least approximated estimates of the underlying parameters. The precision of these methods is, as usual in this field of statistical analysis, related to the sample size and the existence of overlapping components in the data.
Keywords : Stellar kinematics - Gaussian mixture - Data analysis -

1. Introduction

De nos jours, on note la popularité croissante des méthodes graphiques dans le cas de l'analyse des données multidimensionnelles où une visualisation a priori demeure difficile. Dans le cas des données univariées, une grande variété de méthodes fournit des procédures exploratoires aisées à mettre en oeuvre et capables de :
- donner des indications confirmant ou infirmant le type de mélange postulé,
- donner des valeurs approchées des paramètres constituant le mélange, valeurs qui pourront être ensuite utilisées soit telles quelles, soit comme valeurs d'initialisation d'un algorithme itératif du maximum de vraisemblance.

Ces méthodes sont démontrées ici en application à l'échantillon d'étoiles proches de type A2V défini dans (Gomez et al 1989). Il sera analysé au regard de la composante U du vecteur vitesse $(l_{II}=0, b_{II}=0)$ dont nous supposerons les erreurs a priori négligeables.

2. Méthodes basées sur la fonction de densité
2.1. Histogramme et multimodalité

La considération de l'histogramme associé aux données traitées (Fig 1) met en évidence un phénomène de multimodalité. Il est clair que ce résultat est sensible à la longueur de l'intervalle de classe choisi pour la représentation (5 km/s pour Fig 1 ; 10 km/s pour Fig 2). S'appuyant sur les données, une coupure drastique en 2 sous-populations définies par :

$$Pop_1 = (U / U < -1 \text{ km/s}) \qquad Pop_2 = (U / U >= -1 \text{ km/s})$$

conduit aux statistiques suivantes :

	moyenne	variance	seuil de significativité du test de normalité de Kolmogorov-Smirnov (Bougeard 1987)
Pop_1	-22.5	147.0	>15%
Pop_2	+10.8	50.2	>15%

et à une première détection approchée d'un mélange à 2 composantes gaussiennes (cf : Bougeard et Grenier 1986).
Note : La fiabilité de ce résultat est sensible au fait que les composants sont assez bien séparés et que la taille de l'échantillon n'est pas trop petite...

2.2 La Méthode de Bhattacharya (1967)

Une méthode plus élaborée, bien qu'également basée sur les données de l'histogramme, a été proposée par Bhattacharya (1967). Si on admet que la variable U étudiée est une variable aléatoire continue dont la densité f(U) vérifie

$$f(U) = \sum_{i=1}^{k} p_i \frac{1}{s_i \sqrt{2\pi}} e^{\left(\frac{-[U - m_i]^2}{2 s_i^2} \right)} \qquad \text{où} \quad \sum_{1}^{k} p_i = 1$$

ce qui correspond à la contribution de k sous-populations (la i-ème étant gaussienne de moyenne m_i, écart type s_i, en proportion p_i), alors U ne sera généralement plus gaussienne (Bougeard 1986). De plus, dans le cas de composantes bien séparées, au voisinage d'un pic de l'histogramme seule une composante donnée, i, joue, et on a

$$\log f(U) \approx \log\left[p_i \frac{1}{s_i \sqrt{2\pi}} \right] - \frac{[U - m_i]^2}{2 s_i^2} \qquad (E1)$$

$$\frac{d}{du} \log f(U) \approx -\frac{U - m_i}{s_i^2} \qquad (E2)$$

On observe que le logarithme de f(U) a une dérivée affine en U de pente négative. Ceci est à la base de la méthode, le terme de gauche dans (E2) étant approché par (1/h) $(\log(n(j+1)) - \log(n(j)))$ où h est l'amplitude constante des classes et n(j) le nombre d'observations dans la classe j.

On trouve Fig 3, le graphique des différences d'ordre un des logarithmes des fréquences observées dans l'histogramme de Fig 1 pour l'échantillon A2V analysé : on y note bien deux droites de pente négative (ajustées à la main pour plus de clarté), chacune correspondant à une potentielle composante gaussienne dans les données.
Note : L'équation (E1) conduit à l'ajustement d'une parabole concave, technique qui a été généralisée récemment par (Postaire et Vasseur, 1981) au cas des reconnaissances de forme sur données multivariées par identification de régions de densité concave.

3. Méthodes basées sur la fonction de répartition : Harding et Cassié

3.1 Nous avons vu que les méthodes basées sur la fonction de densité donnaient des indications sur le nombre éventuel de composants du mélange. L'approche basée sur la fonction de répartition va, elle, nous permettre d'avoir une estimation rapide de la proportion p_i de chaque composant. Plusieurs méthodes sont disponibles (voir Titterington et al 1985 chap.4 pour une revue), nous présentons ici la plus classique dite méthode de Harding (1949) et Cassié (1954) (voir aussi Fowlkes 1979) qui est basée sur l'utilisation de papier gausso-arithmétique (échelle en ordonnée galtonnienne). Ainsi tracée, la fonction de répartition d'une "gaussienne" donne une droite (droite de Henry) tandis qu'un mélange de composantes normales assez séparées conduit à une courbe présentant autant de points d'inflexion qu'il y a de composants (en fait n points d'inflexions pour n+1 composants).

3.2   On trouvera Fig 4 le graphe dans le cas de l'échantillon traité (intervalle de classe : 5 km/s). Il apparaît effectivement un point d'inflexion P d'ordonnée p_1 : cette ordonnée p_1 est égale au pourcentage de la composante 1 (la plus basse sur Fig 4) dans le mélange.

Si on suppose p_1 lu exactement, il suffit de considérer les points Fig 4 d'ordonnée inférieure à p_1 et de multiplier leurs ordonnées par $(100/p_1)$. On remarque que les nouveaux points tracés s'alignent sur une "droite de Henry", confirmant une nature gaussienne de la composante 1 dont il suffit de relever la moyenne et l'écart-type sur le dit graphe. Après élimination de la composante trouvée, le processus est réitéré, ce qui conduit ici aux estimations :

composante 1 : $p_1 \approx 0.65$	$m_1 \approx -22$ km/s	$s_1 \approx 12$ km/s	
composante 2 : $p_2 \approx 0.35$	$m_2 \approx +9$ km/s	$s_2 \approx 7$ km/s	

Plusieurs évaluations de l'ordonnée du point d'inflexion permettent au besoin d'obtenir un encadrement de ces estimateurs.

4.Conclusion

Si les méthodes graphiques connaissent une très forte popularité actuelle dans le cas de données multivariées, leur usage dans le cas univarié semble relativement oublié en dépit de leur pertinence, de leur rapidité de mise en oeuvre et de leur extension au cas log-gaussien (Diamini et al 1984). Nous avons montré ici sur un échantillon d'étoiles A2V, analysé selon la composante de vitesse U, comment les méthodes basées sur la fonction de densité étaient à même de détecter la potentialité d'un mélange à 2 composants. Une analyse plus fine, fournie par la méthode de Harding-Cassié basée sur la fonction de répartition, a confirmé la nature gaussienne de ces composants et a permis d'obtenir des valeurs approchées des paramètres du mélange, estimations précieuses pour d'éventuelles modélisations paramétriques ultérieures, type maximum de vraisemblance.

5. Références

Bhattacharya, 1967 : Biometrics, mars p115-135

Bougeard M.L., 1986 : polycopié cours DEA, Observatoire de Paris

Bougeard M.L., 1987 : Astronomy & Astrophysics 193, 156-166

Bougeard, Grenier, 1986 :"Méthodes graphiques et mélanges gaussiens", unpublished preprint

Bougeard, Arenou, Gomez, 1989 :"Comparative approach of parametric and nonparametric
    classification methods in stellar kinematics, 47ème Int. Stat. Inst. Congress, Paris 1989

Cassié R.M. 1954 : Aust. J. Mat. Freshw. Res. 5, p513-522

Cazes P. 1976 : Revue de Stat. Appl. vol 24 n°1 p63-82

Diamini P. Massé, Aubenque, 1984 : J.SOC. STAT. Paris 125 n°3 p158-163

Harding J.P. 1949 : J. Marine Biol. Ass. 28 p141-153

Gómez, Delhaye, Grenier, Jaschek, Jaschek 1989 : Astr. &Astrop. (soumis)

Postaire J.G., Vasseur C.P. 1981 : I.E.E.E. Trans, PAMI-3, p163_179

Soubiran C. 1988 : DEA, Observatoire de Paris (dir : Bougeard & Gomez)

Titterington, Smith, Makov, 1985 : "Finite mixture of distributions, Wiley & Sons

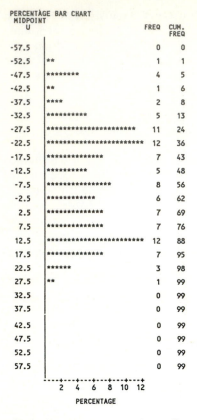

```
PERCENTAGE BAR CHART
MIDPOINT
 U                                  FREQ   CUM.
                                           FREQ
-57.5                                 0      0
-52.5   **                            1      1
-47.5   *******                       4      5
-42.5   **                            1      6
-37.5   ****                          2      8
-32.5   **********                    5     13
-27.5   **********************       11     24
-22.5   ***********************      12     36
-17.5   **************               7     43
-12.5   *********                     5     48
 -7.5   ****************              8     56
 -2.5   ***********                   6     62
  2.5   **************                7     69
  7.5   **************                7     76
 12.5   ************************     12     88
 17.5   **************               7     95
 22.5   ******                        3     98
 27.5   **                            1     99
 32.5                                 0     99
 37.5                                 0     99

 42.5                                 0     99
 47.5                                 0     99
 52.5                                 0     99
 57.5                                 0     99
        +--+--+--+--+--+--+
        2  4  6  8 10 12
          PERCENTAGE
```

Fig 1 : Histogramme par 5 km/S

```
PERCENTAGE BAR CHART
MIDPOINT
 U                                         FREQ
-57.5   *                                    1
-47.5   *****                                4
-37.5   *****                                4
-27.5   **********************               19
-17.5   ******************                   16
 -7.5   *****************                    15
  2.5   ***************                      13
 12.5   **************************           22
 22.5   *******                              5
 32.5                                        0
 42.5                                        0
 52.5                                        0
        +--+--+--+--+--+--+--+
        3  6  9 12 15 18 21
          PERCENTAGE
```

Fig 2 : Histogramme par 10 km/s

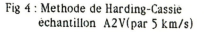
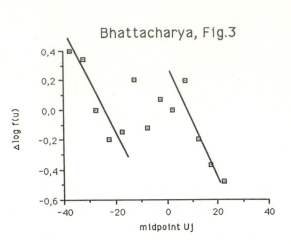

Bhattacharya, Fig.3

Fig 4 : Methode de Harding-Cassie
échantillon A2V(par 5 km/s)

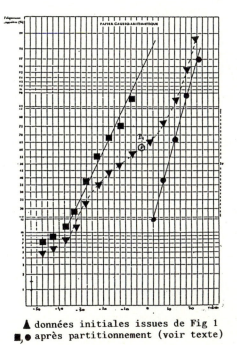

▲ données initiales issues de Fig 1
■,● après partitionnement (voir texte)

Multiple Periodic Analysis of Unevenly Spaced Data

CHEN BING

Purple Mountain Observatory, Nanjing, China

Summary

Because the statistical properties of power spectral analysis are much difficult to establish theoretically, especially to unevenly observed data. In this paper, the multiple periodic analysis methods of irregularly spaced data points are discussed by numerical simulations.

Key Words: Unevenly observed data---Multiple periodic analysis--- Numerical simulations.

1. Introduction.

In this paper, we analyse the spectral analysis methods in details, and point out the spectral analysis techniques developed by Deeming (1975) and Lomb(1976) have a strong limitation on procesing the unevenly observed data, with a noisy signal.

2. Comparative studies of Deeming's and Lomb's analysis techniques

Deeming's periodogram and Lomb's least-squares spectral analysis have a wide application in astronomy. Here, we don't prepare to describe these techniques and their applications, our main aims lie on a comparative studies by numerical simulations. We give two periodic signals which can express as:

$$F(t)=Sin(2\pi \times 11.5t)+Sin(2\pi \times 13.2+\pi/6)$$

and superimposed a noise which satisfies the normally distribution with zero mean and constant variance, and the random numbers distrbuted on the interval $(-1/2,1/2)$ are generated by Monté Carlo method.

We undertake spectral analysis for evenly and unevenly data respectly by the two kinds of techniques. The unevenly spaced data are sampled with data window got by Balona et.al in South Africa and Australia with photometric measurement. Table 1 shows the values of the harmonic contents, according to two different methods, from one result of numerical simulations. From table 1, by comparing the results of two

method \ harmonic content \ data sample	unevenly spaced data			evenly spaced data t = 0.015 n		
	f	A	S	f	A	S
periodogram	11.55	1.37	–0.012	11.50	0.924	0.757
	13.18	1.46	–0.016	13.20	0.937	0.43
least-square spectral analysis	11.51	0.913	0.023	11.50	0.986	0.015
	13.20	0.927	0.549	13.20	0.986	0.523

different methods, we found that the least-square technique is a better method than periodogram in determining multiperiods, and spectral analysis method usually give a erratic spectrum in unevenly data.

3. Validity of observed data

In Sect.2, it has been shown that the leakage and alias will appear during spectral analysis, especially to unevenly observed data. A question is raised naturlly, is it under which condition that the observed data remain good enough to determine the multiperiods by spectral analysis techniques? This problem is too complex to answer on general case. By numerical simulation, we concluded, to get right multiperiods in power spectrum, (1)The length (T) of time series must satisfy T≩30P (2)The observation time must cover at least one fifth of the whole time (T) i.e. t≩T/5.

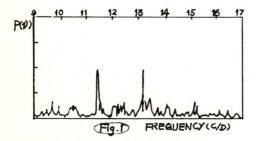

Fig.1--Fig.3 are power spectrum of time series. The time series are got according to the method mentioned in Sect.2, but are sampled with three different obeservation data window.

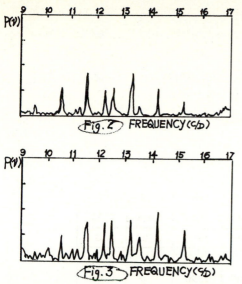

Fig.2   FREQUENCY(c/b)

Fig.3   FREQUENCY(c/b)

4. Prewhiting method

In spectral analysis, we usually determine the multiperiods by prewhiting method. By numerical simulation, it can be found that a noisy signal sampled by unevenly spaced points give a very erratic spectrum. This can be explained as follow: Let a time series concluding 2 periodic signals. They have near frequencies (ν_1, ν_2) and amplitudes (Fig.4,a). Fig.4,b is the window function, the power spectrum is the sum of $f(\nu_1) * \omega$ and $f(\nu_2) * \omega$ (ω represent convolution), the most prominent extremes in power spectrum don't correspounds the frequencies ν_1 and ν_2. A effective method to solve this problem is to add window function, here we can not discuss it in details.

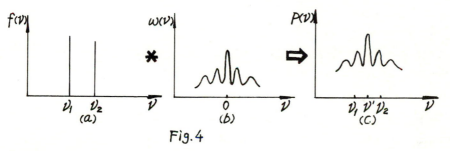

Fig.4

5. Conclusion

Periodogram and least-square spectral analysis are usuful tools in determining multiperiods for unevenly data points, but due to finite observation time and unevenly spaced data, the leakage and alias are very severe if the signal concludes multiperiods and noise. The best way to solve this kind of problem is to present new multiperiods analysis techniques and establish international observation network.

I would like to thank Profs. Xiong Da-run, Zhang He-qi, H.J.Su and J.R.Mo for helpful comments. Some observation data of variables got by Balona et.al are used in this paper. The computing has been carried out on the VAX 11/780 of Purple Mountain Observatory.

References

1. Deeming,T.J. 1975, Fourier analysis with unequally spaced data.
   AP & Space Science (35) 389

2. Lomb,N.R. 1976, Least-square frequency analysis of unequally spaced
   data. AP & Space Science(39) 447

3. Scargle,J.D. 1982, Statistical aspects of spectral analysis of unevenly spaced data. APJ(263) 835

4. Antonello,E. et.al 1986, The two periods of the Cepheid CO Avigue.
   APJ (157) 269-275

5. Balona L.A. et.al Observation of 4 Beta Cephei stars in NGC 6231.
   South Africa Astro. Obs. No.6

6. Heck,A. et.al 1985, On period determination methods.
   A & AP suppl.series (59) 63-72

7. Xu Zen-tao,Chen Bing 1987, The sunpot activity before 17th century.
   Proceedings of the fourth Asian-Pacific regional meeting of the
   International Astronomical Union.(119)

Chen Bing: Purple Mountain Observatory.
            210008 Nanjing,China

Unbiased Estimator of the Inclination Dependence of Galaxy Diameters

JACEK CHOŁONIEWSKI

Astronomical Observatory of the Warsaw University
Aleje Ujazdowskie 4
00-478 Warszawa, Poland
telex: 817063 oauw pl

1 INTRODUCTION AND OVERVIEW

Isophotal diameters of spiral galaxies (which are subject to direct measurements) are function of their inclinations to the line of sight. This dependence is caused by projection effects (which tend to increase diameters) and by extinction on dust (which tends to decrease diameters). There is a number of publications where inclination dependence of galaxy diameters is given (see Tully and Fouque 1985 and references therein), but in none of them the basic property of the used estimators is given: their biases.

In this paper I construct an estimator for determination of inclination dependence of galaxy diameters. I have used for its derivation a method related to the method of moments what guarantees that it is unbiased (at least asymptotically). The results produced by the estimator are presented in Fig.1.

I have used for computations spiral galaxies from CfA catalogue of galaxies (Huchra et al. 1983). The sample is magnitudo-limited what makes data in it censored. This fact is explicitly taken into account in my analysis.

Heidmann et al. (1972) introduced an intuitively correct estimator of inclination dependence of galaxy diameters. My analysis (which will be published elsewhere), based on the formalism presented here, shows that this estimator is biased.

2 NOTATION

A galaxy can be described by its:

> i - inclination
> D - absolute major diameter [kpc]
> a - observed major diameter [arcmin]
> L - luminosity
> f - flux
> r - distance

Moreover:

> subscript "o" - denotes face-on values $(i = 90°)$
> function $\phi(.)$ - denotes selection/censoring-free
> distribution function of its arguments
> (e.g. $\phi(L)$ denotes luminosity function)
> f_{lim} - denotes flux limit of the sample
> $\Theta(f - f_{lim})$ - denotes selection function
> $\Theta = 1$ when $f \geq f_{lim}$
> $\Theta = 0$ when $f < f_{lim}$

3 STATEMENT OF THE PROBLEM

Let us assume that:

$$D = D_o \beta(i) \tag{1}$$
$$L = L_o \alpha(i) \tag{2}$$

what means that the functions $\beta(i)$ and $\alpha(i)$ describe fully inclination dependence of diameter and luminosity respectively.

Our task in this paper is to construct estimator for $\beta(i)$.

4 DISTRIBUTION FUNCTION

We need to construct our estimator a distribution function of (L_o, D_o, r, i) for the sample galaxies: $F(L_o, D_o, r, i)$. This function can be expressed as:

$$F(L_c, D_o, r, i) = \phi(L_o, D_o)\phi(i)r^2\Theta(f - f_{lim}) \tag{3}$$

where $f = L/r^2 = L_o\alpha(i)/r^2$. Eq.3 holds when we assume that (L_o, D_o), i and r are statistically independent and that distribution of galaxies in space is homogeneous.

5 CONSTRUCTION OF THE ESTIMATOR

The expectection value of any function of L_o, D_o, r and i: $Y(L_o, D_o, r, i)$ is:

$$E_i(Y) = \frac{\int_o^\infty dL_o \int_o^\infty dD_o \int_o^\infty dr \ Y(L_o, D_o, r, i)F(L_o, D_o, r, i)}{\int_o^\infty dL_o \int_o^\infty dD_o \int_o^\infty dr \ F(L_o, D_o, r, i)} \tag{4}$$

where index "i" denotes that we take variable "i" as a parameter (according to the nature of our problem).

Using eqs 1, 3 and 4 we have: $E_i(D) = \beta(i)E_i(D_o)$, what gives my estimator for $\beta(i)$:

$$\widehat{\beta(i)} = \langle D\rangle_i / \langle D_o\rangle \tag{5}$$

where instead of $E_i(Y)$, which is unknown, I have used average values computed directly from the sample: $\langle Y\rangle_i$. It is known (Eadie et al. 1982) that $E_i(Y) = \langle Y\rangle_i$ for an infinite sample so my estimator is unbiased only asymptotically.

Results given by the estimator are presented in Fig.1.

6 CONCLUSIONS

I have constructed in this paper, using exact formulae of mathematical statistics, an estimator describing inclination dependence of galaxy isophotal diameters. My analysis takes into account selection (censoring) effects and explicitly expresses all neccesary assumptions.

The main intention of this paper is to convince astronomers that such analysis is possible and should be done also in other astronomical contexts.

ACKNOWLEDGMENT

This paper was partly supported by the project: CPBP 01.11.

REFERENCES

Eadie, W. T., Drijard, D., James, F. E., Roos, M. & Sadoulet, B., 1982. "Statistical Methods in Experimental Physics", North-Holland Publishing Company, Netherlands.

Heidmann, J., Heidmann, N. & de Vaucouleurs, G., 1972. Mem. R. astr. Soc., **75**, 105.

Huchra, J., Davis, M., Latham, D. & Tonry, J., 1983. Astrophys. J. Suppl., **52**, 89.

Tully, R. B. & Fouque, P., 1985. Astrophys. J. Suppl., **58**, 67.

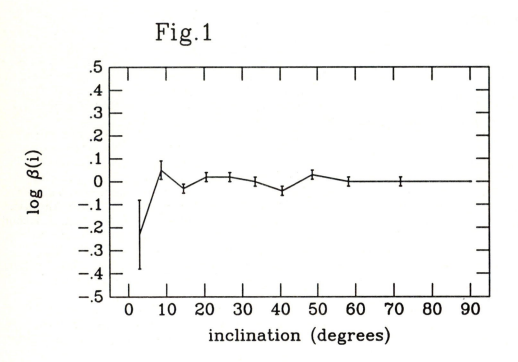

Fig.1

Astrometric catalog error estimates

C. S. COLE

U.S. Naval Observatory – Washington, DC

ABSTRACT: Astrometric catalogs, both observed and compiled, include error estimates for the star parameter estimates. Astronomers who use these catalogs are often unaware of the imperfect nature of the error estimates and consequently may make assumptions which are not justified.

This paper discusses the relative merits of various types of error estimates used in astrometric catalogs. Also discussed are the methods being used to estimate the errors of the Faint Fundamental Extension of the Fifth Fundamental Catalogue (FK5). Finally, comparisons of error estimates with position differences show that published error estimates are generally underestimated.

1. INTRODUCTION

Astrometry, more than any other branch of astronomy, produces reference data which are used by much of the scientific world. Because of this unique role, it is prudent that what is provided to astrometric consumers be examined for consistency with standard statistical practice. Error estimates in various observed and compiled catalogs differ in what is actually estimated and how those estimates are calculated.

Many catalogs published before 1950 quoted estimates for the *probable error*. The probability of a parameter being different from its estimate by more than the probable error is one half. Recent catalogs commonly use estimates of the *mean (or standard) error*, ideally an estimate of the standard deviation of the probability distribution of the parameter estimate. There are two methods currently in use for computing estimates of the mean error. The first method will be called *errors from weights*. This method has a long history and was developed for its computational simplicity and because both very large and very small error estimates are avoided. The second method, which will be called *errors from residuals*, follows the common statistical practice of estimating errors from the residuals of an adjustment process (e.g. least squares).

In discussing these two types of error estimates, this paper will always distinguish between a parameter and its estimate. In general, no particular probability distribution is assumed, however certain practices or estimates implicitly assume a normal distribution.

2. METHODS OF ERROR ESTIMATION

Consider the very simple linear system

$$Y_\iota = \beta + \epsilon_\iota, \quad \iota = 1, 2, 3 \ldots n \qquad (1)$$

where Y_ι is an observed quantity, β is the target parameter and the ϵ_ιs are independent with identical probability distributions. Using the method of least squares one estimates β by

$$b = \frac{1}{n} \sum_{\iota=1}^{n} Y_{\iota}. \qquad (2)$$

The variance of b, σ^2, can be estimated from the law of propagation of errors,

$$\hat{\sigma}^2 = \sum_{\iota=1}^{n} (\frac{\delta b}{\delta Y_{\iota}})^2 \, \hat{\sigma}_{\iota}^2 \qquad (3)$$

where $\hat{\sigma}_{\iota}^2$ is an estimate of the variance of Y_{ι}. This reflects the idea of the first method of estimating the mean error, wherein it is computed from the quality of the input data regardless of how well those input data agree. In practice this is done with a mean error of unit weight, ϵ_0, and a sum of weights

$$\hat{\sigma}^2 = \epsilon_0^2 \, / \sum_{\iota=1}^{n} w_{\iota} \qquad (4)$$

where w_{ι} is the weight of observation ι.
σ^2 can also be estimated by

$$\hat{\sigma}^2 = \frac{1}{n-1} \sum_{\iota=1}^{n} (Y_{\iota} - b)^2. \qquad (5)$$

This is the idea of the second method; the error is estimated from the agreement of the input data. (Weights can also be used with the second method if desired.)

The errors from weights suffer from two drawbacks. First, a single case which has "noisy data" will not have a representative large error estimate and second, it does not follow standard statistical practice. When an estimate of the mean error is stated, the initial assumption is that the error was estimated via the second method. The second method has the drawback that it is possible to estimate the error to be zero. While an individual looking at this error estimate would understand that it is a result of the probability distribution of the parameter estimate, a person using the data in machine readable form might not allow for the possibility of a zero error estimate and assign an infinite weight to one data point in an adjustment.

The method of estimating errors from weights was developed for its computational simplicity. With today's electronic computers, it is actually easier to compute error estimates from residuals than from weights. Many recent catalogs including the AGK$_3$R (Scott and Smith 1971) and the SRS (Smith and Jackson 1985) quote errors estimated from residuals.

3. ERROR ESTIMATES FOR THE FAINT FUNDAMENTAL EXTENSION

The FK5, now being completed at the Astronomisches Rechen-Institut (ARI), will consist of the Basic FK5 and the FK5 Extension. The FK5 Extension will consist of about 1000 stars from the FK4 Sup (ARI 1963) and about 2000 fainter stars. The preparation of the fainter stars, the Faint Fundamental

Extension (FFE), is being carried out jointly by the USNO and the ARI (IAU 1988, Schwan 1988, Corbin 1985a, 1985b).

Positions and proper motions for the FFE are being computed at the USNO with errors estimated from residuals. However, in order to have a uniform system of error estimates for the entire FK5, it will be necessary to estimate errors from weights for the FFE. The planned procedure for estimating errors from weights is basically as follows.

The error from weights, ϵ_{ari} (e.g. $\epsilon_\alpha \cos \delta$ or ϵ_δ), is equal to the mean error of unit weight, ϵ_0, divided by the root of the sum of the weights for that star.

$$\epsilon_{ari} = \epsilon_0 / \sqrt{(\sum_\iota w_\iota)} \tag{6}$$

where w_ι is the weight of catalog ι for a given star. The summation is over all catalogs contributing to the position of an individual star. In order to compute the error estimates in equation (6), one must determine a mean error of unit weight.

To determine ϵ_0, the condition that the average error estimate of the FFE stars should not change is enforced:

$$\sum_{FFE} \epsilon_{usno} = \sum_{FFE} \epsilon_{ari} \tag{7}$$

where ϵ_{usno} is the error estimate computed from residuals and the summation is over all stars in the FFE. Thus ϵ_0 can be found by substituting equation (6) into equation (7) and solving for ϵ_0.

$$\epsilon_0 = \sum_{FFE} \epsilon_{usno} / \sum_{FFE} \sqrt{(1/\sum_\iota w_\iota)} \tag{8}$$

Position errors are estimated using the procedure described above. Proper motion errors are estimated by substituting $w_\iota \tau_\iota^2$ in place of w_ι in equation (6), where τ_ι is the difference (catalog epoch minus mean epoch) for catalog ι and a given FFE star.

A preliminary version of the FFE was constructed with errors from residuals and, using the above procedure, errors from weights were computed. Figure 1 shows histograms of $\epsilon_\alpha \cos \delta$ and ϵ_δ against the number of stars. Note that the errors from weights have a much narrower distribution than the errors from residuals.

4. CATALOG COMPARISONS

Consider a star parameter β_ν (e.g. $\alpha_\nu \cos \delta_\nu$ or δ_ν) and two independent estimates $b_{\nu 1}$ and $b_{\nu 2}$. The variance of $(b_{\nu 1} - b_{\nu 2})$ is

$$\sigma_\nu^2 = \sigma_{\nu 1}^2 + \sigma_{\nu 2}^2 \tag{9}$$

where $\sigma_{\nu 1}^2$ and $\sigma_{\nu 2}^2$ are the variances of $b_{\nu 1}$ and $b_{\nu 2}$ respectively. Two methods for estimating σ_ν^2 are

$$\hat{\sigma}_\nu^2 = (b_{\nu 1} - b_{\nu 2})^2 \text{ and} \tag{10}$$

$$\hat{\sigma}_\nu^2 = \hat{\sigma}_{\nu 1}^2 + \hat{\sigma}_{\nu 2}^2 \tag{11}$$

where $\hat{\sigma}_{\nu 1}^2$ and $\hat{\sigma}_{\nu 2}^2$ are variance estimates of $b_{\nu 1}$ and $b_{\nu 2}$ respectively. For

Figure 1. Distribution of Preliminary FFE Error Estimates

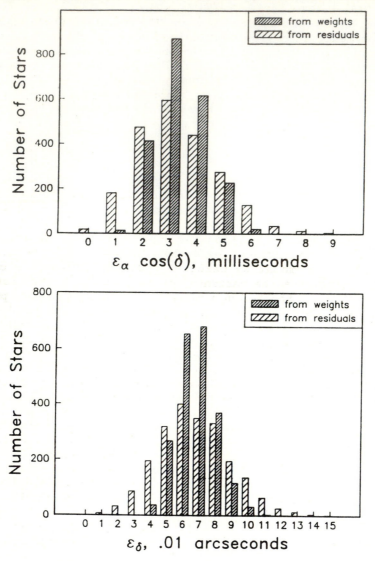

$\varepsilon_\alpha \cos(\delta)$, milliseconds

ε_δ, .01 arcseconds

an individual star, equation (10) gives a poor (but unbiased) estimate of $\hat{\sigma}_\nu^2$. But by comparing

$$\sum_{\nu=1}^{N} (b_{\nu 1} - b_{\nu 2})^2 \quad \text{and} \quad \sum_{\nu=1}^{N} \hat{\sigma}_{\nu 1}^2 + \sum_{\nu=1}^{N} \hat{\sigma}_{\nu 2}^2 \tag{12}$$

where N stars are available for comparison, one can gain an insight into the quality of the various error estimates. This is demonstrated in the following astrometric catalog comparisons.

The positions in several astrometric catalogs were computed for the same epoch and position differences formed. The sums of squares of these position differences were then compared to the sums of squares of the error estimates in the various catalogs. Tables 1 and 2 give these comparisons.

Table 1. Leoncito 50 - FK4 at WL_{50} individual epochs, 70 sub-polar and 4 FK4 orbit star positions removed, N=1115.

	Sums of Squares			
	$\alpha \cos(\delta)$		δ	
	$\overset{s}{.}001^2$	%	$\overset{"}{.}01^2$	%
Position differences	276696		397379	
FK4 error estimates	114019	41.2	175480	44.2
WL_{50} error estimates	15625	5.6	64142	16.1
Unexplained Excess	147052	53.1	157758	39.7

Percentages shown are percentages of the sums of squares of the position differences. The percentages of unexplained excess are interesting for they indicate the amount by which the two catalogs being compared collectively underestimate the variances of the star parameter estimates.

The FK4 (Fricke and Kopff, 1963) positions were computed from published positions, proper motions, parallaxes and radial velocities. FK4 error estimates include individual position and proper motion error estimates and system position and proper motion error estimates. The system error estimates of the FK4 do not include estimates of the mean error of the equinox and equator points. About three quarters of each FK4 error estimate in Table 1 comes from the associated proper motion error estimate due to the difference in mean epoch between the FK4 and the WL_{50} (Smith 1978).

Table 2. SRS - SAOC at SRS individual epochs, N=18963.

	Sums of Squares			
	$\alpha \cos(\delta)$		δ	
	$\overset{s}{.}001^2$	%	$\overset{"}{.}01^2$	%
Position differences	62269276		146751987	
SAOC error estimates	45274238	72.7	111735145	76.1
SRS error estimates	1992344	3.2	2419273	1.6
Unexplained Excess	15002694	24.1	32597569	22.2

SAOC (SAO, 1966) positions were computed using published positions and proper motions. About ninety percent of each SAOC error estimates in Table 2 comes from the associated proper motion error estimate due to the difference in mean epoch between the SAOC and the SRS.

The purpose of these comparisons is to show that published astrometric catalog variances are underestimated by one fourth to one half. The catalogs used in these comparisons were chosen as being representative of recent astrometric products. One notices that most of the error estimates are due to the proper motion error estimates and that the SAOC proper motion error estimates are more realistic than those of the FK4.

The source of the "Unexplained Excess" in Tables 1 and 2 has been discussed elsewhere. It is generally attributed to ambiguous *systematic errors* but has been more accurately defined in terms of *underparameter-ization* (Firneis and Firneis 1975) and *parameter variance* (Eichhorn and Cole 1985, Eichhorn and Williams 1963).

The multiple correlations of the model parameters and the target parameters are included in the target parameter error estimates in a global (or simultaneous) reduction. As global reductions become increasingly practical with the continuing development of computers and software, errors estimated from weights will gradually disappear. It is only a matter of time before normal points and tabular corrections are replaced by global reductions and analytical models.

5. SUMMARY

Recent astrometric catalogs have become standardized by reporting estimates of the mean error, rather than estimates of the probable error, as was common prior to 1950. Two distinct methods of estimating the mean error are currently in use. Recent applications of both of these methods have tended to underestimate the true mean error. The increasing use of reduction procedures that simultaneously estimate both model parameters and target parameters will lead to errors, estimated from residuals, free of the problems of underestimation and estimates of zero error.

The author wishes to thank the Astronomical Data Center at the NASA Goddard Space Flight Center for providing some of the data used in this investigation.

REFERENCES

ARI 1963, *Veröff. Astron. Rechen-Inst. Heidelberg* 11
Corbin, T. 1985a, in *IAU Symposium 111, Calibration of Fundamental Stellar Quantities*, D.S. Hayes et al. (eds.), 53-70
Corbin, T. 1985b, *Cel. Mech.* 37, 285-298
Eichhorn, H. and Cole, C.S. 1985, *Cel. Mech.* 37, 263-275
Eichhorn, H. and Williams, C.A. 1963, *Astron. J.*, **68**, 221-231
Firneis, M.G. and Firneis, F.J. 1975, *Astron. Nachr.* **296**, 95-100
Fricke, W. and Kopff, A. 1963, *Veröff. Astron. Rechen-Inst. Heidelberg* 10
IAU 1988, "Reports on Astronomy", *Trans. IAU XXA*, 29-39
SAO 1966, *Star Catalog: Positions and Proper Motions of 258,997 Stars for the Epoch and Equinox of 1950.0*, Smithsonian Institution, Washington
Schwan, H. 1988, in *IAU Symposium 133, Mapping the Sky*, S. Débarbat et al. (eds.), 151-157
Scott, F.P. and Smith, C.A. 1971, in *Conference on Photographic Astrometric Technique*, H. Eichhorn (ed.), 181-190
Smith, C.A. 1978, in *IAU Colloquium 48, Modern Astrometry*, F.V. Prochazka and R.H. Tucker (eds.), 447-453
Smith, C.A. and Jackson, E. 1985, *Cel. Mech.* 37, 277-284

Analysis of a Stellar Velocity Field through Orthogonal Series of Functions

F. COMERON – J. TORRA

**Departament de Fisica de l'Atmosfera, Astronomia i Astrofisica
Universitat de Barcelona, Barcelona, Spain**

Abstract

Orthogonal series development is applied to a sample of stars in order to find out the characteristic parameters of the local galactic kinematics. This method reveals itself as a useful tool to discern whether or not the proposed kinematic models can explain satisfactorily the observed field of velocities.

----

1 - Objectives and stellar sample

Our objective is to find some of the characteristic parameters of the galactic systematic field of velocities for a sample of young, bright main- sequence stars via the fitting of a theoretical field to the observed one. Several models are proposed in order of growing complexity, and new systematic effects are added as suggested by the O-C residuals.

The analysed sample of O and B stars was compiled by Figueras (1986) from the SAO Catalog with astrophysical data (Ochsenbein, 1980), and contains nearly 3.000 stars. We use only radial velocities, since proper motions are rather meaningless for most of this sample because of the great mean distance of the stars.

2 - Method

The main tool of our work consists of the use of orthogonal series of functions. Essentially, we develop the observed field of radial velocities of stars into a series of orthogonal functions, and obtain a set of "observational coefficients". Then, we also develop the geometry of the kinematic effects of the theoretical field we want to fit and obtain the corresponding field of "theoretical coefficients".

The result of multiplying the theoretical coefficients by different combinations of kinematic parameters should match the observational coefficients, provided that the model is good enough. So, the final part of our work will be solving of a linear system of equations and finding the best set of kinematic parameters.

3 - Orthogonal functions

Orthogonality of the terms of a series development implies low correlations among them. The advantages of such developments have been explored and applied to a variety of astronomical problems. mainly by Brosche and co-workers (see e.g. Brosche (1966), Brosche and Schwan (1981) and Brosche et al. (1989)).

The way of calculating the development coefficients can be either least- squares or numerical integration; in this later case, each coefficient is calculated independently from the others.

Another advantage of using series development is that the information is dramatically compacted: we do not need to solve a system of N equations, one for each of the N stars, but instead we have a system with one equation per coefficient. The number of coefficients needed for an adequate description of the velocity field is not very high- of course, it is much lower than N.

Finally, we remark that the O-C residual coefficients have an immediate geometric interpretation according to the indexes of the functions in which they appear.

Choosing the sets of orthogonal functions for series developments is basically a geometrical matter. The arguments of the functions are the spatial coordinates; in our case, we have restricted to the galactic plane, so our coordinates are projected distance and galactic longitude.

We want the orthogonal set of functions to satisfy the following three conditions:

- To be a complete set of functions, i.e., any function of these arguments may be reproduced by a linear combination of the orthogonal ones.
- To be periodical when one of its coordinates is periodical, to avoid discontinuities at the origin of this coordinate.
- To have a uniform behavior within the desired interval, in order to make the numerical integration which leads to determination of the coefficients easier.

We have chosen for our problem a set consisting of products of Jacobi polynomials of orders (0,1) (Nikiforov and Ouvarov, 1983) and Fourier terms.

In addition to the orthogonal development coefficients, we can also make an estimation of their probable errors due to the observational inaccuracies in the radial velocity, the uncertainties in the determination of the parallax and the cosmic dispersion of stellar velocities. With the calculated coefficients and the estimated errors, we can write a likelihood function whose unknowns are the combinations of kinematic parameters that fit the theoretical model to the observed one. Maximizing of the likelihood function provides the best set of kinematic parameters we are searching for.

4 - Results

We have proposed three successive models enclosing the following effects:

Model A: Solar motion only.
Model B: Solar motion and axisymmetric differential rotation.
Model C: Solar motion, axisymmetric differential rotation and spiral arms.

Aplication of model A leaves a very high residual at the terms with $m = 2$, $q = 0$, which is just the expected effect resulting from differential rotation. Model B solves entirely this problem.

Model B also shows high residuals in some terms, which suggests the need for other systematic terms to be taken into account. We have then tried model C with the result that, although some residuals are improved, no reasonable choice of the number of arms and pitch angle can explain the high discrepancies at some coefficients.

The terms with higher residuals in model C are those with m = 0, i.e., independent of the galactic longitude. This suggests that an expansion of the local system of stars may be taking place, as suggested by some authors (Frogel and Stothers, 1977). Our method is quite sensitive to the existence of such an expansion.

We have thus proposed a fourth theoretical model (model D) which incorporates this expansion to the features of model C, and have tried to adjust the following parameters for this last effect: rate of expansion and rate of decay with distance, distance of the center of the expansion to the Sun and galactic longitude of this center. It is found that the most conflictive residuals are substantially reduced after application of this model.

The numerical results and their physical significance are discussed elsewhere (Comerón, 1989), but we can point out the good agreement between them and those found by other authors using independent techniques. Particularly, the kinematic parameters found for spiral structure are consistent both with the physical constrains of the density waves theory and with most of the observational data which locate the Sun near the inner edge of a spiral arm. The local expansion and decay rates are also consistent with the ages of the young stars which compose our sample. This good agreement strenghtens our confidence in the validity and advantages of the presented method.

Acknowledgements

This work has been supported by the CICYT under contract ESP88-0731.

References

Brosche, P.: 1966, Veröff. des Astronomischen Rechen-Instituts nr. 17, Heidelberg.
Brosche, P. & Schwan, H.: 1981, Astron. Astrophys. 99, 311.
Brosche, P., Wildermann, E. & Geffert, M.: 1989, Astron. Astrophys. 211, 239.
Comerón, F.: 1989, Master Thesis, University of Barcelona (to be published)
Figueras, F.: 1986, Ph. D. Thesis, University of Barcelona.
Frogel, J.A. & Stothers, R.: 1977, Astron. J. 82, 890.
Fuchs, B. & Wielen, R.: 1987, in "The Galaxy", G. Gilmore & B. Carswell (eds.), D. Reidel, Dordrecht.
Nikiforov, A. & Ouvarov, V.: 1983, "Fonctions Spéciales de la Physique Mathématique", Mir, Moscow.
Ochsenbein, F.: 1980, Bull. Inform. CDS no. 19, 74.
Rohlfs, K.: 1977, "Lectures in Density Waves Theory", Springer Verlag, Berlin

ω Orionis : an Illustrative Case of Period Analysis

J. Cuypers
Koninklijke Sterrenwacht van België
Ringlaan 3, B-1180 Brussel, Belgium

L.A. Balona
South African Astronomical Observatory
P.O.Box 9, Observatory, 7935 South Africa

1. Introduction

It is generally known that period analysis of variable stars can lead to erroneous results when only small data samples are available. This is well illustrated in the study of Be stars. Although long-term light variations were often observed in these stars, the short-period light variability of these stars was only recently established (see e.g. Percy(1987) for a review). The determination of the periods remains difficult for several reasons: the time scale of the variations is close to one day, the amplitudes are small, the light curves are non-sinusoidal and sometimes a trend or a long-term variation is present as well. When an object has been intensively observed in order to obtain a reliable period, one can easily illustrate the difficulties and errors that could arise when only limited data samples are available.

2. The period of ω Orionis

ω Orionis was included twice in an intensive multi-site photometric observing campaign (Balona et al. 1987, Balona et al. 1990). Both times observations were made at South African Astronomical Observatory (SAAO) and at the European Southern Observatory (ESO). The period of ω Orionis was found to be very close to one day by Balona et al. (1987). It is almost impossible to obtain good phase coverage from one site. When the star is observed only once a night (fig. 1), the short-period variability can easily be confused with a trend or a longer period, as was done for most of the Be stars in the past. The periodogram (fig. 2) indicates that the frequencies 0.02, 0.98, 1.02, 1.98, 2.02, ... cycles/day are equally probable. As known from time series analysis the significant frequencies can be written as Δf=0.02 cycles/day or n $\pm$ Δf with n=1,2,... cycles/day. Even if the short-period variability is well established by observations over several nights (fig. 3 and 5) the period is not unambiguously determined since the periodogram indicates different values for the most probable period (fig. 4 and 6). Observers from different sites would consider a different period as the real one. Although hints can be found in some case studies (see e.g. Cuypers, 1983), no adequate methods or statistical tests seem to exist to identify the real period among its aliases. Only combining the two observational sets will in this case resolve the ambiguity (fig. 7). Note how the frequency 0.98 cycles/day is not significant at all in the periodogram of the combined data.

An attempt to fit the data with a sinusoidal wave (fig. 8) shows that large residuals are present. This has also been observed in other Be stars and named flickering by Cuypers et al. (1989). A phase diagram with a period of 0.98 days indicates also that still one third of the light curve was not observed (fig. 9). There is no doubt on the reality of the 0.98 days periodicity as seen on the phase diagram (fig. 10) when a few SAAO observations of 1987 were added. It is however not excluded that twice this period is the physically real period of the light variations of this star as is the case in many other Be stars.

3. Conclusions

Analysis of the light variations of the bright Be star ω Orionis showed clearly how erroneous results can occur in period analysis of variable stars. Periods close to one day can only be resolved unambiguously if the stars are intensively observed from more than one site. Short-period variability can easily be confused with a trend or a longer period, as was the case for the Be stars, of which a large fraction, if not all, turned out to be short-period variables. The small and limited experiment described here indicates that it is probable that the periods of a lot of bright stars are still not detected or not correctly identified.

References

Balona, L.A., Marang, F., Monderen, P., Reitermann, A., Zickgraf, F.-J., 1987, Astron. Astrophys. Suppl. Ser. **71**, 11.

Balona, L.A., Cuypers, J., Marang, F., 1990, in preparation.

Cuypers, J., 1983, Astron. Astrophys. **127**, 186.

Cuypers, J., Balona, L.A., Marang, F., 1989, Astron. Astrophys. Suppl. Ser., in press.

Percy, J.R., 1987, *The Physics of Be Stars*, ed. A. Slettebak and T. Snow, Cambridge University Press, Cambridge, p. 44.

Based on observations made at the European Southern Observatory (ESO) and at the South African Astronomical Observatory (SAAO)

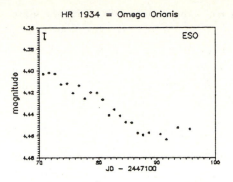

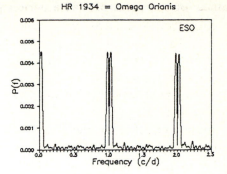

Fig. 1. ESO Data of ω Orionis (one obs./night, 1988).

Fig. 2. Lomb–Scargle Periodogram of the ESO 1988 Data (one obs./night).

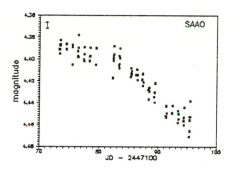

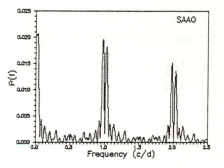

Fig. 3. All SAAO Data of ω Orionis (1988).

Fig. 4. Lomb–Scargle Periodogram of the SAAO 1988 Data.

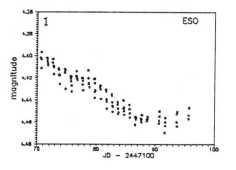

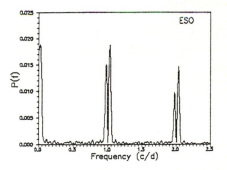

Fig. 5. All ESO Data of ω Orionis (1988).

Fig. 6. Lomb–Scargle Periodogram of the ESO 1988 Data.

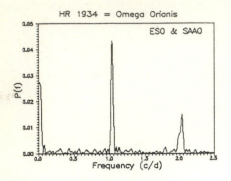

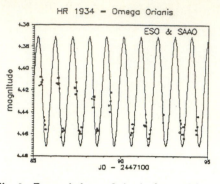

Fig. 7. Lomb-Scargle Periodogram of the combined ESO-SAAO 1988 Data.

Fig. 8. Part of the ω Orionis data with an approximate fit.

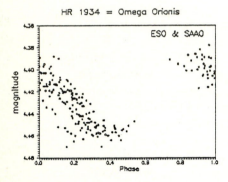

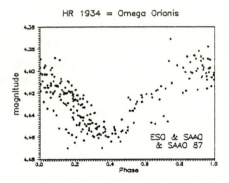

Fig. 9. Phase diagram of ω Orionis with the combined ESO-SAAO 1988 Data.

Fig. 10. Phase diagram of ω Orionis with the ESO-SAAO Data, including observations made at SAAO in 1987.

Detection of time variability from synthetic images

L. DENIZMAN

Observatoire Astronomique, Centre de Données Stellaires, Strasbourg
(France)

Method

Whenever a picture is converted from one form to another, e.g. copied, scanned, transmitted or displayed the quality of the output will be lower than the input. Many enhancement techniques are designed to compensate for the effects of a specific (known or estimated) degradation processes of the images.

It is well known that, because of the optical characteristics of the Schmidt camera system, after a defined angular radius from the telescope axis pixels are geometrically distorted. Moreover another problem is photometric(intensity) registration of the plates which are taken in different observational conditions. This is unavodiable because of the instrumental and observational variations such as emulsion on the plates, variation of the reflectivity of the mirrors, variation of the zenith angle etc.

Thus the problem we face can be defined as follows: For searching the time variability of a set of pixels of the same region (or regions) on the plates how to transport the mentioned pixels to a common reference system by roto-translation and scaling (geometric registration) and lastly by a photometric registration of them, to get a homogeneous data set named data cube. In other words it is necessary to eliminate all the effects which cause variability but the real physical variability between two set of pixels.

In this work a procedure for the simulation of the best algorithm is searched for to automate the detection of time variability of point-like sources, from photographic-like synthetic images in STARLINK (1) and ASTRONET (2) software environments.

A short description of the selected programs used in simulation is given as follows:

- IMAGE:This program generates a stellar field (up to 10 star profiles) with Gaussian stellar profiles, standard normal noise and stationary background (3). Synthetic image is stored in a two dimensional array.

- OPERATOR:In order to get a geometrically distorted image this program is used which simply rotates, translates and/or scales the given coordinates of the objects on the 2-D frame with respect to the axis center. Then program calculates linear transformation coefficients.

These coefficients are tested in the STARLINK by the routines XYKEY, XYLIST and XYFIT

- REGISTER:This program computes the new intensity values of the output picture with the use of previous transformation coefficients. -

Method

Whenever a picture is converted from one form to another, e.g. copied, scanned, transmitted or displayed the quality of the output will be lower than the input. Many enhancement techniques are designed to compensate for the effects of a specific (known or estimated) degradation processes of the images.

It is well known that, because of the optical characteristics of the Schmidt camera system, after a defined angular radius from the telescope axis pixels are geometrically distorted. Moreover another problem is photometric(intensity) registration of the plates which are taken in different observational conditions. This is unavodiable because of the instrumental and observational variations such as emulsion on the plates,variation of the reflectivity of the mirrors, variation of the zenith angle etc.

Thus the problem we face can be defined as follows:For searching the time variability of a set of pixels of the same region (or regions) on the plates how to transport the mentioned pixels to a common reference system by roto-translation and scaling (geometric registration) and lastly by a photometric registration of them, to get a homogeneous data set named data cube. In other words it is necessary to eliminate all the effects which cause variability but the real physical variability between two set of pixels.

In this work a procedure for the simulation of the best algorithm is searched for to automate the detection of time variability of point-like sources, from photographic-like synthetic images in STARLINK (1) and ASTRONET (2) software environments.

A short description of the selected programs used in simulation is given as follows:

- IMAGE:This program generates a stellar field (up to 10 star profiles) with Gaussian stellar profiles,standard normal noise and stationary background (3). Synthetic image is stored in a two dimensional array.

- OPERATOR:In order to get a geometrically distorted image this program is used which simply rotates, translates and/or scales the given coordinates of the objects on the 2-D frame with respect to the axis center. Then program calculates linear transformation coefficients. These coefficients are tested in the STARLINK by the routines XYKEY, XYLIST and XYFIT

- REGISTER:This program computes the new intensity values of the output picture with the use of previous transformation coefficients. -

*Nearest pixel:*Transformed pixel coordinates are generally a real number. In this mode simply, the nearest pixel center or nearest integer value of the input pixel assumed to be the output. So this value is assigned to the output pixel. *-Bilinear interpolation:*This mode scans the input picture with a box of four neighbour pixels. And computes the output value of the pixel with linear interpolation in both X and Y directions, then the average value of these two interpolations is assigned to the output pixel. *-2nd degree polynomial interpolation:*In this mode scanning of the input picture is done with 9 neighbour pixels. And output pixel value is calculated with the 2nd degree Lagrangian interpolation in both X and Y directions, lastly output value is the averege of the interpolations. *-3rd degree polynomial interpolation:*Same as previous mode but 16 neighbour pixels are used for the scanning the image and 3rd degree Lagrangian interpolation applied.

The correspondent of this program in STARLINK is RESAMPLE which uses first two methods of approximation as defined; also *uniform* mode is added which uses a weighted mean of the nearest 9 pixels.

- STARFIT:This STARLINK program is used to find a set of parameters describing a model star image, which can be used when performing stellar photometry. The program combines a number of star images specified by the user and determines a mean seeing disk size, radial fall-of parameter, axis ratio and the axis inclination of a model star image.

- STARMAG:This program is used to determine the integrated brightness and the magnitude of star images by fitting a 2-D surface to the data. The program is intended for the images which have been previously linearised and spatially calibrated, although it is not necessary for the background to have been subtracted. The method assumes that all star images have the same shape.

- ROOTMSQER:This program calculates the root mean square error between two images and generates a percentage error file, an image frame which is the percentage of the intensity difference in registered image.

Conclusions

After the construction of the simulation procedure with the available software environments the detection of the time variability from Schmidt-like synthetic images for point-like sources is performed. For this one template and eight image frames which are containing four standard stars, stationary background and one variable star which has a known form of intrinsic light variation, are generated. **Figure 1 (a,b,c)** presents 3-D graphics of template plate, first plate and registered plate without noise respectively.**Figure 2 (a,b,c)** ,same as previous figure but with S/N= 10. Image frames include photometric and geometric degradation together; so two registration steps used respectively. Images first registered photometrically then this photometrically registered image is geometrically registered onto the template frame.From the tests performed in this simulation final conclusions can be given as follows:

1. Geometric registration causes the biggest intensity loss for images which increases proportionally with the increasing noise.

2. Photometrical registration tests showed that intensity loss and error for estimating the magnitudes are more smaller than the geometric case; i.e. for S/N=infinite Varδm=0.0004 and for S/N=4 Varδm=0.0294 ,for the photometric registration case; and for S/N=infinite Varδm=0.0839 and for S/N=3 Varδm=0.0970 , for the geometric registration case.

3. Polynomial registration algorithms did not give significantly better results corresponding to bilinear cases.

4. Light curve of the star on the synthetic image frames is chosen as a Gaussian. Lastly deviations from this model light curve for every registration algorithm is computed in order to see the limiting magnitude amplitude for detection of time variability which is smaller than 0.1 magnitudes. This result seems good especially for a relatively high noise level such as S/N=10.

References

(1) STARLINK, *User notes*;1984

(2) G. Sedmak,*Data Analysis in Astronomy,***Eds.:** V. Di Gesu, L. Scarsi,P. Crane, J.H Friedman and S. Levialdi;*Plenum Press* ;1985

(3) P. Santin ,*private communication*; 1984

Acknowledgements

This research was proposed and supervised by Prof. G. SEDMAK, director of the *Astronomical Observatory of Trieste (OAT)* and submitted to the *International School for Advanced Studies , Astrophysics section,* Trieste as a *"Magister Philosphiae"* thesis. All the computer tests and simulation is carried on DEC VAX 750, and ARGS system of OAT. I wish to thank to the staff of ASTRONET Trieste pole, technicians of the OAT computer center, and especially to Dr. P. Santin and Dr. F. Pasian for very useful discussions and support which they gave during my stay.

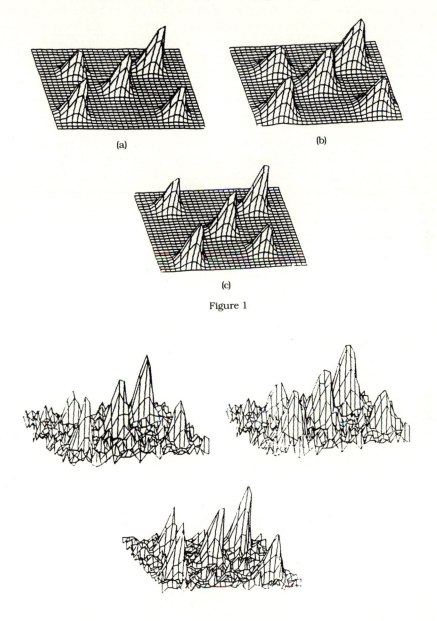

(a)

(b)

(c)

Figure 1

Figure 2

Regularization Parameter Estimation

L.DESBAT

Groupe d'Astrophysique de Grenoble
CERMO BP 53x
38041 Grenoble cedex
France

1 REGULARIZED DECONVOLUTION

We will consider a simple deconvolution problem:

$$y_i = \int_0^1 beam(t_i - u)x^0(u)\,du + e_i, \quad t_i = \frac{i-1}{n}, \quad i = 1,\dots,n$$

where $beam, x^0$ are smoothly periodic functions on $[0,1]$ and the errors e_i's are a "white noise" of known standard deviation σ (in our experiments, σ is one tenth of the maximum of the unnoisy data). We have to estimate x^0, knowing $beam$ and the vector $\mathbf{y} = (y_1,\dots,y_m)^t$ of noisy data . A classical discretization yields $y_i = (B\mathbf{x}^0)_i + e_i, \quad i = 1,\dots,n$, where B is circulant. The straightforward resolution of this least squares problem is often completely meaningless. The regularization method is a well known approach for such a problem (e.g. [Lannes,Roques,Casanove 87]). Thus we introduce a 'regularization matrix' Δ, chosen here to be a circulant discretization of some differentiation operator, and we solve:

$$\min_{\mathbf{x}} ||B\mathbf{x} - \mathbf{y}||^2 + \tau||\Delta\mathbf{x}||^2,$$

($||\cdot||$ denotes the Euclidean norm, $\tau > 0$). As soon as $Ker(B) \cap Ker(\Delta) = \{0\}$, this problem has a unique solution $\mathbf{x}(\tau)$:

$$\mathbf{x}(\tau) = (B^t B + \tau\Delta^t\Delta)^{-1}B^t\mathbf{y}. \tag{1}$$

We use the normalized Discret Fourier Transform: $F = 1/\sqrt{n}\,\Omega$ where $\Omega_{k,j} = \exp\left(-2i\pi(k-1)(j-1)/n\right)$, $k,j = 1\dots n$. Letting $\tilde{\mathbf{x}} = F\mathbf{x}$, $\tilde{\mathbf{y}} = F\mathbf{y}$, and for the matrix $B = \bar{F}diag(d_i)F$, $\Delta = \bar{F}diag(\delta_i)F$, we obtain an expression similar to (1):

$$\tilde{x}(\tau)_i = \frac{\bar{d}_i\tilde{y}_i}{|d_i|^2 + \tau|\delta_i|^2} \quad i = 1,\dots,n. \tag{2}$$

2 THE CHOICE OF THE REGULARIZATION PARAMETER

The crucial point in this approach is the choice of the parameter τ. Usual methods like Generalized Cross-Validation (GCV), Mallows' C_L criterion, are based on the Mean Square Prediction Error: $MSPE(\tau) = E(1/m||B(\mathbf{x}(\tau) - \mathbf{x}^0)||^2)$. In the case of periodic deconvolution, Rice (see [Rice 86]) shows that the use of the Mean Square Estimation Error, $MSEE(\tau) = E(1/n||\mathbf{x}(\tau) - \mathbf{x}^0||^2)$, can give better results than $MSPE(\tau)$. He proposes the following method to estimate from the data, the minimizer of $MSEE(\tau)$: in the expression

$$1/n||\mathbf{x}(\tau) - \mathbf{x}^0||^2 = 1/n(||\mathbf{x}^0||^2 + ||\mathbf{x}(\tau)||^2 - 2 < \mathbf{x}(\tau), \mathbf{x}^0 >),$$

the first term $||\mathbf{x}^0||^2$ does not depend on τ, the second $||\mathbf{x}(\tau)||^2$ can be computed, The third can be estimated by:

$$EPSEE(\tau) = \sum_{i=1}^{n} \frac{|\tilde{y}_i|^2}{|d_i|^2 + \tau|\delta_i|^2} - \sigma^2 \sum_{i=1}^{n} \frac{1}{|d_i|^2 + \tau|\delta_i|^2}. \tag{3}$$

It is indeed easy, from equation (2), to check that $EPSEE$ (Estimator of the inner Product in the mean Square Estimation Error) is an unbiased estimate of $PSEE(\tau) = E(< \mathbf{x}(\tau), \mathbf{x}^0 >)$. Rice proposes then to minimize the following function for choosing τ: $E^*(\tau) = 1/n(||\mathbf{x}(\tau)||^2 - 2EPSEE(\tau))$.

3   REGULARIZED ESTIMATORS

When using $EPSEE$, we have sometimes difficulties. E^* often produce too small parameters, when a lot of the eigenvalues d_i are near 0. The variance of $EPSEE$ can explain this problem:

$$\text{var}(EPSEE(\tau)) = \sum_{i=1}^{n} \frac{|d_i|^2(4\sigma^2|\tilde{x}_i^0|^2 + 2\sigma^4)}{(|d_i|^2 + \tau|\delta_i|^2)^2}. \tag{4}$$

It is a decreasing function of τ. If $\exists i, d_i = 0$, then $\lim_{\tau \to 0_+} \text{var}(EPSEE(\tau)) = +\infty$. We can note that the truncation to the term with non zero d_i does not change the expectation of (3), so that $GEPSEE$ is an unbiased estimator of $PSEE$,

$$GEPSEE(\tau) = \sum_{i=1, d_i \neq 0}^{n} \frac{|\tilde{y}_i|^2}{|d_i|^2 + \tau|\delta_i|^2} - \sigma^2 \sum_{i=1, d_i \neq 0}^{n} \frac{1}{|d_i|^2 + \tau|\delta_i|^2}, \tag{5}$$

better than $EPSEE$ because $\forall \tau > 0$, $\text{var}(GEPSEE(\tau)) \leq \text{var}(EPSEE(\tau))$. For a given n, it theoritically solves the variance problem of $EPSEE$. Sufficient conditions for the existence of these estimators in more general mean squares problems are given in ([Desbat,Girard]).
We can note in (4) that the variance problem comes from the small eigenvalues d_i. This suggests to introduce a bias in the estimation of $PSEE$ and to propose:

$$BEPSEE(\tau) = \sum_{i, |d_i| > \alpha} \frac{|\tilde{y}_i|^2}{|d_i|^2 + \tau|\delta_i|^2} - \sigma^2 \sum_{i, |d_i| > \alpha} \frac{1}{|d_i|^2 + \tau|\delta_i|^2}.$$

We give in ([Desbat,Girard]) a way to choose α from *beam* and an estimation of $\max_i |\tilde{x}_i^0|$. We choose α so that the variance gain, $\text{var}(GEPSEE)-\text{var}(BEPSEE)$, is as biggest as possible under the condition that the introduced bias is relatively small. We propose to minimize $BE^*(\tau) = ||\mathbf{x}(\tau)||^2 - 2BEPSEE(\tau)$. The criterion $E(||\mathbf{x}(\tau) - \mathbf{x}^0||^2)$ has been changed to $E(||\mathbf{x}(\tau) - \mathbf{x}_\alpha^0||^2)$, where $\mathbf{x}_\alpha^0$ is a truncated version of $\mathbf{x}^0$. The truncation is a regularization feature. Hence, it is natural to propose a regularized version of $GEPSEE(\tau)$, and to replace $E(||\mathbf{x}(\tau) - \mathbf{x}^0||^2)$ by $E(||\mathbf{x}(\tau) - R(\mathbf{x}^0)||^2)$ where $R(\mathbf{x}^0)$ is a regularized version of $\mathbf{x}^0$. If we take: $R(\mathbf{x}^0) = \mathbf{x}_{\tau_0}^0 \overset{\text{def}}{=} \arg\left(\min_{\mathbf{x}\in\mathbb{R}^n} ||B(\mathbf{x} - \mathbf{x}^0)||^2 + \tau_0||\Delta\mathbf{x}||^2\right)$, we can show that

$$REPSEE_{\tau_0}(\tau) = <\mathbf{x}(\tau), \mathbf{x}(\tau_0)> -\sigma^2 \sum_{i=1}^n \frac{|d_i|^2}{(|d_i|^2 + \tau_0|\delta_i|^2)(|d_i|^2 + \tau|\delta_i|^2)},$$

is an unbiased estimator of $E(< \mathbf{x}(\tau), \mathbf{x}_{\tau_0}^0 >)$. As for α, we give in ([Desbat,Girard]) a way to automatically produce a τ_0 denoted τ_0^c . Then, we propose to minimize $RE_{\tau_0^c}^*(\tau) = ||\mathbf{x}(\tau)||^2 - 2REPSEE_{\tau_0^c}(\tau)$.

Using the notation $b \overset{\text{def}}{=} E(REPSEE) - PSEE$, $v \overset{\text{def}}{=} \text{var}(REPSEE)$

$$v = \sum_{i=1}^n \frac{|d_i|^4(2\sigma^4 + 4\sigma^2|d_i|^2||\tilde{x}_i^0|^2)}{(|d_i|^2 + \tau|\delta_i|^2)^2(|d_i|^2 + \tau_0|\delta_i|^2)^2} \; ; \; b = \tau_0 \sum_{i=1}^n \frac{|d_i|^2|\delta_i|^2|\tilde{x}_i^0|^2}{(|d_i|^2 + \tau|\delta_i|^2)(|d_i|^2 + \tau_0|\delta_i|^2)}$$

and $c \overset{\text{def}}{=} E((REPSEE-PSEE)^2) = b^2 + v$, we could try to find τ_0 which minimizes $\max_\tau c(\tau, \tau_0)$. As c is a decreasing function of τ, $\max_\tau c(\tau, \tau_0) = c(0, \tau_0)$. Because of the multiplication by $|\delta_i|^2/(|d_i|^2 + \tau_0|\delta_i|^2)$, the estimation of $c(0, \tau_0)$ is numerically difficult when τ_0 tends to zero.

An heuristic procedure consists in computing $\tau_{0opt}(\tau) = \arg(\min_{\tau_0} c(\tau, \tau_0))$ and then to choose $mopt\tau_0 = \min_\tau \tau_{0opt}(\tau)$. As b is an increasing function of τ_0, this choice ensures a small bias for all value of τ and a variance reduction compared to the one of $GEPSEE$. $mopt\tau_0$ can be estimated (this gives τ_0^e and $RE_{\tau_0^e}^*$) with $ec = eb2 + ev$, an unbiased estimate of c. ev is an unbiased estimator of v obtained by replacing in the expression of v the unknown terms $|d_i\tilde{x}_i^0|^2$ by $|\tilde{y}_i|^2 - \sigma^2$. It can noted that $eb2$ defined by

$$\left(\sum_i a_i|\tilde{y}_i|^2\right)^2 - \sigma^4\left(\sum_i a_i\right)^2 - 2\sigma^2\left(\sum_i a_i^2(2|\tilde{y}_i|^2 - \sigma^2) + \left(\sum_j a_j\right)\left(\sum_i a_i(|\tilde{y}_i|^2 - \sigma^2)\right)\right)$$

with $a_i = \frac{\tau_0|\delta_i|^2}{(|d_i|^2 + \tau|\delta_i|^2)(|d_i|^2 + \tau_0|\delta_i|^2)}$, is an unbiased estimator of b^2. This produces τ_0^e.

The top left figure shows $\mathbf{x}^0, \mathbf{y}$ and *beam*. The top right figure shows the standard deviation gain on the parameter regularization estimator. In the third figure, for a given value of the kernel width $l \in \{.01, .05, .09\}$, the global performance of an estimator M, ($M \in \{GCV, E^*, BE^*, RE_{\tau_0^c}^*, RE_{\tau_0^e}^*, RE_{max(\tau_0^c, \tau_0^e)}^*, RE_{mopt\tau_0}^*\}$), is described by the histogram of the 500 observed values of its efficiency defined by

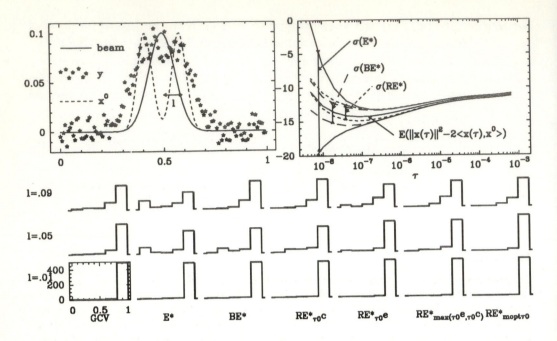

$E_M = \frac{||\mathbf{x}(\tau_{opt}) - \mathbf{x}^0||^2}{||\mathbf{x}(\tau_M) - \mathbf{x}^0||^2}$ ($\tau_{opt} = \arg(\min_\tau ||\mathbf{x}(\tau) - \mathbf{x}^0||^2)$), in 500 experiments which differ only by the simulated noise. We can see in the experiments that, BE^* and RE^* greatly improve E^*. The parameter $mopt\tau_0$, computed with the unknown vector $\mathbf{x}^0$, shows that we can hope to produce a better method than GCV in this case. Using the maximum of τ_0^c and τ_0^e, we build a more efficient method than GCV.

4   CONCLUSION

There is a need to regularize not only the deconvolution procedure, but also some regularization parameter estimators. The parameter estimation regularization should be studied in other inverse problems.

References

[Desbat,Girard] L. Desbat, D. Girard *The 'minimum reconstruction-error' choice of regularization parameters: some effective methods and their application to deconvolution problems.* preprint to appear.

[Lannes,Roques,Casanove 87] A. Lannes, S. Roques, M. J. Casanove. *Stabilized reconstruction in signal processing.* J. of Modern Optics, vol.34, pp. 161-226 (1987).

[Rice 86]   J. A. Rice. *Choice of smoothing parameter in deconvolution problems.* Contemporary Math. Vol.59, pp. 137-151 (1986).

UNCERTAINTIES ON CALIBRATIONS OF ABSOLUTE MAGNITUDES AND THEIR INFLUENCE ON THE STUDY OF THE EXTINCTION LAW

Jorge R. Ducati

Universidade Federal de Rio Grande do Sul, Instituto de Física, departamento de astronomia, Avenida Bento Gonçalves 9500, 90.099 Porto Alegre - RS Brazil

Charles Bonatto

Universidade Federal de Rio Grande do Sul, Instituto de Física, departamento de astronomia, Avenida Bento Gonçalves 9500, 90.099 Porto Alegre - RS Brazil

Jacques Guarinos

CDS, Observatoire de Strasbourg, 11 rue de l'université, 67000 Strasbourg, France

Introduction :

Calibrations of absolute magnitudes are commonly used to calculate spectroscopic parallaxes. It is usually admitted that distances obtained by this method carry an error of 10 to 15 %. Nevertheless, Guarinos et al. (this volume) suggest that such values of absolute magnitude, for any spectral type, are more realistically spread around a probable value, in a gaussian distribution. If the dispersion of the distribution is taken into account, some distances calculated by this method can differ seriously from more frequent values. Consequences can be dramatic for studies of galactic structure. The often used test of similar distances as a criterium for membership in clusters, in this case, is not valid.

Many investigations on the spiral structure, or on the distribution of the interstellar medium, rely heavily on spectroscopic parallaxes calculated for thousands of stars. Errors in distances introduced by the dispersion in luminosity calibrations can, even in large samples, mask the effects or structures searched. This work tries to estimate the importance of errors from these calibrations, comparing a simulated sample with real observational data.

The method :

The first step to construct our simulated sample was to define its luminosity function. One of the largest mass of observational data usefull for the calculus of stellar distances presently available is the catalogue described by Guarinos et al. (1989). Accordingly, we defined a sample of 45000 stars following, approximately, the frequencies in luminosity classes and spectral types found in that catalogue. The same was done regarding the distance range allowed for each group of stars, considering that, beyond a certain distance, the equivalent observed data lacked good coverage. This implied, for example, that for A0 V stars the simulation put 1200 stars regularly spaced between zero and 500 parsecs. The complete simulated sample had 45000 stars to distances up to 2000 pc.

The second step was to associate to each star a value of interstellar extinction, assuming some general reddening law. We used the value 1 mag/kpc.

The third step was to introduce a perturbation on the distances. We supposed that each original simulated distance was derived from the calibrated absolute magnitude of the respective spectral type. But each absolute magnitude in a calibration, as said above, is in fact the more probable value in a gaussian distribution. Other values of M_v can also occur for the same spectral type. We accordingly applied on each distance a perturbation which produced the new value :

$$d' = d \cdot 10 \exp(MM/5) ,$$

where MM = M - M', M being the calibrated absolute magnitude (more probable value) and M' its new value, extracted from the generation of random numbers obeying a gaussian law, centered at M and with a dispersion depending on the spectral type (Guarinos et al., this volume). This perturbation produced a list where distances are not sorted out, but still associated with extinctions arranged in ascending order.

The fourth step was to recover the ascending order in distances, which had the effect of destroying the order of extinctions. Finally, we calculated mean values of extinction within concentric rings of a certain thickness. These results could be compared with observational data.

Results :

In fig. 1 (simulated sample), we present the mean A_v for rings 25 pc thick. The M_v calibration used was that from Corbally and Garrison (1983). This can be compared with observational data (fig. 2) from the catalogue of Guarinos et al. concerning the same calibration. Nevertheless, one has to keep in mind that for our catalogue, the reddening law adopted was different $(A_v = 3.E(B - V))$.

In both diagrams, it is showed that the catalogue used is not complete at large distances. This was expectable since only the early type stars can be visible at large distances. With the simulated sample, (fig. 1), the perturbation on distances has no effect up to 1000 pc. Beyond this distance the curve becomes less steep, and beyond 1500 pc a large dispersion around the mean curve appears, increasing with the distance, while with the observational sample there is a stage with a large dispersion from 1500 pc.

Figure 1 shows perfectly the statistical effect of incertitudes of absolute magnitudes on the estimation of distances : At small ranges, the resultant mean effect is null, while at large ranges (beyond 1500 pc), where the sample becomes poor and uncomplete in late type stars, a large dispersion appears. This goes with a bending of the curve. Both these phenomenons are visible in figure 2 which was obtained from the real sample .

In figure 3 we present the same diagram as in figure 2, but using the Schmidt-Kaler calibration (1982) : The same tendancies are shown.

Conclusion :

Our conclusion is that a large part of the dispersion in the $A_v(distance)$ diagrams for distances beyond 1500 pc comes from the combination of 2 effects : the incertitudes of absolute magnitudes and the uncompleteness of the sample, which prevents the effects of the incertitudes on M_v regarding different spectral types from cancelling each other out.

Considering the whole available observational data (with good MK types and UBV photometry), one can say that it is a difficult task to study the extinction law beyond 1500 pc.

The next step of our study will be to test different laws of reddening in various directions carefully selected, using our catalogue and the incertitudes of absolute magnitudes.

References :

- Corbally, Garrison (1983) : The MK Process and Stellar Classification. Proceedings of a workshop, Toronto, 1983, p.277, Garrison ed.
- Guarinos, García, Ducati (1989) : Bull. Inform. CDS $n°$ 37.
- Schmidt-Kaler (1982) : Landölt Bornstein, vol. 1, Springer Verlag.

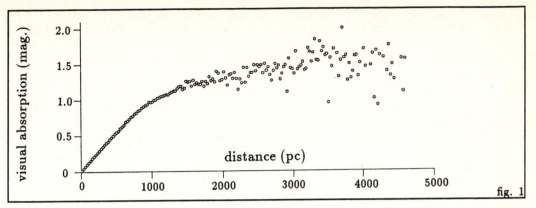

Diagram showing the mean values of extinction within concentric rings 25 pc thick for the simulated sample of 45000 stars. Absolute magnitudes come from corbally and Garrison calibration (1983).

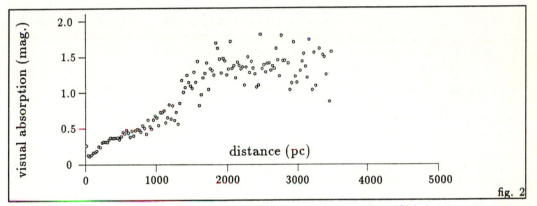

Diagram showing the mean values of extinction within concentric rings 25 pc thick for the real sample of 20000 stars. Absolute magnitudes come from corbally and Garrison calibration (1983).

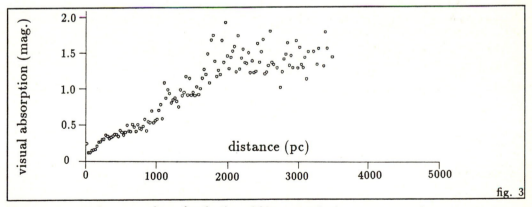

Diagram showing the mean values of extinction within concentric rings 25 pc thick for the real sample of 20000 stars. Absolute magnitudes come from Schmidt-Kaler (1982).

Global methods of reduction in photographic astrometry

C. DUCOURANT, M. RAPAPORT

Observatoire de l'Université de Bordeaux I,   B.P. 89, 33270 FLOIRAC-FRANCE

1 INTRODUCTION

In photographic astrometry, the determination of the celestial coordinates of a star is based on the rectangular coordinates (x,y) of the image of the star on the plate. Its standard coordinates (X,Y) are the coordinates of the direction of his projected position on the tangent plane to the celestial sphere. If the telescope was perfect, it would be possible to write (with well selected units) : $x = X$, $y = Y$. In fact, the distortions due to the telescope, the atmosphere, the plate imply a more complex relation between standard coordinates and measurements. The whole instrumental distortions are represented by a set of parameters (plate constants)

$$X = A_1 + A_2x + A_3y + A_4x^2 + A_5y^2 + A_6xy + ... \qquad (1)$$
$$Y = B_1 + B_2x + B_3y +...$$

The degree of this polynomial expansion depends on the chosen model. In the case of Schmidt plates, the order 3 is requested.

Classically, these parameters are determined, using a set of reference stars, by a least square treatment of (1). For these reference stars, the standard coordinates are assumed perfectly known.

Once (A,B) are known, one can determine the celestial positions of objects from their measured positions on the plate. This is a very classical procedure often used in astrometry. But in photographic astrometry, in most cases, it is not possible to consider the standard coordinates of reference stars as known quantities. As for the plates that use in Bordeaux, the errors the reference catalogue are of the same order as those due to measurements. So one must consider that in equation (1), (x, y) and (X, Y) are observations with noise.

2. THE METHODS OF REDUCTION

A similar problem is studied by Eichhorn (1985). In this work, Eichhorn considers data coming from catalogues (celestial coordinates of reference stars) as the result of measurements subject to a random error (mean zero and known sigma). He linearizes the equations of condition (1) around an aproximate solution and takes into account the catalogue errors by adding to the system of linearized equations, some equations of probabilistic constraint of the type :

$\Delta\alpha \, Cos\delta = 0$
$\Delta\delta \qquad = 0$

This problem is essentially non linear and the determination of the plate constants (A,B) is thus the generalization of the problem of fitting a straight line in a cloud of points whose both coordinates are noisy. This can easily be seen in the case of an artificial model with 2 constants :

$x = A1 + A2X$

This specific case is treated by Fasano and Vio by iteratives techniques (1988).

We have used a method derived from the work of Jefferys (1980, 1981) to solve the problem of the reduction of a plate taking into account the errors on the catalogue. The least square treatment is then adapted to the non linear case which is solved by successive approximations of the Newton-Raphson type. The tvery interesting point in this technique is the possibility to access to the covariance matrix of the estimated parameters.

2.1 Reduction of a single plate

The astrometric reduction of a plate considers as observed quantities :
- the measurements on the plate
- the catalogue positions of reference stars

These quantities are considered as being affected by a random error of mean zero and known variance. This means that we consider that there are no systematic errors in the catalogue. We need to know the relative precision of the measurements and of the catalogue. Practically, when one reduces data from a single plate with such a technique, the final positions of unknown objects remain the same as the classical solutions. The only difference is a more realistic estimation of the precision of the results (Table 1).

Table 1 : Comparison of global and classical solutions of the reduction of a single plate. The sigma is given by the global solution. The classical RMS is 0.53".

Number	$\alpha_{global} - \alpha_{classical}$	$\sigma\alpha$	$\delta_{global} - \delta_{classical}$	$\sigma\delta$
1	0.001"	0.23"	- 0.001"	0.22"
2	0.001	0.24	0.000	0.24
3	-0.001	0.25	0.002	0.25
4	-0.001	0.25	0.001	0.24
5	-0.001	0.24	0.001	0.24
6	0.001	0.22	-0.001	0.22
7	-0.002	0.23	0.002	0.23

Classical RMS = 0.53"

2.2 Simultaneous reduction of several plates

We had at the Bordeaux Observatory three plates (1891, 1909, 1988) from the Carte du Ciel survey, covering the same sky region and containing 20 AGK3 stars and the quasar 3C273.

A classical, separate reduction of each exposure provides differents positions for 3C273, thus an apparent proper motion of :

μ_α = -0.007"/year

μ_δ = -0.010"/year

The differences that we observe are too large to be completely explained by measurements errors -these differences come probably from the errors of AGK3 proper motions- Therefore, we have performed the simultaneous reduction of the plates, where the random errors on reference star's concern :

a) the positions at the epoch of AGK3

b) the AGK3 proper motion

We have solved the complete problem, adding the a-priori information that the quasar is a fixed point. This allows us to center the plates one with respect to the others. Through that procedure, the catalogue data are corrected. In the local referentiel that this adjusted catalogue defines, the quasar has an apparent proper motion of :

μ_α = -0.002"/year

μ_δ = -0.003"/year

Such a motion is now within the error bars of the measurements.

3. CONCLUSION

We have found three points of interest in the application of such global methods to the reduction of plates :

- they allow to take into account the whole information available (errors on measurements and on catalogue data)

- they give a realistic estimation of the variance of the solutions (precision on objects positions, precision on the adjusted catalogue)

- they allow to treat simultaneously several plates and thus increase the precision of the results.

4. REFERENCES

H. Eichhorn
"The direct use of spherical coordinates in focal plane astrometry".
Astron. Astrophys. 150, 251-255 (1985)

W.H. Jefferys
"On the method of least squares".
Astron. Journal 85, n°2, 177-181 (1980)
"On the method of least squares II"
Astron. Journal 86, n° 1, 149-155 (1981)

Fasano, G., Vio, R.
"Fitting a straight line with errors on both coordinates".
Bulletin Information du C.D.S. n° 35 (1988)

COMPARISON OF CALIBRATIONS OF ABSOLUTE MAGNITUDES FOR THE CALCULUS OF STELLAR DISTANCES

Jacques Guarinos

CDS, Observatoire de Strasbourg, 11 rue de l'université, 67000 Strasbourg, France

Beatriz García

CDS, Observatoire de Strasbourg, 11 rue de l'université, 67000 Strasbourg, France
and Universidad Nacional de La Plata, faculdad de Ciencias astronómicas y Geofísicas,
casilla de correo 677, 1900 La Plata, Argentina

Jorge R. Ducati

Universidade Federal de Rio Grande do Sul, Instituto de Física, departamento de astronomia,
Avenida Bento Gonçalves 9500, 90.099 Porto Alegre - RS Brazil

Introduction :

It is well known that the calculus of stellar distances is a crucial problem in astronomy. In most cases, distances are obtained using spectroscopic parallaxes. This method, based on the application of calibrations of absolute magnitudes to stars whose MK type is available has only a statistical significance. Thus, an error of 10 to 15 % can be expected in the value of each of these distances. These errors can alter seriously the results of investigations on the galactic structure or on the interstellar medium.

Moreover, to calculate a distance, one needs to estimate the visual absorption, which is a very difficult task. In most cases, it is assumed that this quantity is a multiple of the intrinsic colour $(B - V)_0$, for which a calibration can be applied, given the MK type (Schmidt-Kaler, 1982). However, the photometric data of the star can provide a second estimation of the intrinsic colour, which can differ seriously from the first one. This is an other cause of the large errors in the calculus of stellar distances. In this case, since both distances and visual absorptions are ill-estimated, it is dangerous to intend to derive a relationship between these two quantities.

Aim of this study and method used :

The aim of this work is to estimate the effect of the errors brought by the automatic use of calibrations of absolute magnitudes and intrinsic colours on the calculus of stellar distances. Such distances are compared to "individual" distances whose calculus is allowed by :
  - the use of the colour indices of each star;
  - a simulation of errors on the absolute magnitudes M_v provided by the calibrations, using a gaussian distribution centered at M_v. Its dispersion depends on the photometric data of the star, its luminosity class, and the calibration adopted.

In the "individual distances", the colour indices are used for the calculus of the dispersion of the gaussian distribution as explained below, but also to estimate the visual absorption through the calculus of the intrinsic colour by the Q method.

Four different calibrations of absolute magnitudes are dealt with, and compared when applied to the same sample of stars.

This is only the first part of this study since we only took into account one cause of the dispersion of the absolute magnitudes, which is the discrepancy between the MK classification and the Q method (based on photometric data) regarding the value of the absolute magnitude.

The data :

The data come from our catalogue containing 20000 stars. The way we constructed this catalogue is explained in a previous paper (Guarinos, García, Ducati 1989).

For each star this catalogue provides the following data : Identification, MK type in a standard format, galactic coordinates, UBV photometry with incertitudes and number of measurements, and the parameters used for the computation of a distance.

Our program can give a visual absorption and a distance in different ways : The intrinsic colour can be either extracted from the Schmidt-Kaler calibration (1982) given the MK type, or computed using the Q method (García, Claria and Levato, 1988). However this method concerns only the early stars (from O to A2) and cannot be applied to the supergiants. Thus, the computation of the distance depends on three parameters which are fixed according to the user's preference:
- the calibration for the absolute magnitude, chosen among the following ones :
    - Corbally and Garrison, 1978 and 1983;
    - Schmidt-Kaler, 1982;
    - Grenier, Gómez, Jaschek C., Jaschek M., Heck for magnitude- limited samples, 1985;
    - Grenier, Gómez, Jaschek C., Jaschek M., Heck for distance- limited samples, 1985.
- the source of the intrinsic colour;
- a parameter which decides whether the program will generate a simulation of the errors in the estimation of the absolute magnitude or not.

Simulation of the errors :

We computed the dispersion of the absolute magnitudes for the stars belonging to bins whose membership depends on the value of the intrinsic colour according to the Q method, the luminosity class and the calibration.
This quantity was assumed to be the dispersion of the gaussian distribution of the actual value of the absolute magnitude around the value provided by the calibration. A random numbers generator obeying a gaussian law was used for the calculus of the distance of each star of our catalogue. Consequently, the value of the absolute magnitude depended on the intrinsic colour computed from the observed colours, while it was assigned according to the spectral type as it appears in the MK type catalogue. The dispersion of the absolute magnitudes was calculated for each spectral type. Thus, the calibration providing values of the absolute magnitudes closest to the ones calculated from the photometry will be considered as the most suitable for our sample of stars. We only kept the bins large enough for having a statistical significance.

Results :

In the Grenier et al. calibration, the stars earlier than B5 are not concerned, and we will have to keep this in mind for interpreting the diagrams.

Figure 1 and 2 show both the average value of the absolute magnitude and the error computed as explained above, according to the intrinsic colour. These H-R diagrams were used to assign the dispersion of the gaussian distribution representing the error in the simulation. Figure 1 concerns the main sequence stars for the Corbally and Garrison calibration while figure 2 concerns the main sequence stars for the Grenier et al. calibration. In the latter calibration, it is seen that the lack of stars earlier than B5 alters the statistics. Nevertheless, one can notice that the incertitudes, thus the size of the simulated errors, are slightly smaller than for the Corbally and Garrison calibration in the reliable part of the diagram. That means that the Grenier et al. calibration seems to correspond slightly better to the photometric data of the stars belonging to our catalogue, between $(B-V)_o =$ -0.18 and $(B-V)_o =$ -0.02 . A similar diagram for the Schmidt-Kaler calibration shows the same characteristics as for the corbally and Garrison calibration.

Figures 3 and 4 show the difference between the distance computed from the calibration and the distance obtained using the photometric data and the simulation of the errors described above, according to the spectral type. Figure 3 concerns the main sequence stars in the Corbally and Garrison calibration. It is seen that the two different ways of computing the distance can give very different results for the hottest stars. Figure 4 deals with the main sequence stars in the Grenier et al. calibration. As expected (see fig. 2), the agreement between the two calculus of the distance is slightly better.

Conclusion :

As it was expected, there can be a very large discrepancy between a distance calculated from a spectroscopic parallax and a distance calculated taking into account the photometric measurements of the star. This is particularly true for the earlier stars which can reach large distances (see fig. 3). The larger part of this discrepancy comes from our generation of a gaussian distribution around the value of the absolute magnitude provided by the calibration.

But the Grenier et al. calibration for magnitude-limited samples, mainly based on Strömgren photometry, shows a rather smaller discrepancy with the Johnson's photometry data of the stars of our catalogue, especially for the dwarfs. The absolute magnitudes obeying a gaussian law show a smaller dispersion around the value given by this calibration, compared to the other calibrations. Consequently, the "individual distances" described above better fit the distances obtained directly from the calibration in the case of the Grenier et al. calibration. Unfortunately, the hottest stars cannot be assigned an absolute magnitude in this calibration.

References :

- Corbally, Garrison (1983) : The MK Process and Stellar Classification. Proceedings of a workshop, Toronto, 1983, p.277, Garrison ed.
- García, Claria, Levato (1988) : Astrophysics and Space Science 143, 317.
- Grenier, Gómez, Jaschek C., Jaschek M., Heck (1985) : Astron. Astrophys. 145, 331.
- Guarinos, García, Ducati (1989) : Bull. Inform. CDS $n°$ 37.
- Schmidt-Kaler (1982) : Landölt Bornstein, vol. 1, Springer Verlag.

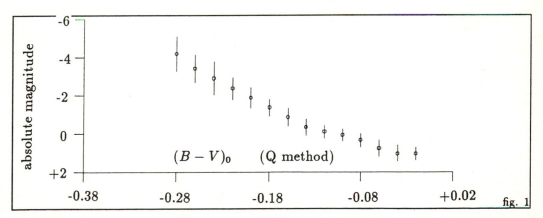

fig. 1

Diagram showing the average values of the absolute magnitudes and their incertitudes calculated as explained in the text, for the Corbally and Garrison calibration.

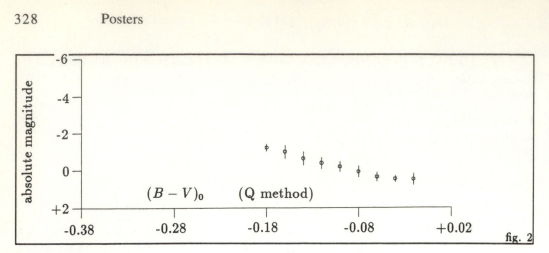

fig. 2

Diagram showing the average values of the absolute magnitudes and their incertitudes calculated as explained in the text, for the Grenier et al. calibration.

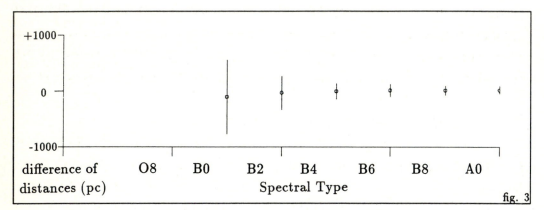

fig. 3

Difference between a distance calculated from a spectroscopic parallax, and a distance calculated taking into account the photometric data of each star. This diagram concerns the Corbally and Garrison calibration.

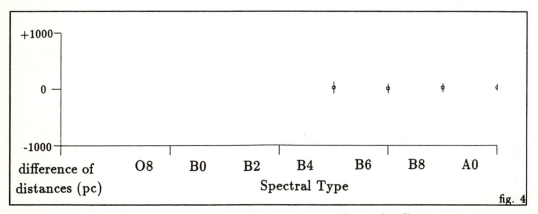

fig. 4

Difference between a distance calculated from a spectroscopic parallax, and a distance calculated taking into account the photometric data of each star. This diagram concerns the Grenier et al. calibration.

The main component analysis of the longitudinal distribution of solar activity

L. HEJNA

Charles University, Prague

Abstract

In this contribution, preliminary results of the main component analysis of Bartels diagram of time series of daily values of sunspot group numbers for solar cycles 18, 19 and 20 are presented. Obtained results suggest that the most significant feature in the longitudinal distribution of sunspot activity is the existence of preferred solar hemispheres alternating with mean period 2.5 Bartels rotations.

1. INTRODUCTION AND DATA PROCESSING

The issue of the preferred (active) longitudes in the solar activity distribution into heliographic longitudes still belongs to open problems of the present-day solar astronomy. The existence of active longitudes could have a significance of primary order both for the problems of short-term and medium-term solar activity forecasts and for the theory of the solar dynamo. The results obtained up to now prove the existence of that phenomenon. However, a sufficiently complete and compact description is still missing (See, for instance Gaizauskas 1985). That is why it is sensible to study these phenomena on the basis of other data sets, using the classical as well as quite new methods for their analysis.

The main aim of this contribution has been an attempt to study the longitudinal distribution of sunspot activity using the expansion of the time series of daily values of sunspot group numbers (for eleven-year cycles 18, 19 and 20) arranged in the shape of Bartels diagram into the system of its proper orthogonal functions (See for instance Vertlib et al. 1971). Under the Bartels' diagram we understand an arrangement of the time series into the form of the matrix, where each row represents a single 27 day Bartels rotation. The detailed discussion of the significance of 27 day period in behaviour of similar solar indexes was given by Balthasar and Schüssler (1983).

If we comprehend each row of Bartels' diagram as a concrete realization of some general longitudinal distribution (affected by an influence of evolutionary processes and so on), we may look for a development of this distribution into a system of its proper orthogonal functions (in the main component analysis sense) and the shape of these functions will describe single independent components of the distribution, we are looking for.

It may be shown that the proper functions, we are looking for, must be identical to the eigenfunctions of the covariation matrix.

Since the covariation matrix is real and symmetric, the solution of the corresponding equation makes no difficulties. Even the eigenvalues will be real and non-negative. These eigenvalues describe a part of total dispersion explained by relevant proper function.

In the case of our analysis the rows of processed matrix have been reduced to zero mean and for the calculation of eigenvectors and eigenvalues the well known Jacobi method has been used.

2. RESULTS AND CONCLUSIONS

The most important of obtained proper functions (for normalised eigenvalues = 0.247, 0.210 and 0.130) and are in graphical form presented in Fig. 1. Some of main features of obtained results are:

1) The most significant component in the longitudinal distribution of sunspot activity (explains approximately 1/4 of total dispersion) may be recognized as the situation when one of the solar hemispheres is preferred (its activity is above - average) whereas the second is damped. The boundaries between these two hemispheres are stable during the complete investigated time interval (three cycles of solar activity) and lie at longitudes corresponding to Bartels days 9 and 23. The signs of expansion coefficients determine which of both hemispheres will be preferred in concrete Bartels rotation.

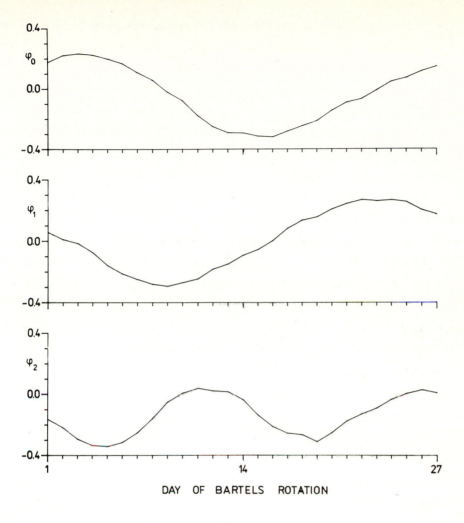

Fig. 1

2) Behaviour of expansion coefficients in time seems to be at first glance rather chaotic, but the run test for testing whether the sequence of their signs is random or not (Sachs, 1982) shows that for the first two proper functions the behaviour of corresponding coefficients (for level of significance $p = 0.05$) is not random. The mean length of iterations is roughly 2.6 Bartels rotations. The periods of the above-average activity on one hemisphere are therefore replaced by periods of reduced activity with just this time period.

3) Strongly smoothed (for whole solar cycles) values of expansion coefficients at the most significant proper function are equal

to -1.291 (for solar cycle no 18), -0.732 (19) and 0.349 (20). It means that during the cycles number 18 and 19 rather the hemisphere corresponding to the second half of Bartels rotation was generally preferred, whereas during the cycle no 20 we can observe opposite situation. This fact is in good agreement with results obtained by Balthasar and Schüssler (1984) and with their conception of solar "memory". But we must note that identically smoothed expansion coefficients at the second proper function behaves similarly but probably with the phase shift of one solar cycle.

Generally we can conclude that the method of main orthogonal component applied to data sets of daily values of selected solar activity indexes may be a very interesting and effective instrument for solar activity longitudinal distribution study.

References

Balthasar, H., Schüssler, M.: 1983, Solar Phys. **87**, 23.

Balthasar, H., Schüssler, M.: 1984, Solar Phys. **93**, 177.

Gaizauskas, V.: 1985, Inv. Rev. Present to Solar-Terr. Prediction Workshop, Meudon, 18-20 June, 1984.

Sachs, L.: 1982, "Applied Statistics", Springer-Verlag Publ. Co., New York, p. 375.

Vertlib, A.B., Kopecký, M., Kuklin, G.V.: 1971, Issled. Geomagn. Aeron. i Fiz. Solntsa **2**, 194.

L. Hejna
Dept. of Computer Technique, Fac. of Math. and Phys., Charles University
Ke Karlovu 3, 120 00 Prague 2, Czechoslovakia

High Precision Wavelength Calibration of Echelle CCD Spectra[*]

H. HENSBERGE
Koninklijke Sterrenwacht van België, Brussels, Belgium
W. VERSCHUEREN
Theoretical Mechanics and Astrophysics, University of Antwerp (RUCA), Belgium

SUMMARY

Systematic errors in wavelength calibration of CCD echelle spectra may arise from unrecognized blending in rich laboratory spectra and from the use of inappropriate methods to determine line centre in the case of undersampled lines. A global calibration based on selected lines is found appropriate.

1 INTRODUCTION

This study is based on data obtained with CASPEC and CCD#3 on the E.S.O. 3.6m telescope. A differential study of Th-Ar wavelength calibration frames taken before and after every stellar exposure gives evidence for systematic effects in the residuals to the adopted calibration. After allowance was made for a small shift of the spectrum over the CCD (typically few 10^{-2} pixel), an internal consistency of roughly 0.03 pixel r.m.s. was observed, in agreement with the prediction of Monte Carlo simulations taking into account read-out and photon noise. However, using standard MIDAS procedures, an r.m.s. of about 0.15 pixel emerged for the residuals, in accordance with D'Odorico and Ponz (1984). Comparison of the residuals of a given line on various frames exposes the presence of systematic effects. This contribution discusses their origin and how to avoid them.

2 LINE BLENDING IN CALIBRATION SPECTRA

The discussion refers directly to our Th-Ar spectra, but the basic arguments are more generally valid for rich spectra. In order to evaluate the influence of line blending in the thorium spectrum, we convolved the laboratory spectrum (Palmer and Engleman, 1983) with a PSF representative for our data. Subsequent gaussian line fitting over 5 pixels illustrates the displacement of the blended feature relative to the laboratory wavelength of the considered isolated line. Effects of the order of 0.1 pixel are commonly caused by spectral lines one order of intensity fainter than the line they disturb. The distribution of these displacements for thorium lines of measurable strength on our CCD spectra (fig. 1) shows a pronounced core of almost

---

[*] Based on observations collected at the European Southern Observatory, La Silla, Chile

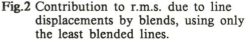

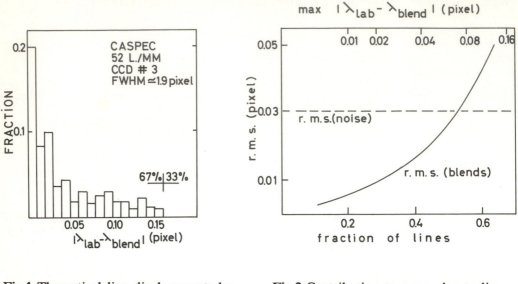

Fig.1 Theoretical line displacements by blends for lines in our calibration spectra $\lambda\lambda$ 3700–4700Å.

Fig.2 Contribution to r.m.s. due to line displacements by blends, using only the least blended lines.

unblended lines superposed on a shallow, continuous tail. The inclusion of blended lines will in general both distort the calibration solution and alter the meaning of the residuals. The distortion generally grows with the addition of degrees of freedom above what is needed physically; order per order independent polynomial fits fall into this category. The contribution of the line displacement by blends, shown in fig. 1, to the r.m.s. computed with respect to the undistorted solution is given in fig. 2 for the case of our CASPEC data as a function of the maximum displacement tolerated to include a line in the calibration. From this figure it is evident that the measured r.m.s. will be determined by random noise effects only if the larger part of the lines is either rejected or when their laboratory wavelength is properly corrected for the influence of blending fainter lines.

3 AN IMPROVED GLOBAL WAVELENGTH CALIBRATION

Using our CASPEC calibration frames, we have checked the applicability of a global calibration procedure. First we defined for each line two quantitative parameters describing the expected displacement due to blending and its sensitivity to the reduction parameters. The theoretical and empirical evidence considered to define them fully independently from r.m.s. considerations is given in detail in Hensberge and Verschueren (1989), in a paper oriented more specifically to MIDAS. Using the expected wavelengths of selected fairly unblended features rather than the laboratory wavelength of the principal component, the succes of several types of calibration formulae has been explored. The main conclusions are

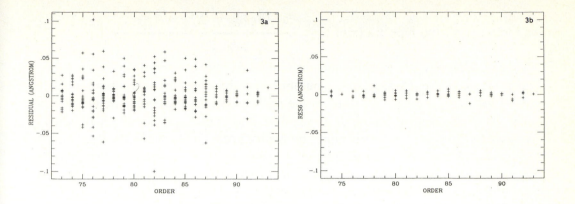

Fig.3 Residuals in the wavelength calibration for the standard MIDAS procedure (third degree polynomial order per order) with gaussian fitting on all detected lines (3a) and in case of the proposed global fit using selected lines (3b)

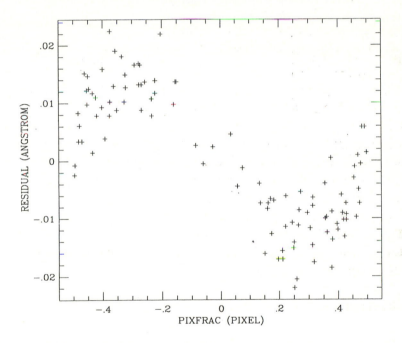

Fig.4 Influence of the method 'gravity' on the residuals in the wavelength calibration (global fit using selected lines). PIXFRAC is the 'gravity' computed pixel fraction position of line centre.

a A global fit of the type : $\lambda(x,m^*) = \dfrac{a_0 + (a_1 + b_1 m^*)x + a_2 x^2 + a_3 x^3}{1 + c_1 m^*}$

(where x denotes the position in pixel units in the extracted order and $m^* =$ const.-m, m is the physical order number and the constant is defined such that $m^* = 1$ for the highest m) reproduces the positions of 85 relatively unblended lines over 20 orders with a r.m.s. of about 0.04 pixel. The factor $(1 + c_1 m^*)^{-1}$, with c_1 roughly 0.01, is essential and cannot be reproduced satisfactorily by a polynomial of degree lower than 8, explaining the reduced succes of bivariate polynomial fits.

b Order per order third degree polynomial fits to all detected lines except the most obvious blends give calibration solutions distorted at many locations by 2 km s^{-1} or more, while the r.m.s. of almost unblended lines relative to the global fit is only 0.3 km s^{-1}. The need for unblended lines and for a high ratio of number of lines to number of free parameters necessitates the global calibration.

c The residuals relative to an order per order fit using nearly all detected lines are dominated by blending effects (fig. 3).

4 NOTE ON THE DETERMINATION OF LINE CENTRE

Some software packages use simpler algorithms than gaussian fits for determining the position of line centre in the case that the profile is poorly sampled. In particular, a method named 'gravity' defines line centre as the centre of gravity of the two brightest pixels relative to the third brightest. Although less susceptible to wide blends (it uses 3 pixels only), it is easy shown that the computed position contains a systematic error depending on the fractional position on the pixel of true line centre (Hensberge and Verschueren, 1989). The amplitude maximum of the error is of order 10^{-2} pixel only for gaussians with FWHM $= 1$ but of order 10^{-1} pixel for FWHM $= 2$. The errors introduced by 'gravity' thus may dominate the other contributions to the r.m.s. (fig. 4) and may distort the solution itself, mainly in the case of fits to single orders.

Acknowledgements

The Belgian Nationaal Fonds voor Wetenschappelijk Onderzoek is acknowledged for a grant in connection to this work (nr. S2.0091.88)

References

D'Odorico S., Ponz D. 1984, The Messenger 37, 24
Hensberge H., Verschueren, W. 1989, The Messenger 57 (in press)
Palmer B.A., Engleman R. Jr. 1983, Atlas of the Thorium Spectrum, Los Alamos National Laboratory, ed. H. Sinoradzky

Kinematic Study from Proper Motions using Spherical Harmonics

M. HERNANDEZ-PAJARES – J. NUÑEZ

Dept Fisica de l'Atmósfera, Astronomia i Astrofisica, Universitat de Barcelona Barcelona, Spain

SUMMARY.

In this paper we relate two methods of analyzing the kinematic parameters of the local macroscopic motions of the Galaxy from proper motions. First, the Ogorodnikov-Milne model which consists in the 3-D Taylor expansion of the mean velocity field. Second, the 2-D spherical harmonic development of the velocity components. Only the hypothesis of separability of the stellar density function of the sample into angular and radial parts is needed. Finally, we apply the relations to 4700 A-M stars.

1. INTRODUCTION AND PURPOSES.

We can consider two methods for the analysis of the mean stellar velocity field of the Solar Neighbourhood:

i) The Ogorodnikov-Milne model, Ogorodnikov (1932) and Milne (1935), which consist in a general 3-D Taylor expansion of the velocity field. Edmondson (1937) extended it up to 2nd order (hereinafter OM). Some modern literature about this method is in 1st order Núñez (1981); in 2nd order Figueras (1986), Figueras & Núñez (1989) and Hernández & Núñez (1989a); and in 3rd order Avadisova & Palous (1989).

ii) The 2-D development into spherical harmonics (hereinafter SH). Some important references about this topic are Brosche (1966) for a general scalar field, and Brosche and Schwan (1981) for the radial velocity and proper motions, after taking out the standard motions.

What are the advantages and disadvantages using OM or SH analysis? For the OM model, the main advantage is the intrinsic physical meaning of the parameters. However it presents two disadvantages: the non-orthogonality of Taylor terms and the difficulties in the interpretation of the significant dependences. The use of expansions into spherical harmonics has more or less opposite properties. It has the advantages of high orthogonality in a quasi-uniform angular distribution, and the facilities in the representation of unexpected systematic variations. Further, it does not need the individual distances to work. But, it presents a problem: the difficulties in the interpretation of the significant dependences. In view of this important matter, we have studied and developed the meaning of SH coefficients in terms of the OM-ones: In Hernández-Pajares & Núñez (1989b) we have done this for the radial velocity and galactic velocity components, v_r, U, V, W,

and in this paper we present the study for the proper motions using the notation instroduced in this reference. Therefore now, only a hypothesis about the distribution of stars is needed: the separability of the function "density of the number of stars" into the radial and angular part. Finally, we have applied the theoretical relations obtained between OM and SH coefficients for the proper motions to one sample of A-M stars contained in the Figueras (1986) catalogue (see Hernández-Pajares, 1989c for the detailed equations).

2. PROJECTION ON THE SPHERE OF THE PROPER MOTION COMPONENTS.

Let $v_r(r,l,b)$ be a scalar field, with 'r' being the heliocentric distance and l,b the galactic longitude and latitude; let $n(r,l,b)$ be the star-number density; they are defined in a region between r=R1 and r=R2. For us, v_r is the radial velocity, or some other velocity component that we could consider. When we analyze the distribution of v_r by spherical harmonics we do not take into account the distance r. This is equivalent to analyzing the projection of v_r on the sphere. Then, in order to connect the OM and SH analysis, we can do the following: i) Develop $v_r(r,l,b)$ into the explicit dependences in r,l,b using the 2nd order OM model. ii) Try to express the angular dependences of $<v_r>(l,b)$ in terms of the K_{nml}'s. We have found that this is possible for U,V,W and v_r (see Hernández-Pajares & Núñez, 1989b). For the proper motion components it is necessary to multiply them by cosb. iii) The projection will not change the expression of v_r in terms of the spherical harmonics if it affects only the radial coordinate. Indeed, it is easy to prove that this happens if the 3-D density of number of stars is separable into radial and angular parts $n(r,l,b) = A(r) \cdot B(l,b)$.

3. CALCULATIONS AND RESULTS.

After checking the relations obtained with numerical experiments, we have made calculations with a real sample. It contains 4732 stars from the S.A.O. catalogue with Astrophysical Data (Ochsenbein, 1980) with spectral type A-M, photometric distance greater than 70 pc and residual velocity less than 65 Km/s and 2.5 σ; σ is the standard deviation estimated by Figueras (1986), with the OM2 model. This sample was used in Hernández-Pajares & Núñez (1989b). In this paper we proved that it fulfils the hypothesis of separability of the density function of stars into angular and radial parts. Then, we have developed $kr\mu_l \cos^2 b$ and $kr\mu_b \cos b$, the spatial velocity components associated at the galactic longitude and latitude proper motions multiplied by the cosinus of the galactic latitud, for the A-M sample into spherical harmonics series up to order n=8 (81 terms).

The method used to solve the least squares adjustment is the same that Brosche (1966), the Gram-Schmidt orthonormalization procedure with a significance limit of 5% for the F-test, and with a weight given by the cosmic dispersion and the experimental error (see for instance Figueras and Núñez, 1989). We take the FK4 mean experimental errors for $\mu_\alpha \cos\delta$ (0.24 arcsec/century) and μ_δ (0.18) following Fricke, 1967. We obtain 18 and 15 significant terms for the $kr\mu_l \cos^2 b$ and $kr\mu_b \cos b$ respectively, with coefficients up to order $n=8$

The corresponding OM terms in the galactic heliocentric frame are in tables 1a-1b for the longitude and latitude proper motions. From table 1a we observe that the first and second component of the Solar Motion, not taking into account the 2nd order correction, are very similar to the Figueras (1986) results. They are calculated for the same sample with the general second order OM model giving values of $U_\odot=8.83$, $V_\odot=11.3$, $W_\odot=6.64$ Km/s. The Oort' constant $A \equiv S_{12}$ is in agreement but slightly higher than previous determinations from proper motions (Kerr & Lynden-Bell, 1986).

OM parameters from $kr\mu_l \cos^2 b$	Values obtained	nml
ω_3	-5.5 ± 2.3	001
$U_\odot + O(r^2)$	8.89± 0.60	110
$V_\odot + O(r^2)$	11.0 ± 0.6	111
$S_{23} - \omega_1$	-13.6 ± 5.6	211
S_{12}	17.5 ± 3.2	221
$\Omega_{13} + \Omega_{31} + (S_{112} + S_{222} - 4S_{233})/8$	31. ±31.	311

OM parameters from $kr\mu_b \cos b$	Values obtained	nml
$W_\odot + O(r^2)$	6.94± 0.38	001
$S_{33} - (S_{11}+S_{22})/2$	2.8 ± 4.7	101
$S_{23} + 5\omega_1/3$	30.0 ± 3.5	110
$S_{13} - 5\omega_2/3$	0.0 ± 3.9	111
$W_\odot + O(r^2)$	7.92± 0.69	201
$V_\odot + O(r^2)$	13.8 ± 1.0	210
$U_\odot + O(r^2)$	9.7 ± 1.1	211
$7(\Omega_{11}-\Omega_{22})+4S_{123}$	-290 ±100	220
$-7(\Omega_{12}+\Omega_{21})+2(S_{113}-S_{223})$	-120 ±100	221
$S_{222}+S_{112}-4S_{233}$	-350 ±140	410
$S_{223}-S_{113}$	-118 ±55.	421

Tables 1a-1b: Significant Ogorodnikov-Milne parameters from the $kr\mu_l \cos^2 b$ and $kr\mu_b \cos b$ spherical harmonic coefficients (orden nml). The units are Km/s, Km/s/Kpc and Km/s/Kpc^2 for the zero, first and second order OM terms respectively (for more details see for instance Hernández-Pajares and Núñez, 1989b).

Also we can obtain the Oort' constant $B \equiv \Omega_3$ from ω_3, see Eq.(1), taking out the standard values of the precession correction: $\Delta n=0.44\pm0.04$, $\Delta k=-0.19\pm0.06$ arcsec/cty (Fricke, 1977). The value is $B=-8.6$ Km/s/Kpc in agreement but slightly lower than the ones listed in Kerr & Lynden-Bell (1986).

In table 1b the Oort' constant A does not appear as significant, and has a value of 15.8±6.9 Km/s/Kpc associated to the 320 spherical harmonic coefficient with $\zeta=6.3\%$. Also the Solar Motion components, not taking into account the 2nd order correction, do not agree well with those of

Figueras (1986). Then, we can say that the OM results are of poor quality if they come from the $kr\mu_b cosb$ analysis. From tables 1a and 1b we can see that many combinations of second order OM terms appear significant as the 220 term for $kr\mu_b cosb$.

4. CONCLUSIONS.

With this work we have completed the link between the 2nd order Ogorodnikov-Milne and Spherical Harmonic models, extending the study to proper motions. The link, which needs the separability of the density function into angular and radial parts to work, was performed for the radial velocity and U,V,W components in Hernández-Pajares and Núñez (1989b). All these relations together (see Hernández-Pajares, 1989c for the proper motion equations) give kinematic meaning to the SH coefficients and convert this model into a more powerful tool for the stellar kinematic analysis.

As a second part we have applied the resultant relations to a sample of 4700 A-M stars. From the galactic longitude proper motions we obtain, in general, concordant results with previous literature for the Solar Motion and A and B Oort' constants. However the OM-results are of poor quality when we use the galactic latitude proper motions. Some combinations of 2nd order OM terms appear significant.

4. ACKNOWLEDGEMENTS.

We thank F.Figueras for the kind helpful and assistance, and Robin Rycroft of the Language Advisory Service of the Barcelona University for help in correcting the manuscript. This work has been supported by a research grant of the "Ministerio de Educación y Ciencia" of Spain and by the "Comisión Interministerial de Ciencia y Tecnología" under contract PB85-0017.

5. REFERENCES.

Avedisova,V.S., Palous,J.: 1989, Bull. Astron. Inst. Czechosl. <u>40</u>, 42.
Brosche,P.: 1966, Veröff. Astron. Rechen-Inst. Heidelberg No.17.
Brosche,P., Schwan,H.: 1981, Astron. Astrophys. <u>99</u>, 311.
Edmonson,F.K.: 1937, Monthly Notices Roy. Astron. Soc. <u>97</u>, 473.
Figueras,F.: 1986, Ph.D. Thesis, Un. Barcelona.
Figueras,F., Núñez,J.: 1989, in "Fundamentals of Astrometry", I.A.U. Colloq. 100, to be published in Celest. Mech.
Fricke,W.: 1967, Astron. J. <u>72</u>, 1368.
Fricke,W.: 1977, Veröff. Astron. Rechen-Inst. Heidelberg No.28.
Hernández,M., Núñez,J.: 1989a, Astrophys. Space Science <u>156</u>, 3.
Hernández-Pajares,M., Núñez,J.: 1989b, to be published in Astrophys. Space Science.
Hernández-Pajares,M.: 1989c Ph.D. Thesis, Un. Barcelona, in preparation.
Kerr,F.J., Lynden-Bell,D.: 1986, Monthly Notices Roy. Astron. Soc. <u>221</u>, 1023.
Milne,E.A.: 1935, Monthly Notices Roy. Astron. Soc. <u>95</u>, 560.
Núñez,J.: 1981, Ph.D. Thesis, Un. Barcelona.
Ochsenbein,M.: 1980, CDS Inf. Bull. <u>19</u>, 74.
Ogorodnikov,K.F.: 1932, Z. Astrophys. <u>4</u>, 190.

Application of Stepwise Regression in Photometric Astrometry

S. HIRTE – W.R. DICK – E. SCHILBACH – R.-D. SCHOLZ
Zentralinstitut für Astrophysik, Potsdam

ABSTRACT

A special method of stepwise regression applied to the selection of reduction models is described. Some examples illustrate the application of the method to the fitting procedure on plates taken with the Tautenburg Schmidt telescope.

1 INTRODUCTION

One of the most difficult problems in photographic astrometry is to find an appropriate model for the relationship between a reference catalogue and a photographic plate (or between different plates). Generally, a pair of polynomials in
powers of measured coordinates and sometimes of magnitudes
is postulated to represent this relationship. Usually the
form of polynomials is chosen empirically in accordance with
preceding extensive tests, but it is not always possible to
find a reduction model which can be accepted for each case.

By means of a strict mathematical method we tried to establish a reduction model which exhibits the smallest residual
variance among all possible polynomials with only significant terms.

2 THE METHOD OF STEPWISE REGRESSION

A starting reduction model can be represented by the linear
polynomial

$$y = \beta_0 + \sum_{j=1}^{m} \beta_j z_j + \varepsilon \tag{1}$$

with y – dependent variable, z_j ($j=1\ldots m$) – terms constructed from some independent variables x_i ($i=1\ldots p$), e.g. pow-

er, logarithm, exponential, etc., β_j (j=0...m) – regression coefficients to be determined, ε – residual with mean value 0 and variance σ^2.

The stepwise regression allows us to find a model with only significant terms $\beta_j z_j$ in (1). For the estimation of the goodness of the model we use the "prediction determination" B^V of Enderlein (1971) as statistical decision criterion:

$$B^V = 1 - \frac{(n-h-1)\ s_h^2 + (2h+2)\ s_m^2}{(n-1)\ s_y^2 + 2\ s_m^2} \qquad (2)$$

with h($\leq$m) – number of terms in the examined model, n – number of equations, s_y^2 – variance of the dependent variable, s_m^2 – residual variance of the complete starting model, s_h^2 – residual variance of the examined model. The model with the greatest value of B^V is the best one, because B^V increases if a significant term is added and decreases by adding a redundant term. To establish the "best" model two heuristic strategies, foreward selection and reduction, are used: in each step one term $\beta_j z_j$ is added to or deleted from the model, respectively. The two strategies may lead to different models. In this case the model with the greater B^V is accepted. The numerical solution is carried out by applying the sweep operator of Wiezorke (1967).

A Monte-Carlo simulation showed that in many cases the use of stepwise regression for reduction models with powers of rectangular coordinates in z_j reduced the number of terms $\beta_j z_j$ considerably. As a rule the mean square error is smaller than that of the full model (1). Therefore the method of stepwise regression may be especially useful if the number of reference points available is not very high but nonlinear terms z_j in the relationship (1) are suggested.

3 APPLICATION OF THE METHOD OF STEPWISE REGRESSION

The method described was applied to investigate the astrometric properties of Tautenburg Schmidt plates by use of measurements of AGK3 stars (Dick and Hirte, 1989) and to establish the form of the relationship between measured coordinates of objects appearing on different plates. Magnitude dependent terms were partly included in the reduction models (Dick, 1988) but will not be considered here.

Usually 50 to 100 AGK3 stars are available on a Tautenburg Schmidt plate. An example of a reduction polynomial obtained by stepwise regression from measurements of all AGK3 stars on a Tautenburg plate is given in (Dick and Hirte, 1989). Generally, only those terms are left in the "best" model wich are also classified as significant by the t-test. According to the t-test the critical values of the term-to-error ratios are for our cases (degree of freedom > 40) about 1.6 for significance level α = 0.05. In all cases considered the term-to-error ratios in the "best" models were never less than 1.4.

In order to estimate the systematic errors which might be introduced by reduction models, 716 galaxies on two plates nearly completely overlapping were measured. The galaxies were divided into 10 groups according to their magnitudes. Plate-to-plate solution was carried out for all galaxies and for each group separately (Table 1).

Table 1: Term-to-error ratios in x-coordinate with different groups of reference galaxies (N = number of galaxies, s = mean error of unit weight)

group	1	2	3	4	5	6	7	8	9	10	all
N	67	91	112	86	66	59	43	52	85	55	716
s/μm	4.4	5.5	4.2	4.7	4.0	5.1	4.3	5.1	4.2	4.1	4.7
x^2	2.0		2.4	2.0				1.9			3.0
xy	3.7	6.2	2.7	3.4	3.8		2.6		2.2	3.7	9.0
y^2									1.6		
x^3			2.3								1.5
$x^2 y$				2.0		2.0			1.8	1.5	2.3
xy^2	1.5		1.5						2.1		
y^3			2.1				2.2			1.8	1.7

Provided the plate-to-plate geometry represented by galaxy
positions is constant, the "best" polynomials of stepwise
regression should include nearly the same terms with the
different galaxy groups. However, the form of the resulting
polynomials is influenced by the measuring accuracy of single objects in a group, their distribution on the plate,
possible emulsion shifts on the plates and others. It might
also depend on the magnitude of the objects.

4 GENERAL REMARKS

The analysis of measurements on Tautenburg Schmidt plates by
stepwise regression shows that each field or even each plate
has to be treated with its individual reduction polynomial.
The reality of terms on the boundary of significance which
nevertheless are included in the reduction model by stepwise
regression is doubtful. It seems that the terms with term-
to-error ratios less than 1.7 to 1.8 should not be regarded
as real ones. The fitting should be repeated with a fixed
polynomial omitting doubtful terms. This approach is arbitrary, indeed.

The model of stepwise regression is the "best" one only
within the framework of the chosen form and degree of the
starting polynomial. We use only reduction polynomials up to
third order of the measured coordinates x and y because due
to physical reasons higher order terms seem to be not real
even if they formally appear as highly significant.

Forthcoming results in the investigation of the method of
stepwise regression and its application to astrometric reduction of measurements of Schmidt plates are being expected
from further Monte-Carlo simulations now under way.

REFERENCES

Dick, W.R.: 1988, Dissertation (A), Potsdam.
Dick, W.R., Hirte, S.: 1989, Mitt. Lohrmann-Obs. TU Dresden,
                    Nr. 56, 16.
Enderlein, G.: 1971, Biometrische Z.- Berlin, **13**, 130.
Wiezorke, B.: 1967, Metrika.- Wien, **12**, 68.

Binary Star Statistics: the Matter of Sampling

S. J. HOGEVEEN

Astronomical Institute "Anton Pannekoek", Roetersstraat 15, 1018 WB
Amsterdam, The Netherlands

ABSTRACT

When mass-ratio distributions of binary stars are derived from different binary star
catalogues, the results obtained are often widely different. In this paper a model for
'biased sampling' is presented, which may account for the observed differences.

1   MASS-RATIO DISTRIBUTIONS FROM CATALOGUES

Mass-ratio distributions have been presented by (among many others) Trimble and
Walker (1986) and Vereshchagin et al. (1987) for visual binary stars, and by Staniucha (1979) for spectroscopic binaries.

Trimble and Walker investigated the mass-ratio distribution of the systems in
the *Fourth Catalogue of Orbits of Visual Binary Stars* (OVB) compiled by Worley
and Heintz (1983). Vereshchagin et al. have determined the mass-ratio distribution
of the systems in *The Washington Double Star Catalogue* (WDS) compiled by Worley
and Douglass. The mass-ratio distributions found by these authors are displayed in
figure 1.a. The distributions are markedly different: the mass-ratio distribution of
the systems in the OVB peaks strongly toward mass-ratios $q = 1$ ($q = M_{\mathrm{sec}}/M_{\mathrm{prim}}$),
while the q–distribution found for the systems in the WDS indicates that secondary
masses are distributed according to the Initial Mass Function (IMF) for single stars,
at least for mass-ratios $q > 0.4$ (the IMF indicates that masses of single stars are
distributed according to the power law $M^{-2.7}$; see Scalo, 1979).

Staniucha (1979) repeated earlier work of Trimble (1974) and Kraicheva et al.
(1979) when the *Seventh Catalogue of the Orbital Elements of Spectroscopic Binary
Systems* (DAO) compiled by Batten, Fletcher and Mann (1978) became available.
The DAO contains two observational catagories of binaries: the *single lined* (SBI)
and the *double lined* (SBII) spectroscopic binaries. Figure 1.b shows the mass-ratio
distribution derived for each category. Again there is a marked difference between

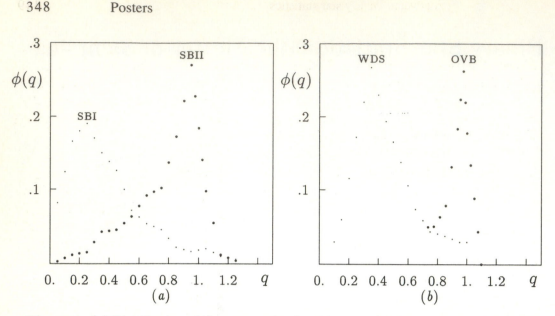

Figure 1: (*a*) Distribution of the mass ratios for 653 SBI and 325 SBII systems in the DAO. $\phi(q)$ represents the fraction of the total number of systems with mass ratio q in the interval Δq. (*b*) Distribution of the mass ratios for the systems in the WDS and in the OVB. For OVB systems only the peak for $q > 0.72$ is shown here. The diagrams have been reconstructed from graphs presented by Staniucha (1979), Trimble and Walker (1986) and Vereshchagin et al. (1987).

the distributions: the SBII systems peak towards $q = 1$, and the q–distribution of the SBI systems can be fitted with an IMF for the secondaries from $q > 0.25$.

2   BIASSED SAMPLING

When we look at the observed mass-ratio distributions presented in section 1, the question arises whether the *observational* categories of binary stars indeed represent *physically* different categories of binary systems, or if they all adhere to a common distribution, which for each observational category is obscured by selection effects. The selection effects governing the observations of binary stars are investigated in two papers, one about the mass-ratio distribution of visual binary stars (Hogeveen 1989, submitted) and one about the q–distribution of spectroscopic binaries (Hogeveen 1989, in preparation). The papers show that it is quite feasable to determine the selection effects affecting the visual binary stars in the WDS, and the single lined (SBI) systems in the WDS. However, it turned out very hard to determine and model the selection effects for the visual binaries in the OVB and the SBII systems in the DAO. However, it is possible to obtain useful results if we make some crude assumptions about when the conditions are 'favourable' for a system to determine its orbit.

For visual binaries the periods of the systems may not exceed the time-span of the observations very much. This in practice means that the separations of the systems cannot be very large, so they will on average have small angular separations. When the systems have small separations the magnitudes of the components may not differ too much in order to be properly observable. Thus 'favourable' systems will be those with nearly equal magnitudes.

The mass-ratio for double lined spectroscopic binaries systems is determined from the radial velocity amplitudes of the components K_{prim} and K_{sec} as: $q = M_{\mathrm{sec}}/M_{\mathrm{prim}} = K_{\mathrm{prim}}/K_{\mathrm{sec}}$. The radial velocities are measured from spectra, and in order to make accurate measurements, the lines of both components have to be visible properly. This in practice means again that the magnitudes of the components cannot differ too much, thus 'favourable' SBII systems are also those with nearly equal magnitudes.

If we assume a Main Sequence mass–luminosity relation, then 'nearly equal' magnitudes also mean nearly equal masses, or a preference for mass-ratios close to $q = 1$.

2.1   The Model

Let us assume that all binaries adhere to an IMF distribution for secondary masses. The observed binaries, and especially those for which orbits have been determined, comprise (in numbers) only a limited sample of the entire population of binary stars. Let us also assume that the systems in the OVB and the SBII systems in the DAO have been selected from this population according to the preferences described in the previous section.

Let us assume a population of $\mathcal{N}$ elements. A sample of N 'favourable' elements can be drawn from this population by dividing the population into N sub-samples (i.e. just as many sub-samples as the number of elements in the final sample) of $n = \mathcal{N}/N$ elements, and choosing from each sub-sample the 'most favourable' element. We have then drawn a sample of N 'favourable' elements from the population, rather then the N very best, because the 'most favourable' element in one sub-sample may be worse than the second, third or even fourth best element in an other sub-sample.

Figure 2 shows what happens if we apply this method of 'biassed selection' to a population of 30,000 elements, distributed according to an IMF-like power law, to obtain a sample of 1,000 elements. A 'favourable' element is here defined as one which has its value close to 1. The figure shows that, although the parent population has a distribution according to the IMF, the elements in the sample have an entirely different distribution, which in fact does resemble the q–distributions derived from the OVB and the SBII systems in the DAO.

In fact, it is possible with this method to derive a sample of $N < \mathcal{N}$ systems from $\mathcal{N}$ systems in the WDS which exhibits the q–distribution of the systems in the OVB (Hogeveen 1989, submitted).

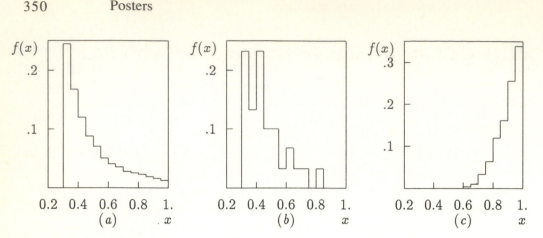

Figure 2: (*a*) Distribution of parent population ($\mathcal{N} = 30{,}000$ elements; $f(x)$ is fraction of elements with value x in interval Δx). (*b*) Arbitrary sub-sample of $n = \mathcal{N}/N = 30$ elements. (*c*) Sample of 1,000 elements, selected as the element with its value closest to $x = 1$ from each sub-sample.

3   CONCLUSIONS

From the above we may conclude that, if the assumptions regarding the selection effects are valid, the mass-ratio distributions of binary stars, derived from the OVB and the SBII systems in the DAO, are heavily affected by the way these samples were selected, and that the distributions are not incompatible with an IMF distribution of secondary masses for all binary stars [the samples in the OVB and the DAO comprise mainly (A,) F, G, and K–type stars].

Because the main 'selection effect' is the decision of the observer to determine an orbit, it is not possible to assess this effect more accurately than in a crude way as employed above.

REFERENCES

Batten, A. H., Fletcher, J. M., Mann, P. J.: 1978, *Seventh Catalogue of the Orbital Elements of Spectroscopic Binary Sytems*, Dominion Astrophysical Observatory, Victoria, B.C.

Kraicheva, Z. T., Popova, E. I., Tutukov, A. V., Yungel'son, L. R.: 1979, *Sovjet Astronomy* **23**, 290

Miller, G. A., Scalo, J. M.: 1979, *Astron. J. Suppl. Ser.* **41**, 513

Staniucha, M.: 1979, *Acta Astron.* **29**, 587

Trimble, V.: 1974, *Astron. J.* **79**, 967

Trimble, V., Walker, D.: 1986, *Astrophys. Sp. Sc.* **126**, 243

Vereshchagin, S. V., Kraicheva, Z. T., Popova, E. I., Tutukov, A. V., Yungel'son, L. R.: 1987, *Sovjet Astron. Lett.* **13**, 26

Worley, C. E., Heintz, W.: 1983, *Fourth Catalogue of Orbits of Visual Binary Stars*, U. S. Naval Observatory, Washington

Outlying points in multivariate solar activity data

M. JAKIMIEC

Astronomical Institute, University of Wroclaw, Poland

1. INTRODUCTION

Usually, a set of predicting variables (X) describing solar active region features for the given day is the basis in short-term (one day ahead) solar flare activity predictions (see e.g. Sawyer et al, 1986), where the predicted variables (Y) characterize solar flare activity for the next day. The predicting algorithms are constructed for the multivariate data sets forming the matrix with n rows and p+q collumns, where n is the number of the data vectors (the size of the sample), p and q are the numbers of the predicting and predicted variables, respectively. An accurate knowledge of the interrelations between X and Y variables is necessary to the construction of the statistical models which, in turn, may be used in the prediction procedure. The short-term predictions are often based on a regression equation of the type:

$$y = b_0 + \sum_{i=1}^{p} b_i x_i + e$$

where x_i (i=1,...,p) are the predicting variables, b_0 and b_i are the estimated regression coefficients and e is an error. Sometimes, the set of the p+q variables contains outlying values and some data vectors with such values reveal non-typical interrelations unlike to the general relations between the appropriate variables.

There is the question of the predicting function stability, i.e. how much the interrelation structure of the variables, and consequently, the obtained predicting algorithms are disturbed by these atypical data vectors. If the data set contains outlying, atypical data vectors, we should moreover examinate whether a more appropriate to the bulk of considered data points model may be obtained by more modern robust methods.

2. DISCUSSION

Now, we will discuss the results of studies of the observational data gathered for the year 1979 from Solar Geophysical Data. A complex of p=14 predicting variables characterizing sunspot groups of the D, E, F Zurich classes was investigated. They are: McIntosh class, sunspot group area, calcium plage area and intensity, magnetic class, magnetic field strength, magnetic field index and seven X-ray flare activity characteristics. All these variables are more comprehensively described by Jakimiec and Wanke-Jakubowska (1988).

In papers by Bartkowiak and Jakimiec (1989a) and Jakimiec and Bartkowiak (1989b) the data characterizing sunspot groups in the decay and in the increase phase of the evolution were analysed separately. Authors applied two methods: the Chi2-plots constructed from Mahalanobis distances (method described by Gnanadesican and Kettenring, 1972), and scatterdiagrams constructed from the first two and last two principal components (method described by Morrison, 1967). They found several unquestionably atypical data vectors. The notypicalness is related either to very strongly flaring sunspot groups or to sunspot groups with non-typical relationships of the characteristics.

To discover some common factors explaining the interrelation structure of the considered variables Jakimiec and Bartkowiak (1989b) employed the factor analysis method described e.g. by Morrison (1967). Authors compared the common factor structures obtained twice for the entire data sets and for the data set after removing of the atypical data vectors. The comparison showed that there are no significant changes of the common factor structures. So we might conclude that intrinsic structure of the sunspot groups characteristics is sufficiently stable despite some outlying data vectors contained in the data set. Moreover, Jakimiec and Bartkowiak (1989c), constructing dendrogrammes examinated the "similarities" between the considered variables. They found, similarly to the preceding results, that the atypical data vectors do not influence significantly the interrelations of the variables.

Investigating the stability of the regression function Jakimiec and Bartkowiak (1989a) applied four statistical methods showing the impact of single data vector on the stability of the regression equation: the diagonal elements of the HAT matrix, the externally studentized residuals (Cook and Weisberg, 1980), the statistic DFFITS introduced by Belsley et al. (1980) and the mean percent of distortion of the regression coefficients. The authors extended the considered previously variables with two predicted variables characterizing X-ray flare activity the next day. They found a number of influential data vectors which, in most cases, were already revealed as outliers. It was found, however, that such individual influential data vectors, although can change in some measure particular estimations of the regression function parameters, and yet can not disturb strongly the whole regression function.

3. CONCLUSIONS

The stability of the regression despite some atypical, outlying data vectors contained in the data set allow us to conclude that the interrelationships of the variables is specified by the bulk of the data vectors. The regression functions estimated for a main part of the data set, should reflect true physical relations between the sunspot group characteristics. As a consequence, we should use for the building of the predicting functions either the data set after removing of the outliers or the data set with weighted data vectors.

Bartkowiak and Jakimiec (1989b) employed the α-trimmed robust method allowing to remove authomatically from the data set such outlying data vectors which give the great residuals (forecasted minus observed values) in the prediction. The data vectors trimmed off by this method correspond to the previously revealed atypical data vectors. These vectors seem to be related to sunspot groups with sudden changes of flare activity, e.g. to a newly emerged magnetic flux. The regression functions obtained by use of the α-trimmed method describe slowly changing process of flare activity defined, in essence, by the evolution of active regions.

Bartkowiak and Jakimiec (1989c) applied also the robust regression method, so called "weighted regression with Huber weights", ascribing to the outlying data vectors the smaller weights the greater are the residuals. The obtained regression functions are very similar to those obtained by α-trimmed regression.

REFERENCES

Bartkowiak, A. and Jakimiec, M., 1989a, Acta Astr., vol. 39, 85.

Bartkowiak, A. and Jakimiec, M., 1989b, submitted to Acta Astronomica.

Bartkowiak, A. and Jakimiec, M., 1989c, paper send to Acta Astr.

Belsley, D. A., Kuh, E. and Welsch, R. A., 1980, Regression Diagnostics: Identifying Influential Data and Sources of Collinearity, (Wiley, New York).

Cook, R. D. and Weisberg, S., 1980, Technometrics, vol. 22, p. 495.

Gnanadesikan, R. and Kettenring, J. R., 1972, Biometrics, vol. 28, p. 81.

Jakimiec, M. and Bartkowiak, A., 1989a, Acta Astr., vol. 39, in print.

Jakimiec, M. and Bartkowiak, A., 1989b, submitted to Acta Astronomica.

Jakimiec, M. and Bartkowiak, A., 1989c, submitted to Acta Astronomica.

Jakimiec, M. and Wanke-Jakubowska, M., 1988, Acta Astr., vol. 38, p. 431.

Morrison, D. F., 1967, Multivariate Statistical Methods (McGraw-Hill, New York).

Sawyer, C., Warwick, J. W. and Dennett, J. T., 1986, Solar Flare Prediction (Colorado Associated University Press, Boulder, USA).

Hierarchical Cluster Analysis of Emission Line Spectra of Planetary Nebulae

G. JASNIEWICZ [1] - **G. STASINSKA** [2]

1.) Observatoire Astronomique, Centre de Données Stellaires, Strasbourg, France
2) Observatoire de Meudon, Meudon, France

It is a widespread custom in observational studies of planetary nebulae to assign an "excitation class" to each object according to the characteristics of its emission-line spectrum (Webster, 1975, Morgan, 1984, Aller and Keyes, 1987). This implicitely assumes that planetary nebulae are a one parameter family. Some authors use the excitation class as an indicator of the star effective temperature (e.g. Zijlstra and Pottasch, 1989), while there is obviously at least one important parameter which governs the emission-line spectrum of planetary nebulae, namely the ionization parameter (cf. e.g. Stasińska. 1989), not to mention the optical thickness of the nebular gas and the electron density. There also exists the idea that high excitation class planetary nebulae have more massive nuclei. and that low excitation classes are associated with low mass nuclei (e.g. Webster. 1988), although photoionization models for planetary nebulae in expansion around evolving nuclei of different masses show that this is not true (Stasińska, 1989).

In fact. planetary nebulae form at least a two parameter family, the basic parameters being the mass of the central star and the age of the nebula. And indeed. the determination of the central stars masses is the subject of many studies and speculations. Here. we wish to address this problem from a statistical point of view. In a first step. we will classify the spectra of planetary nebulae using an objective method. The next (and more difficult) step would be to interpret these classes using theoretical spectra produced by photoionization models for nebulae of different ages around stars of various masses.

In order to perform such a classification, we must select a set of line ratios which are not too much affected by the variations of the chemical composition of the nebular gas from object to object. The number of lines should be greater than 2, and the lines should potentially provide complementary information. Having this in mind. we have chosen HeII4686 / Hβ wich is mainly (but not only) dependent on the effective temperature, [OIII] λ5007 / Hβ, which is also strongly dependent on the ionization parameter and [OII] λ3727 / Hβ, which depends also on whether the nebula is density-bounded.

The sample of planetary nebulae contains 128 galactic objects, whose dereddened spectra heve been taken from Peimbert and Torres-Peimbert (1977), Barker (1978), Aller and Czyzak (1979, 1983), Aller and Keyes (1987), and Webster (1988). We have not included objects for which the [OII] line was not seen, since this absence was interpreted as an effect of high reddening, and not as a real lack of [OII] emission.

We have performed on this sample a principal component analysis (PCA), using the option on the correlation matrix. The eigenvalues of this matrix are 0.44, 0.32 and 0.24 relatively to the unit. Thus, the chosen line ratios are not correlated. We have then classified the objects following the Ward's method of hierarchical clustering, using the projected data upon the eigenvectors defined by the PCA. Six classes have been found, which can roughly be characterized as indicated in the table below. The last column of the table gives the number of objects in each class. Figures 1 to 3 show how the nebulae are distributed in the [OIII] / Hβ vs [OII] / Hβ, [OIII] / Hβ vs He II / Hβ, and [OII] / Hβ vs He II / Hβ planes, each class being represented by a different symbol.

class	[OIII] / Hβ	[OII] / Hβ	HeII / Hβ	nb of objects
1	0. - 10.	0. - 1.2	0. - 0.25	39
2	10.- 20.	0. - 0.8	0. - 0.5	42
3	0. - 8.	1.2 - 5.	0. - 0.5	10
4	3. - 15.	0. - 1.2	0.5 - 1.2	18
5	7. - 15.	3. - 6.	0. - 0.8	6
6	10. - 20.	0.5 - 2.	0.25 - 0.8	13

References

Aller,L.H.,Czyzak,S.J.: 1979, *Astrophys. Space Sci.* **62**,397

Aller,L.H.,Czyzak,S.J.: 1983, *Astrophys. J. Suppl.* **51**,211

Aller,L.H.,Keyes, C.D.: 1987, *Astrophys. J.* **65**,405

Barker,T.:1978, *Astrophys. J.* **219**, 914

Morgan,D.H., *Monthly Notices Roy. Astron. Soc.* **208**, 633

Peimbert,M., Torres-Peimbert,S.: 1977, *Rev. Mex. Astron. Astrophys.* **2**, 181

Stasińska,G.:1989, *Astron. Astrophys.* **213**, 274

Webster, B.L.: 1975, *Monthly Notices Roy. Astron. Soc.* **173**, 437

Webster, B.L.: 1988, *Monthly Notices Roy. Astron. Soc.* **230**, 377

Zijlstra, A.A., Pottasch, S.R., 1989, *Astron. Astrophys.* **216**, 245

Figure 1

Figure 2

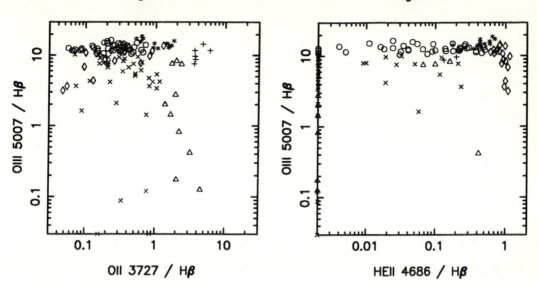

Figure 3

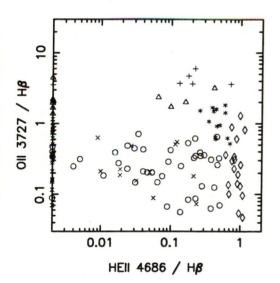

Meaning of symbols:

x   Class 1

o   Class 2

△   Class 3

◊   Class 4

+   Class 5

*   Class 6

Periodicity Probability
Are these flares periodic ?

TAO KIANG

Dunsink Observatory, Dublin 15 (Ireland)

ABSTRACT Four flares at intervals of 49, 51, 44 min were
observed in a star. The probability of a more regular
spacing by chance is shown to be 0.0047 on a Poisson process
approach, and 0.0069 on assuming uniform distribution
between fixed ends. Corresponding formulae for 5 flares are
given, showing how they can be further extended.

On 6 March 1988, astronomers of Armagh Observatory observed four flares
in the eclipsing binary YY Gem within a period of 144 minutes. The
flares were spaced at intervals of 49, 51 and 44 minutes, and each
interval contained a dead period of 10-15 min. the question was, are
these flares significantly periodic ?

Whether or not there might be flares in the dead periods, the
statistical probelm is the same, namely, given a smallish number (n'') of
events with more or less regular spacing, what is the probability of a
more regular spacing arising by chance ?

There are two approaches to this problem depending on what we take as
the random background. In one, we consider the flares to form a Poisson
process on a scale equal to the mean value of the observed intervals. In
the other, we consider the two end flares to be fixed, and each of the
other flares to be a uniform random variable over the fixed span. In
either case, the number of degrees of freedom is $n = n''-2$, and the task
is to find a statistic (function of observations) which can reasonably
be taken as a measure of departure from strict periodicity and then find
its distribution function.

A. Poisson Process on Fixed Scale

In a Poisson process, each interval (inter-arrival time) s follows
independently the exponential distribution with scale parameter μ, say.
The standardized interval $x = s/\mu$ then follows the standardized
distribution,

$$P(dx) = e^{-x} dx, \quad (0 \leq x \leq \infty), \tag{1}$$

with mean and standard deviation both equal to 1. Strict periodicity
means all observed x's equal to 1. So we might be tempted to use the
quantity $\Sigma(x_t-1)^2/1^2 = \Sigma(x_t-1)^2$ as the test statistic. But this is no
good, for the distribution (1) is so highly non-normal that this
quantity will not approach even approximately the known χ^2-distribution
for the small number of degrees of freedom n.

A statistic with a determinable distribution is the range r, the
difference between the largest and smallest x. But since our definition
of x imples the mean x is fixed at 1, the distribution we want is not
the unconditional distribution of r, rather, it is the conditional
distribution of r, conditional on the mean being fixed at 1 ($\bar{x}=1$).

In the present example, if we first ignore the dead periods, we have
$n''=4$ flares and $n'=3$ intervals. The mean interval is $144/3 = 48$, hence

The standardized intervals are, when ordered according to size, 0.917, 1.021 and 1.062, and the range is $r = 1.062-0.917 = 0.145$.

It can shown that the distribution of the range in 3 observations with the mean fixed at 1 is the triangular distribution over the interval $(0, 3)$:

$$P(dr;\ n'=3, \overline{x}=1) = (4/9)\ r\,dr, \qquad\qquad (\ 0\ \le r \le 3/2),$$
$$= (4/9)(3-r)\ dr, \qquad (3/2 \le r \le\ 3\ ). \qquad\qquad (2)$$

Hence, the probability of $r \le 0.145$ arising by chance in the present case is

$$P(r \le 0.145;\ n'=3,\ \overline{x}=1) = (4/9) \int_0^{0.145} r\,dr = 0.0047. \qquad (3)$$

This value seems to agree more or less with what one might have guessed from the three numbers, 49, 51, 44.

Now, suppose there was an extra flare in the first dead period, thus splitting the first interval into two, 25 and 24 min, say. Then we would have $n'' = 5$ flares and $n' = 4$ intervals. The mean interval would be be $144/4 = 36$ min and the range would be $r = (51-24)/36 = 0.75$.

Again, it can be shown that the conditional distribution of r for $n' = 4$ and $\overline{x} = 1$ consists of three parabolic arcs over the interval $(0,4)$:

$$P(dr;\ \overline{x}=1, n'=4) = (9/32)\ r^2\,dr, \qquad\qquad (\ 0\ \le r \le 4/3),$$
$$= (9/64)[-16 + 24\ r - 7\ r^2]\ dr, \quad (4/3 \le r \le\ 2\ ),$$
$$= (9/64)(4-r)^2, \qquad\qquad\quad (\ 2\ \le r \le\ 4\ ). \qquad (4)$$

So, for the hypothetical case of four intervals, 25, 24, 51, 44, the probability of a more regular division of the given span arising by chance is

$$P(r \le 0.75;\ n'=4, \overline{x}=1) = (9/32) \int_0^{0.75} r^2\,dr = 0.040\ . \qquad (5)$$

thus, even if there was a flare in the first dead period, there would still be a sigificant periodicity, "at the 5% level"; of course, if there was not, the significance would be much higher, at the 0.5% level.

B. Uniform Distribution between Fixed Ends

When we have $n''=4$ flares, we re-zero and re-scale the time axis so that the first and the last occur at times 0 and 3, respectively. The other two are then supposed to be uniformly distributed in the interval $0 \le t \le 3$. Let t_1 and t_2 be the times of the later and the earlier of these two. Their joint distribution is a uniform distribution over the region $3 \ge t_1 \ge t_2 \ge 0$ in the (t_1, t_2) plane, that is, the triangle ABC in Fig. 1.

Strict periodicity is represented by the point $O(2, 1)$, and the observed configuration by the point $P(t_1, t_2)$. The question is, what region of the traingle should we take as being closer to O than the point P is? It seems that a reasonable choice would be the triangle A'B'C', "homothetic" to ABC and passing through P. the probability p we seek, then, is just the ratio of the areas of the two traingles, A'B'C' to ABC. Suppose P falls in the subtriangle OBC (and hence, by

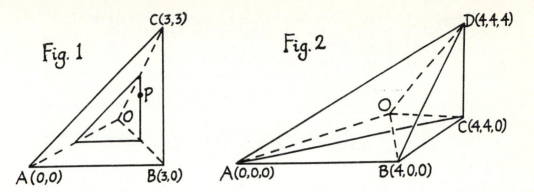

Fig. 1

Fig. 2

definition, on B'C'). Then we have $p = \beta^2$, where β is the relative height of the triangles OB'C' to OBC. It can be shown that, for each of the three possibilities, β is equal to the scalar product of the vector OP and the "relative normal vector" given in the last line of TABLE 1.

To find in which of the three sub-triangles, OAB, OBC, OCA, the observed point P falls, it is convenient to use coordinates referred to O: $x = t_1 - 2$, $y = t_2 - 1$. We evaluate, at P, the three functions f_{OA}, f_{OB}, f_{OC}, defined in TABLE 1, then the signs of the two of the values will decide the issue according to the scheme tabulated (the third sign is irrelevant).

TABLE 1 Signatures and Relative Normal Vectors ($n = 2$)

function	OAB	OBC signature	OCA
$f_{OA} = \quad x - 2y$	+		−
$f_{OB} = \quad x + \ y$	−	+	
$f_{OC} = -2x + \ y$		−	+
relative normal	(0, 1)	(1, 0)	(−1, 1)

In the present example, we have

$x = t_1 - 2 = 3 \times (49+51)/144 - 2 = 0.083,$
$y = t_2 - 1 = 3 \times 49/144 = 0.021,$

and $f_{OB} = +$, $f_{OB} = -$, hence, according to the signatures, P falls in the sub-traingle OBC, for which the relative normal vector is (1, 0). Hence $\beta = 0.083 \times 1 + 0.021 \times 0 = 0.083$ and

$$p = \beta^2 = 0.0069 . \tag{6}$$

This is somewhat larger than the value of 0.0047 given at (3) from the first approach. There is, of course, no reason why they should agree closely. Indeed, considering their completely different character, we should remark on their similarity rather than their difference. The similarity suggests that either statistic can be regarded as a reasonable measure of departure from strict periodicity. My own preference is for the second approach, because the conditional distribution of r of the fist approach is a marginal distribution, obtained after integrating out $n'-2$ variables, whereas the evaluation of p in the second approach does not involve any such "slurring-over". In other words, the first approach ignores part of the information contained in the intermediate valued intervals, while the second approach is sensitive to all the fine details subject to the stated constraint.

Again, if we suppose there was an extra flare in the first dead period, then we would have $n''=5$ events, and we would be dealing with the tetrahedron $4 \geq t_1 \geq t_2 \geq t_3 \geq 0$ in the three dimensional (t_1, t_2, t_3) space. The four vertices of the tetrahedron are $A(0, 0, 0)$, $B(4, 0, 0)$, $C(4, 4, 0)$ and $D(4, 4, 4)$, (see Fig. 2), and strict periodicity is represented by the point $O(3, 2, 1)$. Analogous to TABLE 1, we now have TABLE 2.

TABLE 2   Signatures and Relative Normal Vectors ($n = 3$)

function	OABC	ODAB	ODBC	ODCA
		signature		
$f_{OAB} = -y + 2z$	−	+		
$f_{OBC} = x + z$	−		+	
$f_{OCA} = -x + y + z$	−			+
$f_{ODA} = x - 2y + z$		+		−
$f_{ODB} = x + y - z$		−	+	
$f_{ODC} = -2x + y$			−	+
relative normal	$(0, 0, -1)$	$(0, -1, 1)$	$(1, 0, 0)$	$(-1, 1, 0)$

Again, suppose the extra flare occurred at 25 min after the first one. We then have

$x = t_1 - 3 = (100/144) \times 4 - 3 = -0.222,$
$y = t_2 - 2 = (49/144) \times 4 - 2 = -0.639,$
$z = t_3 - 1 = (25/144) \times 4 - 1 = -0.306,$

and we find $f_{OAB} = +$, $f_{ODA} = +$, $f_{ODB} = -$, hence, according to the signatures shown in TABLE 2, the observation falls in the tetrahedron ODAB, for which the relative normal vector is $(0, -1, 1)$. Hence we have $\beta = 0.639 - 0.306 = 0.333$ and

$$p = \beta^3 = 0.037. \tag{7}$$

Note the exponent of β is 3 here; in general the exponent is equal to the number of free variables, n, or number of flares *minus* 2. The near agreement between (7) and (5) is satisfactory.

Generalization to larger values of n will not be difficult if one simply follows the analytic procedure and not trying to visualize the shape of objects in more than three dimensions.

The Systematic Effect of Shape on Galaxy Magnitudes

M. J. KURTZ – E. E. FALCO

Harvard–Smithsonian Center for Astrophysics, Cambridge, U.S.A.

We have used the FOCAS (Jarvis and Tyson, 1981; Valdes, 1982) software to test for any systematic effects on automated galaxy photometry. We have used a simple observational model universe (Falco and Kurtz, 1988) to generate synthetic CCD frames. Preliminary results show no large systematics save for the two expected: proximity to another galaxy and mean surface brightness (shape). He we begin to quantify the latter.

The SIMULA program (Falco and Kurtz, 1988) currently generates a clustered distribution of galaxies in a euclidean space using a modified version of the Soniera and Peebles (1979) method. This distribution has a correlation function with a slope of 1.8, and the same luminosity function as found by de Lapparent *et al.* in the CfA survey. The limiting apparent magnitude of the simulation is 25.

SIMULA then creates a simulated CCD frame of the model sky. In this version all galaxies are pure deVaucouleur law ellipticals, with a distribution of core radii which matches that found by Kent (1985) for his sample. The ellipticity is random.

The fields are created with several combinations of observational parameters. In this poster we will describe results from simulating a one hour exposure with a low-noise (10 electron read out noise) CCD taken with a telescope with a 1 square meter aperture. The pixels and the seeing are one arc second. The simulated frame is essentially perfect, with no flat field problems, no cosmic rays, no sky variations, and no stars!

We next reduce the simulated data using FOCAS, using the known point spread function in the place of using the FOCAS routines to find it, as they must fail due to the non-existance of stars in the field. For this poster we will only consider those FOCAS did not attempt to split into multiple objects, in an attempt to separate effects.

Figure 1 shows the difference between the FOCAS derived magnitudes and the known total magnitudes, plotted against the total magnitude. All errors above 1.7 are due to a faint object center being closer to the FOCAS derived center than the brighter object which FOCAS actually found. This cannot be considered a real error.

Note that the zero point of the FOCAS magnitudes is arbitrary. The mean scatter about the optimal zero point for the FOCAS mags is 0.13 mag. The actual scatter would be greater than this, and would depend on the calibration methods. It can be seen that the scatter of the fainter objects is not symmetric about the value defined by the brighter objects.

Figure 2 shows the magnitude difference plotted against the mean surface brightness, calculated by using the area which contains 99.5% of the light. This is essentially a measure of the shape of the galaxy light distribution. There is a substantial correlation; clearly errors in the determination of magnitudes are strongly correlated with the shapes of the galaxy light distributions.

The direction of this error is that low surface brightness objects seem fainter than they should (using the bright, high surface brightness objects for the calibration). Because we are comparing an isophotal magnitude (FOCAS) with a total magnitude this is essentially what one should expect. Indeed for these objects FOCAS finds an isophotal area which is a smaller fraction of the 99.5% area than for higher surface brightness objects. Figure 3 shows the log of the ratio of areas (99.5% / FOCAS) versus the magnitude error. The correlation is remarkable.

The extent of the effect in our simulation is 0.5 magnitudes between the brightest and faintest surface brightness objects. After developing such niceties in our model as non-euclidean geometry, and spiral galaxies, we intend to develop shape dependant corrections which should be capable of obtaining improved magnitudes.□

Simula1L mag error vs diff in areas

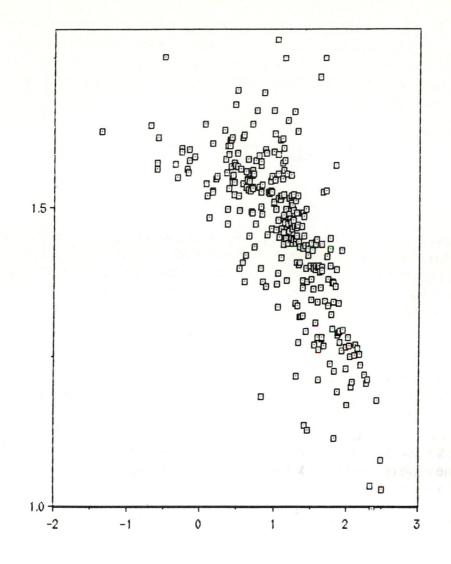

adiff

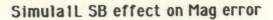

SimulaiL SB effect on Mag error

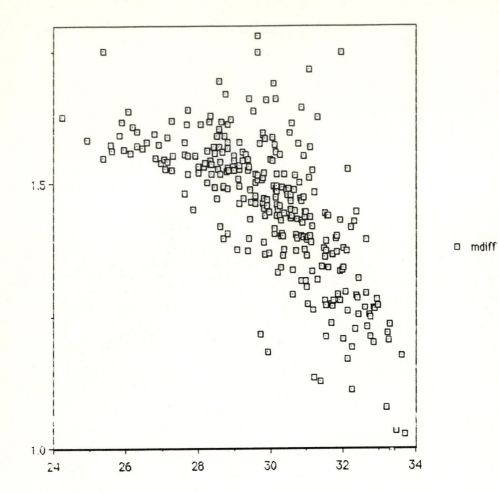

Simula1L Mag Error ve Mag

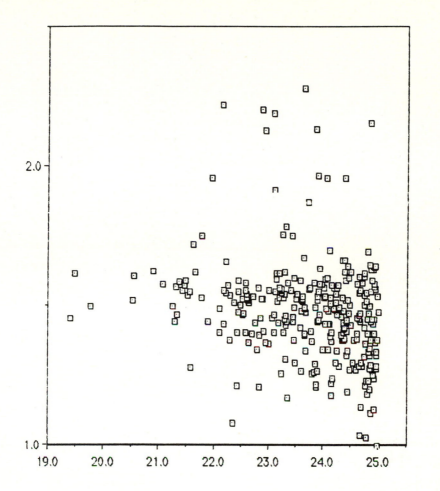

□ m(T)

m(T)

AUTOMATED PROCESS FOR IMAGES DETECTION IN ASTROMETRIC PLATES.

A. López García[1], J. A. López Ortí[1,2],R. López Machí[1,2].

[1] Observatorio Astronómico. Departamento de Matemática Aplicada y Astronomía.
Universidad de Valencia. Spain.

[2] Colegio Universitario de Castellón. Universidad de Valencia. Spain.

ABSTRACT

The automatic process of searching, centering and measuring of an stellar image in astrographic plates is described.

A detailed study about the different alternatives which arise in order to get information about the image transparency, by means of a photometer or a CCD, is done.

1 INTRODUCTION.

In our semiautomatic measuring process of astrographic plates, the images position are got in a sequential way through a route that includes the asteroid. The centering of each stellar image is got from the adjust of an elliptical paraboloid to the plate transparency in a neighbourhood of the centre of the image (Centrado and Refinc, Fig. 1). To do this, sixteen lectures are made with a photoelectric photometer coupled to the measuring engine eye-piece. In a 4x4 points net a quadratic polynomial function of six coefficients is adjusted.

The reduction of each exposure includes a previous adjust that allows to select the stars that will be included in the final adjust (made by polynomic or dependences method).

2 ALTERNATIVE MEASURING PROCESS.

An alternative process, that make use of a CCD camera plus a digitizer board, has been put in practice recently, and we want to expose the following:

1.-   The background plate determination in a neighbourhood of each image is delicate, and this problem has been analyzed by other authors. In our case we have just made a "lecture" of it in a squared zone around each image.

2.-   The centre of each image has been got as the baricentre of the light distribution once the background was eliminated (Fig. 2).

3.-   The light distribution of the asteroid and the stars of one plate have been represented, and a gaussian-like surface has later been adjusted with success (Fig. 3 to 7).

4.-   The differential correction of the CCD image baricentre has been added (3 and 4 measures) to the positions got by means of the photometer lectures (1 and 2 measures).

5.-   The same reduction process (Ajuplc) has been employed in the coupled measures 1,2 and 3,4, taking, in both previous and final adjust, all the stars measured (Tables 1 and 2). The results, got in the polynomical adjust, so as in the previous one, comparison allow us to affirm that the stellar images position got by means of the digitizer board improve the results in both adjusts (lower maximum and mean internal errors).

3   CONCLUSIONS

The positions obtained from the transparency lectures in a squared 4x4 net with the photometric equipment, and the later paraboloid adjust and centering, provide a good enough plate reduction in comparison with the more sophisticated process involving the CCD camera and the digitizer board, that requires a more complex equipment and software as so as a greater time spending.

So we consider that this process is only necessary in the more unfavourable eventualities.

4   REFERENCES.

Lindegren, L. 1978. "Photoelectric Astrometry. A comparison of methods for precise image location". In MODERN ASTROMETRY, IAU Col. nº 48.

López García, A. *et al.* "Astrometría de Asteroides en el Observatorio Astronómico de Valencia. II.- Medida de las placas. Dispositivo mecánico y electrónico". Bol. Astron. del Observatorio de Madrid. (In press).

López García, A. *et al.* "Astrometría de Asteroides en el Observatorio Astronómico de Valencia. III.- Algoritmos y programas". Bol. Astron. del Observatorio de Madrid. (In press).

Sedmak, G. *et al.* 1979. INTERNATIONAL WORKSHOP ON IMAGE PROCESSING IN ASTROMETRY.

TABLE 1.- Internal Errors.

	Method				
	Photoelectric			CCD	
	Measures				
	n° 1	n° 2		n° 3	n° 4
Star n°	Residuals (")				
2	1.1	0.6		0.7	0.7
3	1.2	0.6		0.7	0.5
4	1.0	1.0		0.7	0.3
5	0.7	0.9		0.6	0.2
6	2.1	2.4		1.8	0.7
7	0.9	1.2		0.8	0.3
8	1.3	1.4		0.8	0.3
9	0.5	0.5		0.6	0.7
10	0.6	0.4		0.2	0.2
11	0.1	0.0		0.0	0.0

TABLE 2.- REDUCTION RESULTS.

Method	Asteroid position		Dispersion	
	RA(1950.0)	DEC(1950.0)		
Paraboloid	$1^h\,50^m\,32^s.79$	$4°\,45'\,25".2$	$0^s.00$	$0".5$
CCD	$1^h\,50^m\,32^s.82$	$4°\,45'\,25".9$	$0^s.02$	$0".5$

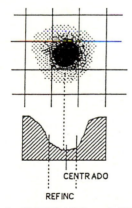

260.5983   278.6549

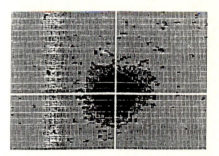

FIG. 1.- CENTRADO and REFINC schematic routines work.

FIG. 2.- Digitized stellar image and relative baricentre coordinates.

372   Posters

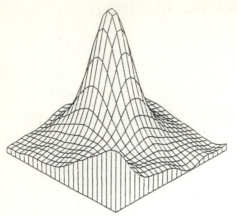

FIG. 3.- Light distribution surface for
the star #4.

FIG. 4.-Topographic map of light distribution
surface for the star #4.

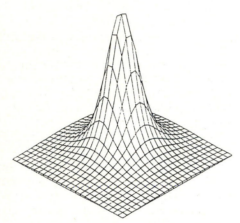

FIG. 5.- Gaussian-like surface for the light
distribution for the star #4.

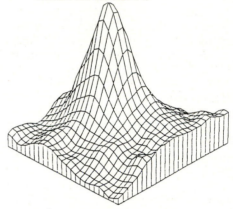

FIG. 6.- Light distribution surface for
the asteroid.

FIG. 7.- Topographic map of light distribution
surface for the asteroid.

Uncertainties in empirical mass-luminosity relation and local minima on the initial mass function of stars

O. Yu. MALKOV

Observatoire Astronomique, Centre de Données Stellaires, Strasbourg, France

In a number of recent papers it has been shown that the initial mass function (IMF) of the field stars is not monotonous but has one or more minima, which do not change its general shape nevertheless. For example Piskunov e.a. (1979) point out the minimum at about $m = 3 - 4$ (hereafter stellar mass is given in solar units) and do not exclude the other minima at $m = 1.5$ and $m = 10$; Vereshchagin (1982) finds a local minimum at $m = 1.5 - 2.5$ and, less confidently, at $m = 5 - 6$. Mermilliod (1976) in his study of open clusters finds the same minimum, too. More complete list of such findings is given by Vereshchagin and Piskunov (1984) who point out that most of investigators find the deficiency of stars at the following m: $1.5 - 2.0$, $3.0 - 4.5$ and 15. The first minimum has a characteristic depth of about 0.5 in logarithm and is detected especially confidently. Another minimum near $m = 3$ has approximately the same depth. It should be noted that different authors give different depths.

§A number of scenaria has been proposed to explain this deficiency: e.g. shortage of stars of given mass; superposition of different IMFs which gives a multimodal observed mass function; bursts of star formation etc. It should be kept in a mind, however, that the IMF could not be derived directly from observations, since data on stellar masses are not very abundant. So normally it is necessary to know the mass-luminosity relation (MLR) to obtain the IMF from luminosity function (LF):

$$f(\log m) = |dM_V/d\log m| \; g(M_V) \qquad (1)$$

Here $f(\log m)$ is the IMF, $g(M_V)$ is surface initial LF, which can be found from the observational LF if we take into account already died stars and height scales.

Let us suppose that LF is known very accurately, and let us investigate how strongly fluctuations in the MLR would change the IMF. Let us disturb the MLR increasing M_V for some $\log m$. Then the derivative in (1) increases in absolute value. Since the LF increases with M_V in this region, a rise of M_V leads an increase of $g(M_V)$ too. Estimates show that when varying M_V by $0.^m6$ at $\log m = 0.7$ (and still being within the observational band, which according to Popper's (1980) data is about 1^m wide in this region) we increase the derivative by factor 3.4 and the LF by 1.6. So the value of $f(\log m)$ changes by factor 6 (or by 0.8 in log f). Another example: if we change M_V by $+0.^m3$ for $\log m = 0.3$, we increase the corresponding values by factor 1.9 and 1.3, and by 2.5 in total (or by 0.4 in log f).

Note, that "negative" MLR disturbances (decreasing of M_V at given $\log m$) would not lead to the desirable effect. The $g(M_V)$ value for the same $\log m$ will be decreased, in this case, which should be compensated by the essential growth of the derivative. In general, the IMF feature (the local maximum or minimum) produced by the MLR variation depends both on the sign of the variation and on the position (ascending or descending branch) on the LF.

The above estimates show that the local minima at log m = 0.7 and 0.3 could be rather easily explained in terms of insufficient knowledge of the detailed behavior of the MLR.

Below we shall show that the most pronounced LF feature (so called Wielen dip) can also be explained in the same manner: i.e. we shall show that there exists a monotonous IMF, which can be obtained from the observed LF, and with use of a MLR being confined in the observational strip.

After rewriting the Eq.(1) in the form of

$$M_V'(\log m) = - [f(\log m) / g(M_V)] \tag{2}$$

we shall obtain a differential equation for the MLR; setting boundary conditions we shall be able to solve the Cauchy problem. We used the LF from Scalo (1986) and PDMF, derived from the IMF, by corrections for stellar life-times according to (Vereshchagin and Kisselyova, 1987). Both a power-law

$$\log F(\log m) = const - C_1 \log m \tag{3}$$

and a lognormal

$$\log F(\log m) = const - C_2 \log m - C_3 \log^2 m \tag{4}$$

IMFs were used. The equation was solved numerically, in the fourth order Runge-Kutte scheme. For comparison, Popper's (1980) observational data on masses and absolute magnitudes of binaries were used. The region containing the most reliable data is limited by -0.3 and +1.2 in log m.

First, the influence of boundary conditions was investigated. It was found that the differences in boundary conditions (if, of course, those are chosen within reasonable limits) are essentially unimportant for log m < < 1.0. Fixing the boundary at $M_V(1.8) = -7^m$ and varying constants $const$, C_1, C_2 and C_3 we investigated the dependence of the calculated MLR on the IMF parameters. It was found that in the case of power-law function (3) with the Salpeter's (1955) coefficient C_1 = 1.35 (or about 1.35) we can obtain the MLR, which confirms satisfactorily the observational data, with the normalization factor $const$ being as 1.65 - 1.80. The influence of variations of $const$ and C_1 on the derived MLR are shown in Fig. 1.

As to the lognormal IMF (4), already three parameters were varied to obtain the best fit to observations. A number of results is presented in Fig. 2. It was found that the parameters are in the following limits: C_2: 1.0 - 1.4, C_3: 0.5 - 0.7, and $const$ is approximately the same as it was in the case of the power-law IMF.

Thus, monotonous IMFs would produce nonmonotonous features (local minima and maxima) on the observational LF due to nonmonotonous behavior of mass-luminosity relation. Applying to the LF an empirical MLR which is usually smoothed out, we unavoidably will find nonmonotonous features in the IMF. Note, that the aim of this paper was not a search for the best MLR, satisfying both the monotonous IMF, and binary stars data. We also did not consider reasons which can be responsible for the MLR fluctuations. It was only shown that the maxima and minima on the IMF perhaps do not reflect a real situation, but they could be explained by insufficiently correctly defined MLR.

References

Mermilliod J.-C., 1976, Astron. Astrophys., 53, 289.
Piskunov A.E., Tutukov A.V., Yungelson L.R., 1979, Pis'ma Astr. Zh. 5, 81.
Popper D.M., 1980, Annu. Rev. Astron. Astrophys., 18, 115.

Salpeter E.E., 1955, Astrophys. J., 121, 161.
Scalo J.M., 1986, Fund. Cosm. Phys., 11, 1.
Vereshchagin S.V., 1982, Pis'ma Astron. Zh., 8, 546.
Vereshchagin S.V., Piskunov A.E., 1984, Nauch. Inform., 57, 76.
Vereshchagin S.V., Kisselyova N.A., 1987, Astron. Zh., 64, 980.

Fig. 1. Numerical solutions of eq.(2) for power-law IMF.
    1 - log F (logm) = 1.65 - 1.35 logm,
    2 - log F (logm) = 1.80 - 1.35 logm.
Observational data from Popper (1980) are indicated by diamonds.

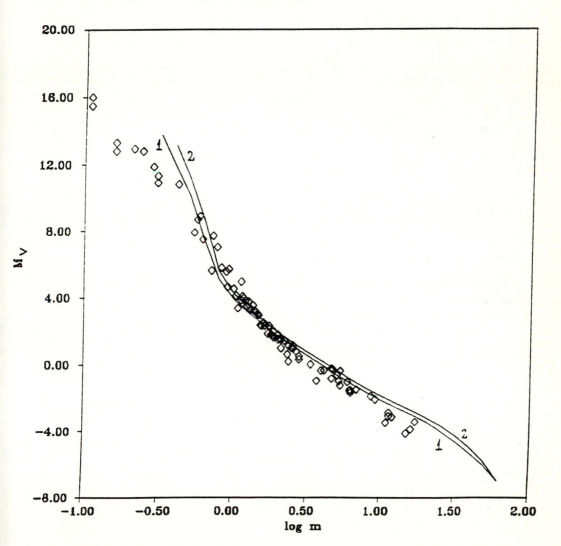

Fig. 2. Numerical solutions of eq.(2) for lognormal IMF.
    1 - log F (logm) = 1.65 - logm - 0.5 log^{2}m,
    2 - log F (logm) = 1.80 - logm - 0.5 log^{2}m.
Observational data from Popper (1980) are indicated by diamonds.

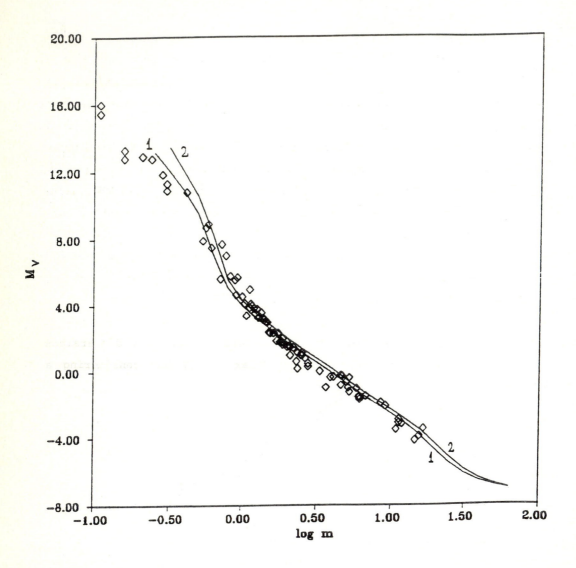

Gradients of the Residual Stellar Velocity Dispersions

M. MORENO – J. TORRA

Departament de Fisica de l'Atmosfera, Astronomia i Astrofisica
Universitat de Barcelona, Barcelona, Spain

1 INTRODUCTION

If in the dynamical theory of a stationary galaxy we introduce an
ellipsoidal velocity distribution, one of the consequences is that the
ellipsoids of homogeneous and identical star systems are independent of
their spatial positions. The velocity mean field and the residual
velocity distribution deviates from this theoretical behaviour.
Second-order momenta irregularities and their dependence with the
position have been found by several authors (Rudnicki, 1969; Shatsova,
1970; Peralta, 1979).Mayor (1974) and Erickson (1975) studied some
particular cases. The most extensive work on this subject is the one of
Oblak (1983), who calculated the gradients of the second-order momenta
in the galactic plane. Torra (1984) and Figueras (1986) determined
gradients of the second-order momenta through linear regression.

Our aim is to determine gradients of the residual velocity dispersions
in a way similar to the one used by Oblak (1983) but considering a
tridimensional field of velocities.

2 METHOD

For each one of the stars in a sample of N, with known radial
velocities, proper motions and parallax, we can write:

$$
\begin{aligned}
v_{ri} &= f_{ri}(X_j, r_i, l_i, b_i) + v\,res_{ri} \\
\mu_{li} \cdot \cos b_i &= f_{li}(X_j, r_i, l_i, b_i) + v\,res_{bi}/k \cdot r_i \\
\mu_{bi} &= f_{bi}(X_j, r_i, l_i, b_i) + v\,res_{bi}/k \cdot r_i
\end{aligned}
\tag{1}
$$

where: l_i, b_i, r_i stand for the usual galactic coordinates and distance.

f_{ri}, f_{li}, f_{bi}: describe the systematic terms of the velocity field. The unknown parameters X_j are: the components of the Sun peculiar motion, gradients of the velocity field and precessional corrections (j=1...12). v res$_{ri}$, v res$_{li}$, v res$_{bi}$: the residual velocity components which constitute the random part of the velocity field.

As observed variance σ^2 we take: $(v - f)^2 = v \, res^2 = \sigma^2$.

If we call $\phi(v_r, \mu_l \cdot \cos b, \mu_b)$ the observed velocity distribution function and $F(v \, res_r, v \, res_l, v \, res_b)$ the residual velocity distribution, the relations between the second-order momenta of ϕ and those of F are found in Trumpler (1962) and Ogorodnikov (1965):

$$\sigma_r^2 = \alpha_{11}^2 \mu_{200} + \alpha_{12}^2 \mu_{020} + \alpha_{13}^2 \mu_{002} + 2\alpha_{11}\alpha_{12}\mu_{110} + 2\alpha_{11}\alpha_{13}\mu_{101} + 2\alpha_{12}\alpha_{13}\mu_{011}$$

$$\sigma_l^2 = \alpha_{21}^2 \mu_{200} + \alpha_{22}^2 \mu_{020} + \alpha_{23}^2 \mu_{002} + 2\alpha_{21}\alpha_{22}\mu_{110} + 2\alpha_{21}\alpha_{23}\mu_{101} + 2\alpha_{22}\alpha_{23}\mu_{011}$$

$$\sigma_b^2 = \alpha_{31}^2 \mu_{200} + \alpha_{32}^2 \mu_{020} + \alpha_{33}^2 \mu_{002} + 2\alpha_{31}\alpha_{32}\mu_{110} + 2\alpha_{31}\alpha_{33}\mu_{101} + 2\alpha_{32}\alpha_{33}\mu_{011}$$

where μ_{ijk} are the central second-order momenta of the residual velocity distribution function F and σ_r^2, σ_l^2, σ_b^2 are the observed variances.

We assume a first order linear expansion around the Sun:

$$\mu_{ijk}(r, l, b) = (\mu_{ijk})_o + \xi \cdot \partial\mu_{ijk}/\partial\xi + \eta \cdot \partial\mu_{ijk}/\partial\eta + z \cdot \partial\mu_{ijk}/\partial z$$

where the axes: ξ, η, z, are directed to the galactic center, direction of galactic rotation and north galactic pole, respectively.

Substituting μ_{ijk} in equations (2) by its linear development, we will have a system of 3N equations for 24 unknowns: the six central second-order momenta and their gradients in the indicated directions. Equations (1) are solved using least-squares as described by Mennessier and Crézé (1969) with the hypothesis that the random part comes from the velocity ellipsoid. The normal equations arising from (1) were weighted by $1/\sigma^2$. Those arising from equations (2), which give the second-order momenta of F and their gradients, were weighted by $1/\sigma^4$.

The system (1) + (2) should be solved as a whole. In practice σ^2 is introduced into (1) as a weight and the two systems can be separated. We used an iterative method, the solutions of (1) coming into (2) and vice-versa. The Q parameter is used as a quantitative measure for the goodness of fit (Press W. H. et al., 1986).

3 RESULTS

We selected from Figueras (1986) two samples (V_{res} < 65 km/s):

-sample (a): 1012 [A0-F5] main sequence stars with 70 pc < r < 200 pc

-sample (b): 1558 [G5-M8] giant stars with 70 pc < r < 1000 pc.

3.1 Central second-order momenta gradients of the residual velocity distribution (in $km^2/s^2 \cdot pc$)

	sample(a)	Figueras	Oblak	sample(b)	Figueras
$\partial\mu_{200}/\partial\xi$	.28±.10	-.18±.04	-.79±.40	.24±.07	.16±.12
$\partial\mu_{020}/\partial\xi$	-.17±.07	.13±.12	.18±.29	-.19±.05	-.10±.06
$\partial\mu_{002}/\partial\xi$	.03±.03	.05±.08	.05±.16	-.02±.03	-.16±.04
$\partial\mu_{110}/\partial\xi$	-.16±.07	-.47±.07	.36±.26	-.04±.05	-.38±.06
$\partial\mu_{101}/\partial\xi$	.62±.06	.18±.04		-.27±.06	-.17±.09
$\partial\mu_{011}/\partial\xi$		-.02±.01			.09±.01
$\partial\mu_{200}/\partial\eta$	-.58±.11	-.18±.12	1.21±.46	.30±.07	.20±.08
$\partial\mu_{020}/\partial\eta$	.16±.08	.05±.04	-.56±.28	.11±.05	-.42±.15
$\partial\mu_{002}/\partial\eta$	-.12±.03	-.14±.07	-.14±.19	.22±.03	.10±.02
$\partial\mu_{110}/\partial\eta$	-.27±.08	-.25±.06	-.24±.26	.04±.05	-.18±.02
$\partial\mu_{101}/\partial\eta$	.26±.08	.09±.02		-.19±.08	.02±.04
$\partial\mu_{011}/\partial\eta$	.02±.06	.11±.06		.21±.05	-.03±.11
$\partial\mu_{200}/\partial z$	.03±.11	-.60±.08		.21±.09	-.44±.27
$\partial\mu_{020}/\partial z$	-.40±.08	-.10±.01		.01±.06	.17±.08
$\partial\mu_{002}/\partial z$	-.07±.04	.38±.14		.13±.03	.51±.17
$\partial\mu_{110}/\partial z$	.13±.08	.10±.04		.12±.06	.31±.14
$\partial\mu_{101}/\partial z$	.29±.06	.14±.04		.04±.06	.07±.05
$\partial\mu_{011}/\partial z$	.13±.06	-.26±.04		.13±.05	.20±.09
	(Q = 0.13)			(Q = 0.07)	

4 CONCLUSIONS

The method developed will allow us to obtain simultaneously the velocity field, the central second-order momenta and their gradients in the directions of the galactic center, galactic rotation and north galactic pole. A very high value is obtained for Oort's A constant in sample (a).

Our determinations agree with the previous ones of Figueras (1986): 12 gradients show the same sign for both samples. The values for $\partial\mu_{110}/\partial\eta$, $\partial\mu_{002}/\partial\eta$ and $\partial\mu_{002}/\partial\xi$ for the sample (a) agree very well with the Figueras (1986) and Oblak (1983) one's. According with García (1978) a decreasement of μ_{110} in the direction of galactic center is obtained. Our results are in agreement with theoretical limit values obtained by Huang (1979) and Zeng (1984) for the variations of μ_{200} and μ_{002}.

We think that the use of improved stellar samples will confirm these first results.

REFERENCES

Erickson,R.R. (1975): Astr. J., **195**, 343.

Figueras,F. (1986): PhD thesis, Univ. Barcelona.

García,L. (1978): Tesi de Llicenciatura, Univ. Barcelona.

Huang Ke-Liang et al. (1979): Ac. Astr. Sinica, **20**, 3.

Mayor,M. (1974): Astr. Astrophys., **32**, 321.

Mennessier,M.O., Crézé,M. (1969): Astr. Astrophys., **2**,355

Oblak,E. (1983): PhD thesis, Univ. Franche-Comté.

Ogorodnikov,K.F. (1965): Dynamics of Stellar Systems, Pergamon, London.

Peralta,J., Peralta, M. (1979): Astr. Astrophys. **74**, 121.

Press,W.H. et al. (1986): Numerical Recipes, Cambridge Un. Press.

Rudnicki,K. (1969): Vistas in Astronomy **11**, 173.

Shatsova,R.B. (1970): IAU Symp. 38, Reidel, Dordrecht.

Torra,J. (1984): PhD thesis, Univ. Barcelona.

Trumpler,R.J., Weaver,H.F. (1962): Statistical Astronomy, Dover, N.York.

Zheng Xue-Tang et al. (1984): Ac. Astr. Sinica, **25**, 2.

Robust Techniques in Astronomy

H. L. MORRISON[1] AND A. H. WELSH[2]

[1]Observatories of the Carnegie Institution of Washington

[2]Department of Statistics, The Faculties, Australian National University

SUMMARY

We describe robust estimators of mean velocity and velocity dispersion which were used in a study of the kinematics of two of the Galaxy's populations — the disk and the halo. The classical estimators of these quantities perform very well if the underlying distribution is Gaussian, but quite small deviations from Gaussian shape cause large errors in the estimates. Since very large samples are needed before the form of the underlying distribution can be established securely, this causes real problems for the small samples which are typical of the galactic halo (and many other areas in astronomy). We describe two robust estimators (the 5% trimmed mean and standard deviation) which are simple, easy to compute, and insensitive to outliers.

1 INTRODUCTION

The work we will describe in this paper was developed as part of a study of the stellar populations in our own Galaxy (Morrison, Flynn and Freeman 1989). This study investigated the kinematics of disk and halo of the Galaxy, and their possible overlap. The extreme rarity of stars from the galactic halo means that most samples of stars from the halo will be small ones, and this poses particular problems for inference.

The classical statistical techniques which are used to investigate stellar kinematics in our Galaxy are the sample mean and standard deviation (to estimate mean velocity and velocity dispersion). While these estimators perform very well if the underlying distribution of the data is Gaussian, their properties are less desirable when the underlying distribution shows even small deviations from Gaussian shape. The most common deviation from Gaussian shape is one where the distribution has longer tails (Huber 1981); and outliers (observations from the tails of the distribution) have a particularly undesirable effect on estimates of velocity dispersion.

The problem, particularly for small samples, lies in the fact that it takes a large number of observations to establish the form of the underlying distribution securely. Thus if we have a small sample and wish to use classical techniques, we have to take the Gaussian nature of the data on trust, knowing that even a small deviation from Gaussian shape can adversely affect the accuracy of the kinematical estimates. Norris and Ryan (1989) and Fuchs and Weilen (1987) give examples of non-Gaussian velocity distributions in the field of Galactic structure.

The field of robust statistics has grown up in reponse to the sensitivity of the classical estimators to variations from Gaussian shape. A robust estimator is one which is not susceptible to small departures from the assumptions — in this case, the assumption of a Gaussian distribution. In this paper, we are only concerned with the situation where the possible departures are not extreme — for example, the velocity distributions are not severely skewed. Examples of robust estimators proposed to replace the classical mean are the trimmed mean (Brieman 1973), and the biweight (Mosteller and Tukey 1977). Statistical research on robust estimators to replace the classical standard deviation is less common. Eddington (1914) proposed using $\sum |V|$ rather than $\sum V^2$ in estimating velocity dispersion, and commented: "This is contrary to the advice of most textbooks, but can be shown to be true". Nowadays, there are several statistical textbooks of which Eddington would approve! While his proposed estimator is robust, it does not make very efficient use of the data. Recent statistical research has identified more efficient estimators. One is the biweight scale estimator discussed in Mosteller and Tukey (1977), and another the trimmed standard deviation, which is described here.

2 TRIMMING THE DATA

These estimators are computationally very simple: one orders the data, removes the highest and lowest 5%, and then calculates the mean and standard deviation with the observations that remain. If there are n observations $x_1, x_2, \ldots, x_n$ and n_{trim} are left after trimming, then the trimmed mean is

$$\overline{x}_t = \frac{\sum_i x_i}{n_{trim}}, \tag{1}$$

and the trimmed standard deviation is

$$\sigma_t = \frac{1}{0.789} \sqrt{\frac{1}{n_{trim} - 1} \sum_i (x_i - \overline{x}_t)^2}, \tag{2}$$

where the index i ranges over all observations which have not been trimmed. The factor of 0.789 in Eq. (2) is needed because the removal of the highest and lowest values biases the estimate. It is calculated so that the classical standard deviation and the trimmed standard deviation estimate the same parameter for pure Gaussian data. Details of these estimators can be found in Breiman (1973; p241ff), Huber (1981) and Welsh and Morrison (1989).

The trimmed estimates require the user to throw away a small amount of data, but in return they provide safety in estimation. Different choices of the amount to trim may be appropriate for different situations: it is necessary to trade off the loss of precision caused by removing some data points against the loss of precision due to suspected outliers.

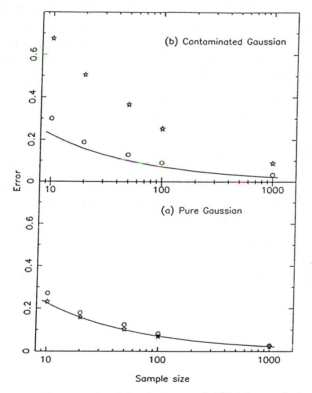

Figure 1 : The error on the standard deviation and 5% trimmed standard deviation estimators. Symbols used are: standard deviation, star; trimmed s.d., circle.

To compare the errors of the classical and trimmed estimators, Monte Carlo simulations were made for the two distributions mentioned above: (a) a Gaussian with $\sigma = 1$, and (b) a Gaussian with $\sigma = 1$ which has been contaminated by choosing 5% of the observations from a Gaussian with $\sigma = 5$. The 'contaminated' sample is used to illustrate the effects of a distribution with larger tails than a Gaussian, the most common deviation from Gaussian shape. Fig. 1 shows the error on the classical standard deviation and the 5% trimmed standard deviation. The solid line shows the $\sigma/\sqrt{2n}$ error commonly quoted for the standard deviation. For the case of a slightly contaminated Gaussian, the errors on the classical standard deviation are much higher, and the trimmed standard deviation is hardly affected; it can be seen that this is the more reliable choice.

The trimmed mean and standard deviation were very useful in our investigation of the kinematics of the Galaxy's populations. Their computational simplicity made it straightforward to extend their use to a more complicated estimation problem. (For more details see Morrison, Flynn and Freeman 1989.) Robust estimation procedures can be applied to many problems, and produce answers which are safer because they are much less dependent on assumptions which cannot be easily verified.

REFERENCES

Breiman, L. (1973). *Statistics with a View toward Applications* (Houghton Mifflin, Boston).

Eddington, A.S. (1914). *Stellar Movements and the Structure of the Universe* (Macmillan, London), p147.

Fuchs, B., and Wielen, R. (1987). In *The Galaxy*, Nato Advanced Study Institute, edited by G. Gilmore and R. Carswell (Reidel, Dordrecht).

Huber, P.J. (1981). *Robust Statistics* (Wiley, New York).

Morrison, H.L., Flynn, C., and Freeman, K.C. (1989). Submitted to Astron. J.

Mosteller, F., and Tukey, J.W. (1977). *Data Analysis and Regression* (Addison-Wesley, Reading).

Norris, J.E., and Ryan, S.G. (1989). Astrophys. J. Letters, **336**, L17.

Welsh, A.H., and Morrison, H.L. (1989). Preprint.

Heather L. Morrison: The Observatories of the Carnegie Institution of Washington, 813 Santa Barbara Street, Pasadena CA 91101, U.S.A.

Alan H. Welsh: Department of Statistics, The Faculties, Australian National University, Acton A.C.T, Australia.

Linear Regression with Errors in Both Variables: a Short Review

Fionn MURTAGH[1]
Space Telescope — European Coordinating Facility
European Southern Observatory
Karl-Schwarzschild-Straße 2
D-8046 Garching bei München

1 THE PROBLEM

In linear regression, we suppose a relationship of the form

$$\mathbf{y} = \alpha + \beta \mathbf{x}$$

where $\mathbf{y}$ and $\mathbf{x}$ are n-valued vectors (containing the n pairs of points in two-dimensional real space), and β and α are, respectively, slope and intercept. Using least squares, values of the slope and intercept are chosen such that

$$\sum_{i=1}^{n}(y_i - \alpha - \beta x_i)^2$$

is minimum. In regressing y on x in this way, we minimize with respect to vertical distances (cf. Fig. 1; see also Fig. 2, p. 1307, of Riggs et al., 1978). We could regress x on y, which would lead to minimization with respect to least squares horizontal distances. We could finally use orthogonal distances, to ensure symmetry in regard to x and y. This latter approach is essentially a first principal components analysis axis, and was initially proposed by Adcock in 1878.

In practice, we may not have error-free measurements. When this applies both to dependent and to independent variables, we have

$$\xi_i = x_i + \delta_i$$

$$\eta_i = y_i + \epsilon_i$$

and we are interested in the linear functional relationship

$$y_i = \alpha + \beta x_i.$$

---

[1] Affiliated to Space Science Dept., Astrophysics Div., European Space Agency.

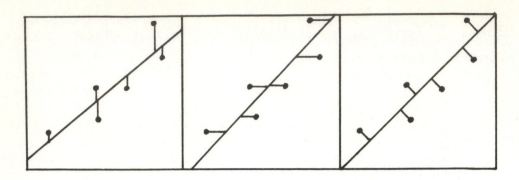

Figure 1: Regression of y on x; regression of x on y; and orthogonal fit.

We suppose that the errors, δ and ϵ, are independent and normally distributed, with expected values 0 and variances σ_{δ_i} and σ_{ϵ_i}. The minimand, ensuing from a maximum likelihood estimation, is

$$\sum_i (x_i - \xi_i)^2 / \sigma_{\delta_i} + \sum_i (y_i - \eta_i)^2 / \sigma_{\epsilon_i}$$

A convenient form for this minimand, which can be obtained from the foregoing, is

$$\sum_i W_i (y_i - \alpha - \beta x_i)^2$$

where

$$W_i = 1/(\beta^2 \sigma_{\delta_i}^2 + \sigma_{\epsilon_i}^2).$$

Minimizing this expression can lead to somewhat different sets of equations to be solved. York (1966) derives a cubic equation; Fasano and Vio (1988) iteratively solve a set of normal equations using gradient descent; and other implementations use variable metric methods (Fletcher-Powell and Marquardt approaches) to solve the normal equations.

The central focus of the paper by Ripley and Thompson is on testing analytical bias, i.e. comparing the best fitting regression line to the line defined by $\alpha = 0$ and $\beta = 1$. Fasano and Vio further discuss two issues relating to bias in the estimated values found. The work of Lybanon, and of Jefferys and collaborators, deals with the errors-in-all-variables curve fitting problem in full generality. The GaussFit symbolic mathematical package implements this work.

It is by now a truism to say that the above solution methods for regression with errors in both variables are not widely known, although they should be. To this day, however, software for the solution of this problem is not widely available. Traditionally, regression problems with analytical solutions were tackled. Although eminently readable, Press et al. (1988, chapter 14) do not address the errors-in-variables problem. Kendall and Stuart (1979, chapter 29) deal with the following cases when both x and y are known only with error: σ_{δ_i} known; σ_{ϵ_i} known; $\sigma_{\delta_i}/\sigma_{\epsilon_i}$ known; and both σ_{δ_i} and σ_{ϵ_i} known. Many other texts have restricted discussion of this issue (Acton, 1966, chapter 5; Draper and Smith, 1981, pp. 122-125; Montgomery and Peck, 1982, pp. 386-389; Seber, 1977, pp. 155-162 and pp. 210-211). Mandel (1984) assumes constant ratios of error variabes. An extensive range of regression methods, which also make this assumption, is to be found in Riggs et al. (1978).

2 RESULTS

The following results were preparatory to implementing a method for the errors-in-both-variables problem in the MIDAS image processing system (MIDAS, 1989). York's (1966) method was implemented by the present author. The results shown below have been given to 4 decimal places (GaussFit outputs results to 15). In some cases, iterative approximation precision can be set by the user, and this might influence exact replicability of the results given here. All implementations were in Fortran, on a VMS machine. Weights are generally (but not necessarily) calculated as the reciprocal of the variance, which is input by the user. For comparison purposes, a robust regression routine based on Kendall's τ (Slepica and Wolf, 1986) was also used. (The latters' algorithm was translated from Pascal into Fortran by the present author; the intercept, not determined in the Slepica and Wolf paper, was calculated as the median of values $y_i - \beta x_i$.) Evidently, errors were not used by this routine. Given the assumptions in effect, it is not suprising that it performs badly compared to the other methods.

Data tables 1 and 2 consist, respectively, of 30 and 37 measurements. They were kindly provided by B.D. Ripley. Data table 3 is that used in York (1966), and further discussed in Lybanon (1984).

Apart from small variations in the error estimates, the results given by all other methods for the slope and intercept are very similar. The error estimates are calculated in an identical fashion by Ripley and Thompson, and by Fasano and Vio. Hence, differences in results here are due to rounding errors. York's errors are approximations, based on ignoring higher order terms in a Taylor expansion.

Table 1		
Ripley and Thompson	Slope:	0.9729
	Intercept:	0.1065
	σ_{sl}:	0.0766
	σ_{in}:	0.0482
Fasano and Vio	Slope:	0.9728
	Intercept:	0.1057
	σ_{sl}:	0.0746
	σ_{in}:	0.0492
York	Slope:	0.9730
	Intercept:	0.1065
	σ_{sl}:	0.0872
	σ_{in}:	0.0576
GaussFit	Slope:	0.9730
	Intercept:	0.1064
	σ_{sl}:	0.0893
	σ_{in}:	0.0562
Slepica/Wolf (Robust)	Slope:	0.7738
	Intercept:	0.6041

Table 2		
Ripley and Thompson	Slope:	1.068
	Intercept:	−0.1607
	σ_{sl}:	0.0082
	σ_{in}:	0.0201
Fasano and Vio	Slope:	1.0677
	Intercept:	−0.1607
	σ_{sl}:	0.0082
	σ_{in}:	0.0201
York	Slope:	1.0677
	Intercept:	−0.1607
	σ_{sl}:	0.0140
	σ_{in}:	0.0345
GaussFit	Slope:	1.0677
	Intercept:	−0.1607
	σ_{sl}:	0.0141
	σ_{in}:	0.0345
Slepica/Wolf (Robust)	Slope:	1.0304
	Intercept:	−0.5188

Table 3		
Ripley and Thompson	Slope:	−0.4806
	Intercept:	5.480
	σ_{sl}:	0.0580
	σ_{in}:	0.2950
Fasano and Vio	Slope:	−0.4807
	Intercept:	5.4806
	σ_{sl}:	0.0582
	σ_{in}:	0.2969
York	Slope:	−0.4805
	Intercept:	5.4799
	σ_{sl}:	0.0710
	σ_{in}:	0.3619
GaussFit	Slope:	−0.4805
	Intercept:	5.4799
	σ_{sl}:	0.0706
	σ_{in}:	0.3592
Slepica/Wolf (Robust)	Slope:	−0.5501
	Intercept:	5.8975

It is planned to alter GaussFit, bringing error calculations into line with Fuller (1987, section 3.2.4).

In conclusion, we have found that the implementations examined gave consistent results. Apart from routines for the errors-in-both-variables problem, what other related routines should be available to the astronomer? We have made reference to one routine for robust regression (robust methods are designed not to be overly influenced by extraneous and outlying points). Robust methods should especially be considered when the data is error-prone, but in a way which is not known precisely to the analyst. Regression using the L_1 metric, rather than least squares, is one widely-used robust approach (see Branham, 1982); Press et al. (1988, section 14.6) briefly discuss others; and the GaussFit package (Jefferys et al., 1988a, 1988b) contains some routines in this area. Beyond robust methods, local fitting methods (see, e.g., Cleveland et al., 1988) are also very important and deserve study.

ACKNOWLEDGEMENTS

For discussion and references, I am grateful to: L. Lucy (ESO, Munich), J. Melnick (ESO, La Silla), R. Ramos de Carvalho (Astronomy Dept., California Institute of Technology), W.H. Jefferys and B.E. MacArthur (Dept. of Astronomy, University of Texas at Austin).

REFERENCES

1. F.S. Acton, *Analysis of Straight-Line Data*, Dover, New York, 1966 (Wiley, New York, 1959).

2. R.L. Branham, Jr., "Alternatives to least squares", *The Astronomical Journal*, **87**, 1982, 928–937.

3. W.S. Cleveland, S.J. Devlin and E. Grosse, "Regression by local fitting: methods, properties, and computational algorithms", *Journal of Econometrics*, **37**, 1988, 87–114.

4. N.R. Draper and H. Smith, *Applied Regression Analysis*, Wiley, New York, 1981.

5. G. Fasano and R. Vio, "Fitting a straight line with errors on both coordinates", *Newsletter of Working Group for Modern Astronomical Methodology*, Issue 7, 1988, 2–7.

6. W.A. Fuller, *Measurement Error Models*, Wiley, New York, 1987.

7. W.H. Jefferys, "On the method of least squares", *The Astronomical Journal*, **85**, 1980, 177–181.

8. W.H. Jefferys, "On the method of least squares. II", *The Astronomical Journal*, **86**, 1981, 149–155.

9. W.H. Jefferys, M.J. Fitzpatrick and B.E. McArthur, "GaussFit — a system for least squares and robust estimation", *Celestial Mechanics*, **41**, 1988a, 39–49.

10. W.H. Jefferys, M.J. Fitzpatrick, B.E. McArthur and J.E. McCartney, "Gauss-Fit: A System for Least Squares and Robust Estimation", User's Manual, Department of Astronomy and McDonald Observatory, Austin, Texas, 1988b, 75 pp.

11. M. Kendall and A. Stuart, *The Advanced Theory of Statistics. Vol. 2. Inference and Relationship*, Charles Griffin, London, 4th ed., 1979.

12. M. Lybanon, "A better least-squares method when both variables have uncertainties", *American Journal of Physics*, **52**, 1984, 22–26.

13. J. Mandel, "Fitting straight lines when both variables are subject to error", *Journal of Quality Technology*, **16**, 1984, 1–14.

14. MIDAS (Munich Image Data Analysis System), User Manual, Vols. 1 and 2, European Southern Observatory, Garching bei München, 1989 (six-monthly releases).

15. D.C. Montgomery and E.A. Peck, *Introduction to Linear Regression Analysis*, Wiley, New York, 1982.

16. W.H. Press, B.P. Flannery, S.A. Teukolsky and W.T. Vetterling, *Numerical Recipes*, Cambridge University Press, Cambridge and New York, 1988.

17. D.S. Riggs, J.A. Guarnieri and S. Addelman, "Fitting straight lines when both variables are subject to error", *Life Sciences*, **22**, 1978, 1305–1360.

18. B.D. Ripley and M. Thompson, "Rregression techniques for the detection of analytical bias", *Analyst*, **112**, 1987, 377–383.

19. G.A.F. Seber, *Linear Regression Analysis*, Wiley, New York, 1977.

20. J.M. Slepica and G.K. Wolf, "A new algorithm and program for robust linear regression", *Statistical Software Newsletter*, **12**, 1986, 80–83.

21. D. York, "Least-squares fitting of a straight line", *Canadian Journal of Physics*, **44**, 1966, 1079–1086.

A BAYESIAN IMAGE RECONSTRUCTION ALGORITHM WITH ENTROPY AS PRIOR INFORMATION

Jorge Núñez[1] and Jorge Llacer[2]

1) Departament de Física de l'Astmosfera, Astronomia i Astrofísica. Universitat de Barcelona and Observatorio Fabra. Av Diagonal 647, 08028 Barcelona. Spain

2) Engineering Division, Lawrence Berkeley Laboratory. Berkeley, CA, 94720, U.S.A.

Abstract
This paper presents the development and testing of a new iterative reconstruction algorithm for Astronomy. We propose a Maximum *a Posteriori* Method of image reconstruction in the Bayesian statistical framework for the Poisson noise case.

1. Fundamental Concepts

Full Bayesian and Maximum entropy solutions are being increasingly used for reconstructing images from noisy and incomplete data in many fields (see, p.e. Skilling and Bryan (1984), Jaynes (1982; 1984)). This paper reports on the development and testing of a new image reconstruction algorithm useful for Astronomy based on Bayesian statistical concepts. The algorithm computes Maximum *a Posteriori* (MAP) images. We also present a comparison of this method with the Maximum Likelihood Estimator Method (MLE) developed by Lucy (1974).

The notation that we are using is the following: p_j $j=1,..,D$ the projection data or the number of counts; a_i, $i=1,..,B$ the object activity or emission density; f_{ji} the Point Spread Function, or probability that an emission in pixel i will be detected in detector j and n_j the noise.

In the image reconstruction problem the object, the projection of the object, and the noise are assumed to be connected by an imaging equation: $f \cdot a + n = p$ or, in discrete version

$$\sum_{i=1}^{B} f_{ji} \, a_i + n_j = p_j \qquad j = 1, \ldots D$$

The application of Bayes' theorem to the image reconstruction problem gives:

$$P(a|p) = P(p|a) \, P(a) \, / \, p(p)$$

The most probable object **a** given the data **p** is obtained by maximizing P(**a**|**p**), or the logarithm of the product P(**p**|**a**) by P(**a**), since P(**p**) is a constant.

We use entropy to define the prior probability P(**a**) in the way proposed by Frieden (1972), called random-grain model by Andrews and Hunt (1977).

The conditional probability P(**p**|**a**) describes the noise in the projections and its possible object dependence. It is fully specified by the likelihood function. We consider that each data element can be assumed to obey Poisson statistics with all data elements statistically independent. We have chosen the Poisson hypothesis because in Astronomy the low-light level detectors obey Poisson statistics. They have sufficient gain to cause the Poisson fluctuations in photon arrivals to dominate over the noise statistics of the detectors. CCDs also fit this situation. In other cases, different choices of the form of the likelihood function like the Gaussian uncorrelated noise should be used (Frieden and Wells (1978)).

2. The FMAPE Method

The reconstruction method we present here is called FMAPE, for Fast Maximum *a Posteriori* with Entropy (Nunez and Llacer (1989)). It will be described here briefly. The FMAPE is based on maximizing the target function

$$BY = - \sum_{i=1}^{B} (a_i/\Delta a) \log(a_i/\Delta a) + \sum_{j=1}^{D} (-h_j + (p_j/\Delta p_j) \log(h_j)) - \mu(\sum_{i=1}^{B} a_i - N) \qquad (1)$$

where

$$h_j = (\sum_{i=1}^{B} f_{ji} a_i)/\Delta p_j \qquad j = 1, \ldots D \qquad (2)$$

The first term is the Shannon entropy with an adjustable parameter, Δa, which controls the relative weight of the entropy vs likelihood. It can be theoretically determined or adjusted to yield reconstructions that converge to feasible images. The second term is the likelihood which contains a vector of projection data p_j incorporating absorption and detector gain corrections and a vector of those corrections Δp_j. The third term insures the conservation of counts and contains one Lagrange multiplier μ. The vector of elements h_j corresponds to a modified projection of the current object a_i and, in fact, prescribes that the corrections to be applied to the matrix elements f_{ji}.

The iterative algorithm that we have devised for the maximization is based on the "successive substitutions" method (Hildebrand (1974)) and is given by

$$a_i^{k+1} = K \, a_i^k \left[\Delta a \sum_{j=1}^{D} [f_{ji}(1/\Delta p_j)(p_j / \sum_{i=1}^{B} f_{ji}a_i^k - 1)]-\log(a_i^k)+C\right]^n \qquad i = 1,\ldots B \qquad (3)$$

There are two constants in (3) whose values are arbitrary. Within the range of values for which the iterative process converges, the convergence point is independent of their value. The first constant is the exponent n which controls the speed of convergence. We have found that for values between 1 and 3, the rate of convergence is roughly proportional to n. The second is the constant C, which insures that no negative values will occur during the iterative process. We use routinely $C = \Delta a$ with no problems. Finally K is computed at the end of each iteration to conserve the number of counts and it is equivalent to calculating the Lagrange multiplier μ by

$$K = 1 / [1 + \Delta a\ \mu - \log(\Delta a) + C]^n$$

3. Preliminary Results

To test the algorithm and to compare it with the MLE algorithm of Lucy (1974), we have done several reconstructions using computed generated data containing one million counts.

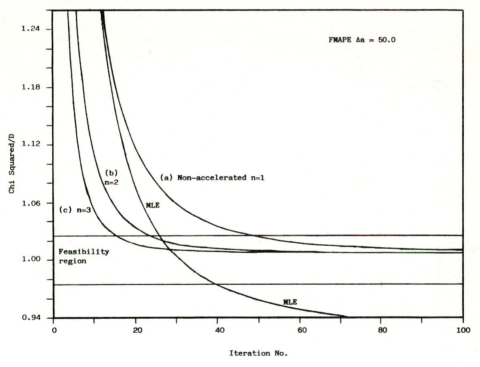

Figure 1

Figure 1 shows the typical behavior of the FMAPE algorithm and compares the values of χ^2/D statistic as a function of iteration number for an MLE reconstruction, and for the FMAPE reconstructions with parameter Δa = 10, 25, and 50.

The path of the MLE algorithm crosses the feasibility region from above in relatively few iterations and decreases continuously below it. By contrast all the FMAPE curves reach a flat path. Reconstructions with Δa

= 10 and Δa = 25 reach convergence above the feasibility region and result in images that have not developed their full contrast at convergence. A value of Δa = 50 yields convergence inside the feasibility region and a visually excellent feasible image. The choice of Δa = 100 (not shown) converges to a non-feasible image, characterized by excessive noise in the regions of high activity. All the reconstructions were carried out with n = 1, i.e., not accelerated.

4. Conclusions

We have proposed a Maximum *a Posteriori* Method of image reconstruction in the Bayesian framework for the Poisson noise case. We have used the entropy to define the prior probability and likelihood to define the conditional probability. The introduction of the parameters Δa and Δp_j j=1,.,D, which can be theoretically computed or adjusted by feasibility, allows us to obtain a Maximum *a Posteriori* Solution without the problem of "grey" reconstructions associated with "pure" Bayesian Methods. We have proposed an iterative algorithm to compute the reconstructions. The algorithm presents the desirable properties of good convergence rate, stability, and positivity, and converges into the feasibility region. The algorithm can be accelerated to be faster than the Maximum Likelihood (MLE) Method.

Acknowledgements

The authors would like to acknowledge valuable discussions with Dr. E. Veklerov from Lawrence Berkeley Laboratory on different aspects of the work presented. This work has been supported, in part, by a grant from the National Institutes of Health (CA-39501) and the U.S. Department of Energy under Contract No. DE-AC03- 76SF00098.

References

Andrews, H.C. and Hunt, B.R.: 1977, *Digital Image Restoration*, Prentice-Hall Inc.

Frieden, B.R.: 1972, *J. Opt. Soc. Amer.*, **62**, 511-518.

Frieden, B.R. and Wells, D.C. 1978, *J. Opt. Soc. Am.*, **68**, No. 1, 93-103.

Hildebrand, F.B.: 1974, *Introduction to Numerical Analysis, 2nd. Ed.*, McGraw-Hill, N.Y.

Jaynes, E.T.: 1982, *Proc. IEEE*, **70**, 939-952.

Jaynes, E.T.: 1984, In *Maximum Entropy and Bayesian Methods in Applied Statistics*. Proc. of the Fourth Maximum Entropy Workshop. Ed. J. H. Justice. Univ. of Calgary. pp. 1-25.

Lucy, L.B.: 1974, *The Astronomical Journal*, **79**, No. 6, 745-754.

Nunez, J. and Llacer, J.: 1989, "A fast Bayesian Reconstruction Algorithm for Emission Tomography with Entropy Prior Converging to Feasible Images" LBL-27074. To be published in *IEEE TMI*.

Skilling, J. and Bryan, R.K.: 1984, *Monthly Notices of the Royal Astronomical Soc.*, **211**, 111-124.

Some Results about the Convergence of Iterative Methods when Solving Singular Least Square Problems

M. RAPAPORT

Observatoire de l'Université de Bordeaux, F-33270 Floirac

When we solve overdetermined systems of condition equations using least square methods, if the design matrix X is of maximal rank, then the matrix of the normal equation, $X^t X$ is symmetric and positive definite. It is well known that the Gauss Seidel iterative method is then converging (Jefferys, 1963).

But we can find in astrometry problems where the design matrix X is singular. The classical theorem about convergence mentionned above cannot longer be used ($X^t X$ is singular). Such a situation was encountered by R. Teixeira and P. Benevides-Soares who treat by a global reduction method two years of meridian observations made with the automatic meridian circle at Bordeaux Observatory. They reduce in a single least squares problem 30 000 observations of 5 000 stars. The dimensions of the design matrix are large ; therefore iterative techniques are of interest in order to solve the normal equations. In the case of the Bordeaux meridian observations, Teixeira uses an iterative method which is equivalent to the Gauss-Seidel algorithm.

In order to give some results concerning the convergence of the algorithm, we will use a theorem which can be found in Young (1971), and which is stated as follows :

$$A (x) = b \text{ is a linear system} \tag{1}$$

A is not necessarily a regular matrix

$$u_{n+1} = Gu_n + k \tag{2}$$

is a linear stationnary method of degree 1 used to solve (1).

We say that :

(2) is weakly convergent if, for all u_0, the sequence $u_1, u_2, ..., u_n$ converges.

(2) is convergent if, for all u_0, the sequence $u_1, u_2, ..., u_n$ converges to a limit independant of u_0.

We will use the following theorem :

The iterative method (2) converges if and only if S(G), the spectral radius of G, is strictly less than 1.

The iterative method (2) is weakly convergent if and only if

- $S(G) \leq 1$
- $(Id - G) u = k$ has solutions
- $\lambda = 1$ is the only eigenvalue of G of modulus unity
- the dimension of the kernel of $(Id - G)$ is equal to the multiplicity of $\lambda = 1$ as an eigenvalue of G.

In order to be able to do easily numerical experiments, we apply the results of this theorem to systems of smaller dimensions than those of the initial problem.

In all cases, we find that the conditions of weak convergence are fulfilled. Furthermore the iterates converge to a solution of (1). We then conjecture that :

The iterative Gauss Seidel technique applied to the singular normal equations arising from meridian observations is weakly convergent.

An other method for solving the overdetermined rank deficient system $X(p) = \ell$ consists in adding a non estimable constraint on the parameters.

The constrained normal equations can then be written following methods developed by Eichhorn et al. (1976), Jefferys (1979), Benevides-Soares (1987). It is to be noticed that the matrix of the constrained normal equations is still not definite positive.

We find in all cases that the matrix G has eigenvalues λ with $|\lambda| = 1$ and $\lambda \neq 1$, so that the conditions of convergence are not fulfilled.

We then conjecture that the Gauss Seidel algorithm applied to the constrained least squares problem arising from the global reduction of meridian observations is not weakly convergent.

References

Benevides-Soares, P. : 1987, private communication

Eichhorn, H.K., and Russel, J. : 1976, *Monthly Notices Roy. Astron. Soc.* **174**, 679

Jefferys : 1963, *Astron. J.* **68**, 111

Jefferys, W.H. : 1979, *Astron. J.* **84**, 1775

Teixeira, R. : 1989, Séminaire de l'Observatoire de Bordeaux

Young, D.M. : 1971, Iterative solutions of large linear systems (A.P. New York)

Centroid Analysis of Space Telescope Wide Field Camera Point Spread Function Images

E. J. SANTORO, R. E. SCHMIDT, and P. K. SEIDELMANN

U.S. Naval Observatory
Washington, D.C.

J. KRISTIAN

Observatories of the Carnegie Institution of Washington
Pasadena, Ca.

ABSTRACT

Noiseless Space Telescope Wide Field Camera point spread function images with known centers were centroided and analyzed. A 2-D Cauchy model proved superior to the Gaussian with a 70% reduction in the largest residual. The Cauchy model produced rms errors of less than 0.32 microns. When noise was added to the images, those images centered between four pixels showed the greatest stability (in terms of centroiding accuracy) with rms errors of less than 0.5 micron for signal-to-noise ratios greater than 5.

1 DESCRIPTION OF IMAGES

For the Wide Field Camera (WFC) of the Hubble Space Telescope (HST) representative point spread function (PSF) images were generated using a combination of the measured WFC PSF from radius 0 to 0.365 arcsec and the predicted PSF at 5755 Angstroms for larger radii. The PSF is assumed to be azimuthally symmetric. A magnified (10×) image has been used to produce 121 normal size images by taking 10×10 block averages at different starting (X,Y) coordinates of the magnified image. In effect, the center of the image reduced this way can be positioned with a 0.1 pixel resolution. The scale for these images is 0.1 arcsec per pixel (15 microns). There is no noise nor sky contained in these images.

2 INITIAL ANALYSIS

The images were centroided using marginal sums and a 1-D Gaussian model with a Taylor series linearization method. In addition to this method, three other programs were tested with the above model. The nonlinear simplex algorithm is derivative free and has slow, but almost sure convergence. The quasi-newton method approximates second derivatives for a much faster rate of convergence, but the region of convergence (in the parameter space) is much smaller than for nonlinear simplex. Lastly, a global minimum algorithm which samples many starting values in the parameter space was used to compare its centroid determination with the other ones. For the 1-D Gaussian model, all of the algorithms produced the same centroids and residuals.

A seven parameter 2-D Gaussian model fit the simulated images better, in the least squares sense, but produced only marginally smaller residuals. In the 1-D model, the absolute value of the residuals ranged from 0 to 1.07 microns and the 2-D Gaussian reduced the maximum centroiding error by only 0.17 microns (i.e., the range was from 0 to 0.90 microns). The fit was dominated by the 3 or 4 pixels near the center because of the extreme kurtosis (peakedness) of the images. Various weighting schemes resulted in spurious solutions at best.

The general 2-D Cauchy model (Rohatgi 1976) has the following form:

$$F(X,Y) = \frac{a}{\left[1 + \left\{\frac{(X-b)}{c}\right\}^2\right]^m \times \left[1 + \left\{\frac{(Y-d)}{e}\right\}^2\right]^m} + f \tag{1}$$

where: a = height
         b = x-center
         c = half-width at half maximum (in x direction)
         d = y-center
         e = half-width at half maximum (in y direction)
         f = background
         m = a positive number greater than or equal to one (shape parameter)

This model has been used previously by Franz (1973) and Harrington (1982). The distribution is similar in shape to the Gaussian but has more pronounced wings. The Cauchy model (using m=2) brought about much smaller centroiding errors (0 to 0.32 microns) than the Gaussian. The decrease in error was over 70% for the largest residual.

Partial tests for m=1,3, and 4 produced slightly larger residuals. Consequently, m=2 is used throughout this report. This effectively reduces the 2-D Cauchy model to six parameters. The parameter m has the effect (along with c and e) of controlling how fast the function decreases in X and Y.

There are nine images with zero residuals in both X and Y coordinates. One image has its center located directly on a pixel. Four images have their centers evenly spaced between four pixels. The other four images have their centers evenly spaced between two pixels either in X or Y. These nine images are referred to as the "best" images in sections 3 and 4 of this report.

3 NOISE ANALYSIS

Random noise was added to the above images in the following way. For each pixel location (i,j) in the image:

$$np(i,j) = p(i,j) + n(0,v)\sqrt{p(i,j)} \tag{2}$$

where: $np(i,j)$ = new pixel intensity
         $p(i,j)$ = noiseless pixel intensity
         $n(0,v)$ = random number generator having Gaussian distribution with mean 0 and variance v. The parameter v was varied to allow for different signal-to-noise ratios.

The signal-to-noise ratio (SNR) for each image is defined as (Gonzalez et al. 1977):

$$SNR = \sqrt{\frac{\sum_i \sum_j p(i,j)^2}{\sum_i \sum_j \left(p(i,j) \times n(0,v)^2\right)}} \tag{3}$$

Statistics for the best images were generated to determine the rate at which the mean error approaches its theoretical limit of 0 at SNR = ∞ . The results are presented in table I below (BMDP Statistical Software 1983). Mean refers to the average value of the centroid residuals in a certain SNR range. For example, the mean error is 0.033 micron for the sixty X residuals in the SNR range from 8 to 12. Standard error of the mean is defined as the standard deviation divided by the square root of the frequency (the number of cases in a specific range of SNR). For XRES (X residuals) and YRES (Y residuals) the units are in microns.

TABLE I.

STATISTICS OF BEST IMAGES FOR LARGE SIGNAL-TO-NOISE RATIOS
(USING A 2-D CAUCHY MODEL)

Variable No. Name	Grouping Variable	Level	Total Frequency	Mean	Standard Deviation	St. Error of mean
3 XRES			532	−0.002	0.280	0.0121
	SNR	8 to 12	60	0.033	0.671	0.0866
		12 to 20	81	−0.011	0.309	0.0344
		20 to 30	89	−0.005	0.237	0.0252
		30 to 40	47	−0.017	0.161	0.0235
		40 to 50	34	−0.014	0.124	0.0213
		50 to 60	19	0.001	0.061	0.0139
		60 to 80	49	0.005	0.069	0.0099
		Over 80	153	−0.005	0.042	0.0034
4 YRES			532	−0.020	0.249	0.0108
	SNR	8 to 12	60	−0.030	0.584	0.0755
		12 to 20	81	−0.051	0.253	0.0281
		20 to 30	89	−0.058	0.245	0.0259
		30 to 40	47	−0.002	0.134	0.0195
		40 to 50	34	0.003	0.135	0.0232
		50 to 60	19	0.000	0.044	0.0102
		60 to 80	49	−0.008	0.081	0.0116
		Over 80	153	0.004	0.040	0.0033

Various regression models were used to fit the absolute value of the residuals to the SNR. All attempts were unsatisfactory because the residual data distribution is also a function of the SNR. In the above table we note that the standard deviation changes with the SNR and this effect cannot be accounted for in a regression model.

The best images were grouped and analyzed by image center location. Statistics for the image centered directly on a pixel and images centered between four pixels are presented in tables II and III below. Residual values are in microns. The addition of noise has the least effect on centroiding accuracy for the images centered between four pixels. In the low SNR range, the mean error is 65% smaller and the standard deviation is 80% smaller for the images centered between four pixels than for the image centered directly on one pixel. The explanation for this difference is not fully understood at this time. Although all the images have the same volume, the image centered on one pixel has an exceptionally sharp peak, whereas the images centered between four pixels have a flatter peak spread evenly over four pixels.

TABLE II.

STATISTICS OF BEST IMAGE CENTERED DIRECTLY ON A PIXEL
(USING A 2-D CAUCHY MODEL)

Variable No. Name	Grouping Variable Level		Total Frequency	Mean	Standard Deviation	St. Error of mean
3 XRES			107	0.011	2.165	0.2093
	SNR	1 to 2	5	−0.309	2.894	1.2943
		2 to 3	28	0.423	2.444	0.4619
		3 to 4	12	0.375	1.883	0.5437
		4 to 5	21	−0.201	2.377	0.5186
		5 to 6	17	−1.074	2.171	0.5265
		6 to 7	18	0.668	1.352	0.3186
		7 to 8	6	−0.535	1.128	0.4606
4 YRES			107	−0.083	2.158	0.2087
	SNR	1 to 2	5	2.721	2.487	1.1124
		2 to 3	28	−0.736	2.543	0.4807
		3 to 4	12	0.175	2.030	0.5860
		4 to 5	21	−0.408	2.222	0.4849
		5 to 6	17	0.424	1.666	0.4040
		6 to 7	18	−0.378	1.375	0.3241
		7 to 8	6	0.700	1.411	0.5760

TABLE III.

STATISTICS OF BEST IMAGES CENTERED BETWEEN FOUR PIXELS
(USING A 2-D CAUCHY MODEL)

Variable No. Name	Grouping Variable Level		Total Frequency	Mean	Standard Deviation	St. Error of mean
3 XRES			160	−0.002	0.713	0.0564
	SNR	1 to 2	3	−0.830	1.264	0.7295
		2 to 3	34	0.105	1.007	0.1726
		3 to 4	53	−0.057	0.750	0.1030
		4 to 5	31	0.034	0.535	0.0961
		5 to 6	22	0.054	0.336	0.0716
		6 to 7	10	0.042	0.305	0.0963
		7 to 8	7	−0.137	0.264	0.1000
4 YRES			160	0.002	0.835	0.0660
	SNR	1 to 2	3	0.340	0.347	0.2002
		2 to 3	34	0.003	1.447	0.2482
		3 to 4	53	0.073	0.654	0.0899
		4 to 5	31	−0.101	0.584	0.1050
		5 to 6	22	0.020	0.514	0.1096
		6 to 7	10	−0.135	0.338	0.1070
		7 to 8	7	−0.096	0.426	0.1610

4 CONCLUSIONS

One hundred twenty-one simulated noiseless HST WFC type images with known centers were centroided and analyzed. A 2-D Cauchy model was clearly superior to the 2-D Gaussian model in terms of centroiding accuracy. The decrease in error was over 70% for the largest residual. Both models produced zero error when centroiding images centered directly on a pixel, centered between two pixels, and centered between four pixels.

The nine best images (zero error cases) were further analyzed in a noisy environment. Various regression models were unsuccessful in fitting SNR vs. residual because the distribution of residuals is also a function of the SNR. Lastly, the images centered between four pixels showed the greatest stability against noise with a mean error less than 0.1 micron and standard deviation less than 0.5 micron for SNR greater than five.

REFERENCES

BMDP Statistical Software (1983). University of California, Berkeley, California.

Franz, O. G. (1973). J. Royal Astr. Soc. Canada 67 (No. 2), 81.

Gonzalez, R. C. and Wintz, P. (1977). Digital Image Processing. Addison-Wesley, Reading, Massachusetts.

Harrington, R. S. (1982). Private communication.

Rohatgi, V. K. (1976). An Introduction to Probability theory and Mathematical Statistics. John Wiley, New York.

ADDRESS

Ernest Santoro, Richard Schmidt, P. Ken Seidelmann
34th and Mass. Ave.
U.S. Naval Observatory
Washington, D.C. 20392

Jerry Kristian
813 Santa Barbara St.
Observatories of the Carnegie Institution of Washington
Pasadena, CA 91101

Analysis of the Mixture of Normal Distributions in Stellar Kinematics : Maximum Likelihood, Bootstrap and Wilks test

Soubiran C.[1], Gómez A.E.[1], Arenou F.[1], Bougeard M.L.[2][3]

(1) URA DO335 CNRS, Observatoire de Paris-Meudon, F-92195 Meudon
(2) Univ. Paris X, IUT, 1 Chemin Desvallieres, F-92410 Ville d'Avray
(3) URA 1125 CNRS, Observatoire de Paris, F-75014 Paris.

ABSTRACT

In this paper, we present two Maximum Likelihood algorithms (EM and SEM) in order to separate gaussian mixtures and estimate the corresponding parameters. An evaluation of the accuracy of the EM estimates is provided by a Bootstrap procedure and the Wilks test is used to verify the number of components of the normal mixture.

These methods have been applied to study the U velocity distribution ($\ell_{II}=0$, $b_{II}=0$) of two samples of A-type Population I stars.

1. INTRODUCTION

Stellar velocity distributions are usually assumed to be gaussian in each component (U, V and W) for population I stars (Mihalas and Binney,1981). In this paper, we analyze the mixture of normal distributions using the maximum likelihood method. The mixture of K normal components is defined by the density function, written in the univariate case:

$$f(x) = \sum_{k=1}^{K} p_k f_k(x) = \sum_{k=1}^{K} p_k \frac{1}{\sqrt{2\pi}\sigma_k} exp\left[\frac{-(x-m_k)^2}{2\sigma_k^2}\right]$$

K is the number of components; p_k, m_k, σ_k and f_k are respectively the proportion, the mean, the standard deviation and the density of the kth normal distribution. The values which maximize LnL (L being the likelihood function of the sample $x_1,\ldots,x_N$) are :

$$p_k = \frac{1}{N}\sum_{i=1}^{N} P(k/x_i), m_k = \frac{\sum_{i=1}^{N} x_i P(k/x_i)}{\sum_{i=1}^{N} P(k/x_i)}, \sigma_k^2 = \frac{\sum_{i=1}^{N}(x_i - m_k)^2 P(k/x_i)}{\sum_{i=1}^{N} P(k/x_i)}$$

where $P(k/x_i)$ is the probability for the ith element to belong to the kth component. The probabilities P being unknown, the latter equations have to be solved iteratively. For this purpose, we present the EM (Dempter et al, 1977) and SEM (Celeux and Diebolt, 1985; Celeux, 1987) algorithms. An evaluation of the accuracy of the EM estimates is provided by a Bootstrap procedure and the Wilks test is used to verify the number of components. These methods are shown to separate the mixture of normal distributions in the U velocity component ($\ell_{II} = 0, b_{II} = 0$) of two samples of A-type dwarf stars (Gómez et al, 1989). One of these samples has also been analysed using graphical methods (Bougeard et al, this issue).

2. THE EM AND SEM ALGORITHMS

2.1 EM

Assuming an initial position $\Phi^0 = (p_k^0, m_k^0, \sigma_k^0, k = 1, \ldots, K)$ of the parameters, the EM iteration uses two steps. The Expectation step which computes the conditional density

$$P^n(k/x_i) = p_k^n f_k^n(x_i) \big/ f^n(x_i)$$

and the Maximization step which chooses as Φ^{n+1} any value of Φ which maximizes LnL(Φ^n). EM needs to assume in advance the number of components of the mixture. It works well and rapidly whenever the components are well separated, the proportions are not too extreme and Φ^0 is not too far from the solution.

2.2. EM with Bootstrap

The accuracy of the EM estimates can be evaluated using a Bootstrap procedure (Efron, 1981). EM estimates are computed for R subsamples, drawn at random from the original sample. The mean and the empirical standard deviation (SD) provide for each parameter a confidence interval [- SD + MEAN ; MEAN + SD] , which measures the degree of overlap of the mixture components and the stability of the results.

2.3. The Wilks test

When the true number of components K is unknown, it is possible to test the hypothesis H_0 (mixture of K components) against the hypothesis H_1 (mixture of K' components, $K' > K$):

H_0 : K components, i.e. 3K-1 unknown parameters
H_1 : K'=(K+ℓ) components, i.e. 3(K+ℓ)-1 unknown parameters

The sample likelihood ratio λ is computed with the two hypothesis. If Ho is true, -2Lnλ follows asymptotically a $\chi_{3\ell}^2$ law. For example, if K'=K+1, -2Lnλ follows a χ_3^2 law and at a 5% level of significance, Ho is rejected if $\lambda \geq 7.81$.

2.4. SEM

The SEM algorithm incorporates a stochastic step between the E-step and the M-step. Due to the S-step, the sequence (Φ^n) generated by SEM is a Markov chain which converges in law to a unique stationary distribution. The empirical mean and standard deviation of this distribution provide a point estimate and a confidence indication of Φ. This algorithm provides the following performances for samples having at least twenty points per component : it always finds K (if the initial K value is greater than the solution value), the results do not depend on starting point which can be drawn at random, it works even when the components are poorly separated and the proportions extreme. But for small samples ($N_k < 20$), the random perturbations of the S-step are too large and it is advisable to run SEM several times and to choose the value of Φ which occurs most often.

3. RESULTS

Figure 1 shows the histograms of two samples of A-type stars (A1V and A2V) located in the close solar neighborhood (distance up to 200 pc). The EM algorithm has been run for both samples assuming one, two and three components. The Wilks test, given in Table 1, found two components in each case, with a 5% level of significance.

Tables 2 and 3 give the EM estimators of the mixture of two components and the corresponding Bootstrap confidence intervals. The EM estimators were obtained after 500 iterations and 100 Bootstrap runs were used for each sample.

The EM estimators were used as initial values for the SEM algorithm application. The SEM estimators obtained after 100 iterations, are given in Table 4. These results show a great stability and confirm the EM estimators.

As a conclusion, under a Gaussian assumption, the existence of two components of different kinematic properties is confirmed for the two samples of A dwarf stars.

We thank Dr. Celeux (INRIA) for supplying the SEM software.

4. REFERENCES

Celeux G.,1987, These d'Etat - Université de Paris IX Dauphine.

Celeux G.,Diebolt J.,1985, *Computational Statistics Quaterly*, **2**,73.

Dempster A.P.,Laird N.M.,Rubin D.B.,1977, *J.R.Statist.soc.*, **B39**,1.

Efron B.,1981, *Biometrika*, **68**,

Gómez A.E.,Delhaye J.,Grenier S.,Jaschek C.,Jaschek M.,1989, *Astron.Astrophys.*(submitted).

Mihalas D.,Binney J.,1981, Galactic Astronomy - W.H. Freeman and Company.

Wilks S.,1963, Mathematical Statistics - Wiley.

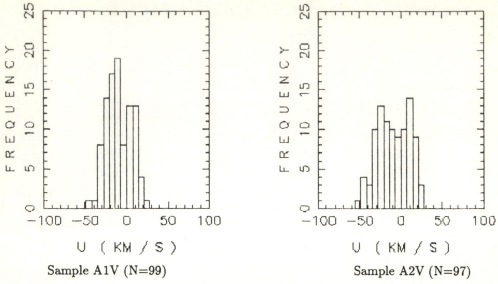

Sample A1V (N=99)   Sample A2V (N=97)

Figure 1 : Histograms

H_0 / H_1	A1V	A2V
K=1 / K'=2	9.04	16.11
K=2 / K'=3	6.97	0.06

Table 1 : WILKS TEST values

A1V (N=99)	$p_1(\%)$	$m_1(km/s)$	$\sigma_1(km/s)$	$p_2(\%)$	$m_2(km/s)$	$\sigma_2(km/s)$
EM	69	-17	10	31	8	6
[;]	[61;77]	[-19;-15]	[4;16]	[23;39]	[6;10]	[2;10]

Table 2 : EM estimates and their BOOTSTRAP confidence interval for the sample A1V

A2V (N=97)	$p_1(\%)$	$m_1(km/s)$	$\sigma_1(km/s)$	$p_2(\%)$	$m_2(km/s)$	$\sigma_2(km/s)$
EM	70	-19	15	30	13	6
[;]	[59;75]	[-22;-16]	[6;22]	[25;41]	[10;14]	[2;11]

Table 3 : EM estimates and their BOOTSTRAP confidence interval for the sample A2V

SEM	p_1	m_1	σ_1	p_2	m_2	σ_2
A1V (N=99)	70	-17	10	30	8	6
A2V (N=97)	70	-19	15	30	13	6

Table 4 : SEM estimates of both samples A1V and A2V

Large voids of clusters of galaxies in the north galactic hemisphere

K.Y. STAVREV

Department of Astronomy and National Astronomical Observatory
Bulgarian Academy of Sciences, Sofia, Bulgaria

ABSTRACT. The results from a search for large voids of clusters of galaxies in the north galactic hemisphere ($b \geq 30°$) based on the analysis of two volume-limited samples ($z \leq 0.14$) of clusters with measured redshifts are presented.

1. INTRODUCTION

The studies of the 3-D distribution of galaxies in the past decade have shown that the near-by Universe ($z \lesssim 0.05$) contains regions with dimensions 20-60 Mpc ($H_o = 100$ km s^{-1} Mpc^{-1}) devoid of galaxies (e.g., de Lapparent et al. 1986). To study the voids in larger volumes than those covered by the redshift surveys of galaxies one may use the clusters of galaxies as tracers of the distribution of galaxies (Bahcall and Soneira 1982a, b; Batuski and Burns 1985). However, in this case the reality of the detected voids is often uncertain mainly because of incompleteness of the samples of clusters used. The inclusion in the analysis of clusters with redshifts estimated from the magnitudes of cluster galaxies in order to achieve sample completeness to larger distance (Batuski and Burns 1985) may lead to detection of false voids and masking of real voids because of large errors in the estimated distances. The purpose of this paper is to determine the locations of the large voids in the spatial distribution of clusters of galaxies on the basis only of clusters with measured redshifts. Some results concerning the voids of rich clusters are given also in Stavrev (1989).

2. OBSERVATIONAL DATA

The search for large voids of clusters of galaxies is based on samples from an updated version of the list of clusters of galaxies

with published redshifts compiled by Lebedev and Lebedeva (1986).
We process first a homogeneous and relatively complete sample of
Abell rich clusters, and second, an inhomogeneous and incomplete
sample of all concentrations of galaxies which lie in the volume
occupied by the first sample in order to get some additional information for the voids. From the surface density distribution of rich
clusters as a function of galactic latitude and the spatial density
distribution of rich clusters as a function of redshift (see Fig. 1
and 2 in Stavrev 1989) the limits of completeness of the sample of
rich clusters are estimated to be $b \geq 40^{\circ}$ and $z \leq 0.08$. To avoid
boundary effects the investigated volume is enlarged to $b \geq 30^{\circ}$ and
$z \leq 0.14$. There are 220 rich clusters (sample 1) within these limits.
The second sample includes 685 clusters and groups of galaxies in
the same volume. In Figure 1 the distributions of the number of
rich clusters ($b \geq 30^{\circ}$) with measured redshifts as a function of
redshift are plotted for Abell distance classes D = 1-6. Relating
these distributions to the 715 unmeasured rich clusters with $b \geq 30^{\circ}$
we find that 51 of them are expected to lie in the volume with
$z \leq 0.08$, and 253 in $z \leq 0.14$, in addition to 86 and 220 clusters in

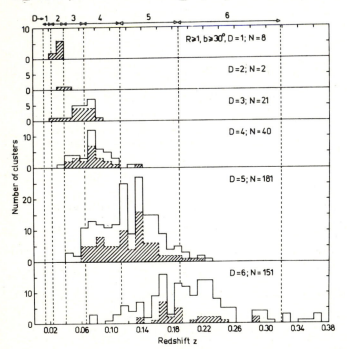

Fig. 1. Distribut-
ion of the number of
rich clusters with
$b \geq 30^{\circ}$ as a function
of redshift for Abell
distance classes D =
1, 2, 3, 4, 5, and 6.
Shaded area indicates
the fraction of clus-
ters with measured
redshifts of more than
one member galaxy. The
dashed vertical lines
delimit the redshift
ranges of the distance
classes defined from
the relation m_{10} =
0.52 + 3.52log cz from
Batuski and Burns
(1985).

sample 1. Hence the completeness of sample 1 with respect to Abell catalogue is estimated as 65% for $z \leq 0.08$ and 45% for $z \leq 0.14$.

3. THE VOID-SEARCHING PROCEDURE

The search for voids of clusters is based on the computation of the distances from the nodes of a cubic grid, built into the investigated volume, to the nearest neighbouring cluster of galaxies. A grid constant $k = 20$ Mpc is chosen. The distances to the clusters are computed from the Hubble law ($H_o = 100$ km s^{-1} Mpc^{-1}). By computing the distances from every grid node to its nearest neighbouring cluster the investigated space is transformed into a discrete 3-D field $d(x,y,z)$, where d is the distance from node (x,y,z) to its nearest neighbouring cluster. The local maxima of this field indicate the voids of clusters. The following criterion for selection of large voids is adopted: a large void is a region indicated by a local maximum $d(x_v, y_v, z_v) \geq 40$ Mpc, which satisfies the condition $d(x_v, y_v, z_v) = \max \left[d(x,y,z) \right]$, where $x = x_v - k$, x_v, $x_v + k$; $y = y_v - k$, y_v, $y_v + k$; $z = z_v - k$, z_v, $z_v + k$. For convenience we call the voids satisfying this criterion "closed" voids, and we may call the empty regions which do not satisfy it "open" voids.

4. RESULTS

Table 1 contains a list of the detected candidates for large voids of clusters of galaxies in the north galactic hemisphere ($b \geq 30°$) to the limiting distance of 420 Mpc ($z \leq 0.14$). The results for sample 1 are given in the first part of the Table, and the second part contains the results for sample 2 (all clusters). The following parameters of the voids are given in columns 2-8: equatorial and galactic coordinates of the void center, distance to the void center, diameter, angular diameter. Column 9 contains the results from the identification of the voids of rich clusters with voids of clusters in the second part of the Table, and with voids detected by Bahcall and Soneira (1982a, b) and Batuski and Burns (1985), marked by "BS" and "BB". The four voids of rich clusters marked by "E" in column 9 have not been identified in part II of the Table because they are populated by poor clusters, i.e. they are not real voids.

TABLE 1. Large Voids of Clusters of Galaxies.

Void No.	RA (h)	DEC (°)	l (°)	b (°)	R (Mpc)	D (Mpc)	A (°)	Identifications
	I. "Closed" Voids of Rich Clusters
1	9.3	55	162	43	262	176	37	II-2, BS VOID, BB 9
2	10.8	29	202	64	245	96	22	E
3	10.9	27	207	64	201	106	29	E
4	11.1	15	236	63	157	88	31	II-4
5	12.2	39	153	76	185	88	27	II-7, (BB 16)
6	13.0	15	315	77	123	116	51	II-9
7	14.5	50	90	60	161	100	35	II-13, BOOTES VOID, BB 18
8	14.6	37	63	66	219	86	22	II-14
9	16.0	58	90	45	368	164	25	II-16
10	16.1	43	68	48	322	154	27	II-18
11	16.4	47	73	44	289	142	28	E, (BB 24)
12	16.9	64	94	37	329	172	29	(E)
	II. "Closed" Voids of Clusters
1	9.5	21	209	44	287	122	24	BS VOID
2	9.7	54	162	47	276	148	30	I-1, BS VOID, BB 9
3	10.1	41	180	54	372	114	17	
4	10.9	8	243	57	166	76	26	I-4
5	12.3	28	207	83	343	160	26	
6	12.3	40	146	76	289	132	26	
7	12.4	31	180	84	181	80	25	I-5, BB 16
8	12.6	38	135	79	305	132	24	
9	12.7	9	297	72	147	102	38	I-6
10	13.7	63	113	53	251	78	18	BB 21?
11	13.8	49	99	65	287	112	22	
12	14.2	43	82	67	368	166	27	
13	14.5	50	90	60	161	100	35	I-7, BOOTES VOID, BB 18
14	14.6	37	63	66	219	84	22	I-8
15	15.3	62	97	48	242	100	23	
16	15.6	63	98	45	398	122	17	(I-9)
17	16.2	34	54	46	249	80	18	
18	16.2	48	75	46	331	90	16	I-10
19	17.2	40	65	35	349	106	17	

REFERENCES

Bahcall, N. A., and Soneira, R. M. 1982a, Ap. J., 258, L17.
_____. 1982b, Ap. J., 262, 419.
Batuski, D. J., and Burns, J. O. 1985, Astr. J., 90, 1413.
de Lapparent, V., Geller, M., and Huchra, J. 1986, Ap. J., 302, L1.
Lebedev, V. S., and Lebedeva, I. A. 1986, Astron. Tsirk., 1469, 4.
Stavrev, K. Y. 1989, in Physics and Evolution of Stars Symposium
   (Visegrad, Hungary), ed. B. Balázs, in press.

Spectral analysis of irregularly observed processes

Toruń Radio Astronomy Observatory, Poland

Abstract

The present work deals with the spectral deconvolution problem pertaining to the spectral analysis of irregularly observed processes. It reveals some new mathematical aspects of the well-known CLEAN algorithm. Further, it develops more general approach to the problem and proposes new L1-norm deconvolution algorithm.

1   Introduction

Spectral analysis is a powerful tool for the detection of periodicities in the data. In astrophysics, there are quite a lot of practical situations in which such analysis is required. For instance, this approach is widely used in the studies of the solar oscillations, the sunspot activity cycles and the variable star light curves. However, the spectral analysis of the astrophysical time series is often difficult or sometimes even impossible as samples are available at irregular intervals for the purely practical reasons. In addition, further difficulties may arise while we are given the data in several segments separated by gaps and/or the resulting signal-to-noise ratio is low. As a matter of fact, an interpretation of the spectrum is highly uncertain and rather difficult in such cases.

New algorithm for the spectral analysis of unevenly sampled time series has recently been proposed (*Roberts et al.,1987*). Their approach is based on the idea incorporated in the CLEAN deconvolution algorithm (*Hogbom, 1974*) which is used routinely in radio astronomical imaging.

This paper presents some new insights to the CLEAN algorithm. More general approach to the spectral deconvolution problem is developed. Finally, new algorithm based on the linear programming technique is proposed.

2   Spectral deconvolution

Let us assume that the continuous function $f(t)$ is sampled unevenly at a finite number of sample points. As a result we get the data set which consists N pairs of values,

$$\{f_n, t_n\} = \{f(t_n), t_n\}, \qquad t_n = 1, 2, \ldots, N \tag{1}$$

The sampling process may be regarded as a multiplication of $f(t)$ with the sampling function $s(t)$ consisting N Dirac delta functions $\delta(t)$. Therefore, one may write

$$f_s(t) = f(t) \cdot s(t) = \frac{1}{N} \sum_{n=1}^{N} f(t)\ \delta(t - t_n) = \frac{1}{N} \sum_{n=1}^{N} f_n\ \delta(t - t_n) \tag{2}$$

Taking the Fourier transform of both sides of Eq. (2) we obtain the spectral convolution equation

$$D(\nu) = F(\nu) \otimes W(\nu) = \int_{-\infty}^{\infty} d\nu'\ F(\nu')W(\nu - \nu') + N(\nu) \tag{3}$$

Since the noise and measurement errors are unavoidable in any real situation we add an extra term in Eq. (3) which represents the background spectrum. Eq. (3) may be written in matrix form as follows

$$d = f \cdot W + n \tag{4}$$

Notice that the Eq. (4) is a complex convolution. Obviously, the true spectrum can be obtain by using an appropriate deconvolution method. The direct inversion cannot be applied here as it does not provide the meaningful solution, so we must to resort to more sophisticated deconvolution procedures.

3   CLEAN algorithm

The CLEAN is an iterative deconvolution based on the assumption that the spectrum consists only a few well separated spectral lines and the background spectrum is negligible. Although a few more refined analysis of the CLEAN have appeared in the literature, its mathematical properties are still poorly understood. One has to point out that CLEAN and its numerous variants have rather been devised on the basis of heuristic arguments. Recent analysis (*Marsh and Richardson, 1987*) hereafter called MR has shed more light on this subject. Namely, they have shown that the CLEAN may be regarded as an approximate method for minimizing an objective function represented by $\sum f_i$, where f_i is the amplitude of ith spectral component, subject to $f_i \geq 0$ and subject to consistency with the data. In other words, the CLEAN algorithm contains a built-in bias towards the solution corresponding to minimum total power. However, these authors did not consider the effect of noise and errors on the final solution. They did not specify explicitly what they mean by "consistency with the data". They have only shown that in the case of data which are consistent with a spectral line, in the absence of noise, the minimization is strictly correct.

4   New spectral deconvolution algorithm

We may rewrite Eq.(4) as follows

$$n = d - W \cdot f \tag{5}$$

Before we discuss the solution of this equation some inadequacies should be mentioned. First of all, the statistics of n component is usually not known. However, we may only assume that n is non-gaussian process. Moreover, the spectral window function has the high sidelobes caused by the sparse sampling. Clearly, this is an example of the so-called ill-posed problem. A solution can be obtained by the regularization procedure. Regularization means adopting some apriori information or constraints, e.g., non-negativity, band-limitidness and imposing some reasonable conditions on the behaviour of the restored function $f(t)$. Various optimization criteria can be invoked. L_2 criterion seems to be not a good choice if we take into account all of the above arguments. Instead, I propose to use the L_1 , or least absolute deviation, criterion to minimize

$$\|n\|_{L_1} + \lambda \|f\|_{L_1} = \sum_i |n_i| + \lambda \sum_i |f_i| \tag{6}$$

where λ is the regularization parameter. This parameter can be calculated for example by the cross-validation technique. Of course, other procedure are available as well. In practice, a weighted form of Eq. (6) may be found more advantageous

$$minimize \qquad \sum_i p_i \, |n_i| + \lambda \sum_i q_i \, |f_i| \qquad (7)$$

Having in mind the findings of MR one can easily recognize that Eq. (7) represents more general approach to the spectral deconvolution problem. It should be pointed out that the present approach allows for the more accurate restoration of an original signal. Equation (7) is often solved by using linear programming (LP) techniques. The details can be found in the literature (*Taylor et al., 1979*). Most often used algorithm to solve the LP problems is the well-known simplex method. Faster algorithms are also available (*Karmarkar, 1984*). The advantage of using the LP to solve Eq. (7) bears on the fact that the optimal and unique solution may be found.

5   References

[1] Hogbom, J.A.,   *1974, Astron. Astrophys. Suppl. 15, 417*
[2] Karmarkar, N.,   *1984, Combinatorica, 4, 373*
[3] Marsh, K.A., Richardson, J.M.,   *1987, Astron. Astrophys. 182, 174*
[4] Roberts, D.H., Lehar, J., Dreher, J.W.,   *1987, Astron. J. 93, 968*
[5] Taylor, H.L., Banks, S.C., McCoy, J.F.,   *1979, Geophysics 44, 39*

Is This Orbit Really Necessary? (II)

CHARLES E. WORLEY

U.S. Naval Observatory
Washington, D.C. 20392
U.S.A.

Summary. This contribution discusses the questionable proliferation of visual double star orbit computations, with examples of some of the associated problems. Possible solutions are proposed.

More than a quarter century ago, W.H. van den Bos (1962) wrote the initial installment of "Is This Orbit Really Necessary?" His main point was that too many orbits were being computed that did not provide us with useful, reliable information. In his opinion, there were two kinds of orbit computation which failed in this respect: (1) recomputation of a definitive orbit and (2) computation of a premature orbit. In the first instance the information was reliable but not useful, and in the second neither reliable nor useful. The time seems ripe for a new look at this problem, as well as for comments on related problems that have since arisen.

The orbit catalog by Worley and Heintz (1983) contained 928 orbits of 847 systems. Since that time orbit computers have continued churning out new orbits at a high rate; no less than 455 have appeared in the literature. There are 111 new systems represented, and recomputations for 244 previously listed, including a substantial number of multiple orbits for the same object. Thus, in approximately seven years, nearly 29% of the systems listed in the catalog ostensibly have had "improved" orbits published. But what is the real situation?

First, let us consider the recomputation of definitive orbits. The orbit catalog lists 62 systems as such; ten have been redone. Four of these, ADS 9744, 12973, 14773, and 15281, are by Tokovinen (1984, 1987), who has used speckle measures exclusively. In no case does he give residuals, but the elements derived are for all intents and purposes virtually identical to previous determinations, with the exception of ADS 12973, where the period has been shortened 0.54 year. However, the speckle measures cover less than half the period, and therefore this result is unconvincing. Also, there are small changes in the values of a for all

these orbits, which may be of significance, but much
subsequent speckle data has accumulated and should be
incorporated before new "definitive" orbits are accepted for
these pairs. Heintz (1984, 1986, 1988a, 1988b) is
responsible for another four orbits, KUI 75, ADS 11046,
11520, and 15972. The orbits of two of these systems, ADS
11046 and 15972, are nearly identical to previous
calculations. In the case of KUI 75, the dynamical elements
are identical within the observational errors; the changes
in the argument of periastron and in the node appear
significant, but can be largely accounted for by their
indeterminacy due to the small eccentricity and inclination.
The two remaining systems are ADS 9716 and 11077. The first
of these, ADS 9716, has been recomputed by Couteau (1988).
It is what I like to call "anticipatory", where small
deviations are corrected before periastron passage.
Unfortunately, almost all such computations have to be
repeated shortly after the periastron! Finally, the solution
for ADS 11077 is a combined solution involving parallax
plates and the visual orbit (Kamper and Beardsley, 1986).
The elements obtained are in fact very similar to those of
the previous determination, and the authors themselves
comment that "the only significant difference in the
residuals is some amelioration of the run in the separations
between 1920 and 1950". So we have ten new definitive
orbits which are reliable, but not useful.

     The second category of unnecessary orbits is those that
are premature or indeterminate. These are the orbits based
on such limited or inexact data that any similarity to the
true elements is entirely accidental. If the period is
short, the pair rapidly deviates from its ephemeris, thus
providing new sport in a few years. A typical example of
this is SCJ 22 (ADS 12469), for which no less than eight
orbits have been computed in the last 30 years, as shown in
Table 1. Listed are the dynamical elements, plus the arc
defined

Table 1.
Orbits of SCJ 22

	Dommanget 1959		Heintz 1961	Heintz 1975	Baize 1984	Heintz 1985	Baize 1987	Heintz 1988
	I	II						
P yr.	142	443	205	172	184	163	159	162
T	1964	1969	2010	1989	1986	1984	1985	1984
a"	2.34	2.16	1.22	1.07	1.08	1.00	0.99	1.00
e	0.93	0.56	0.29	0.48	0.49	0.57	0.56	0.58
arc deg.	70		77	122	172	212	198	239

by the observations at the time of the computation. The
early orbits, based on an obviously insufficient arc, should

never have been attempted. The last four orbits are
"anticipatory", in the sense that the authors should have
exercised restraint and waited until the periastron was well
defined by the measures (even the 1988 orbit by Heintz
already shows small deviations which show that it will have
to be revised). There are literally hundreds of similar
cases in the literature. Of what can the authors be
thinking? Do they believe that observers somehow need
predictions to guide them? Like van den Bos, I assure them
that we do not, as we are as aware of pairs needing
attention as anyone. Unfortunately, I share with van den Bos
the belief that the routine motive is the substitution of
one's own name for that of the previous computer's.

There is another point which causes me considerable
puzzlement. Too many orbits are being computed with
incomplete data sets, with varying effects on the
determinacy of the orbits ranging from minor to disastrous.
In the latter category is the case of ADS 5726. This is a
close, equal-magnitude, pair which has only been observed at
scattered intervals. In such pairs there is always the
problem of quadrant ambiguity, and even in misinterpretation
of the direction of motion. This is exactly what happened in
the recent computation by Docobo and Costa (1984), where
they predict retrograde motion when in fact the motion is
direct. This would not have happened had they consulted the
WDS data base, which contained observations which would have
solved the problem. Again, Baize (1989) has published 12
new orbits. In 11 cases, there exists data in the data base
of which he appears unaware. Since the double star data
basically is complete for all close pairs, and since this
data is supplied free, it is hard to understand why
computers fail to make use of this resource.

There is another annoying, and continuing, problem.
Some computers publish elements with an entirely unwarranted
and misleading profusion of decimals. For example, P and T
are often given to three decimals, implying that we know
these quantities to the day, when in fact we would be very
lucky to know them to the month, or even year! On seeing
such work, I confess to feeling some doubt concerning the
judgement and competence of the author.

The advent of speckle observations, still of widely
varying quality, also leads to problems in their inclusion
in orbits. Some computers apparently assign equal weights
with the visual measures. This is wrong in general. The vast
majority of measures listed by McAlister and Hartkopf (1988)
in their second catalog deserve weights at least 3-5 times
the visual observations, and computers should take special
care to scale their orbits so as to minimize residuals in
the speckle separations. Also, small corrections to

published data are present in this catalog, and these values
should be used in preference to those in the original
publications.

    What of the future? We are in a period of transition in
the observation of visual doubles, where the old methods are
being supplanted by new and very much more accurate
techniques. The rapid spread of economical and powerful
desk-top computers and peripherals, combined with the WDS
data base, makes the computation of orbits an increasingly
trivial pursuit. Indeed, we may be rapidly approaching the
day when something like the theory of sampling, applied to
the upgrading of artificial satellite orbits, will be found
applicable to double star orbits. In that event, addition of
new data to the data base will automatically result in the
computation of an improved orbit.

    Despite the wise words of van den Bos, I believe the
situation to be worse today than it was in 1962. Blame for
this belongs of course to the computers, but it is shared by
editors and referees. They too need to exercise restraint,
and it is they who must insist that the orbit computers use
the full data set and, most important, produce a useful
result. Unfortunately, there has been a tendency on the part
of some editors in recent years to resist the publication of
observations, while encouraging that of orbits. In quoting
Aitken "... an hour in the dome on a good night is more
valuable than half a dozen hours at the desk in daylight",
van den Bos was emphasizing the right approach.

References

Baize, P.   1988   Astron. Astrophys. Suppl. 78, 125.
Bos, W.H. van den   1962   Pub. Astron. Soc. Pacific 74, 297.
Couteau, P.   1988   Circ. Inf. No. 105.
Docobo, J.A.   1984   Circ. Inf. No. 92.
  and Costa, J.M.
Heintz, W.D.   1984   Circ. Inf. No. 93.
                         1986   Astron. Astrophys. Suppl. 65, 411.
                         1988a Ibid.   72, 543.
                         1988b J. Roy. Astron. Soc. 82, 140.
Kamper, K.W.   1986   Astron. J. 91, 419.
  and Beardsley, W.R.
McAlister, H.A.   1988   Second Catalog of Interferometric
  and Hartkopf, W.I.   Measurements of Binary Stars.
                                CHARA Cont. No. 2. Georgia State
                                University.
Tokovinen, A.A.   1984   Letters, Astron. Zhur. 10, 121.
                         1987   Ibid.   13, 1065.
Worley, C.E.   1983   Pub. U.S. Naval Obs., 2nd Ser.,
  and Heintz, W.D.   XXIV, Pt. VII.